AGRO-BIODIVERSITY EXPLORATION

2023

AGRO-BIODIVERSITY EXPLORATION

SB DAS
DEEP PAHALWAN
SB NAHATKAR
DHIRENDRA KHARE

2023

SCIENTIFIC® PUBLISHERS

Print 2026

Edited by

Das SB, Professor and Head, Department of Entomology, College of Agriculture, Jabalpur
Pahalwan Deep, Director Farms, JNKVV, Jabalpur
Nahatkar SB, Director, ABM, College of Agriculture, Jabalpur
Khare Dhirendra, Dean Faculty of Agriculture, JNKVV, Jabalpur

Audit Team Members

Awasthi MK, Professor Department of Soil and Water Engineering, College of Agricultural Engineering, Jabalpur
Barde Vivek, Assistant Professor, Directorate of Instrumentation, JNKVV, Jabalpur
Baghel SS, Associate Professor, Department of Soil Science and Agricultural Chemistry, College of Agriculture, Jabalpur
Bajpai RK, Professor and Head, Department of Forestry, College of Agriculture, Jabalpur
Bhatt Jayant, Professor and Head, Department of Plant Pathology, College of Agriculture, Jabalpur
Bisen BP, Associate Professor, Department of Horticulture, College of Agriculture, Jabalpur
Das S.B, Professor and Head, Department of Entomology, College of Agriculture, Jabalpur
Das Bharati, Senior Scientist, Directorate of Instrumentation, JNKVV, Jabalpur
Dongre RP, Assistant Professor, Department of Horticulture, College of Agriculture, Jabalpur
Jaiswal Ajay, Department of Plant Breeding and Genetics, College of Agriculture, Jabalpur
Mittra NG, Professor and Head Department of Soil Science and Agricultural Chemistry, College of Agriculture, Jabalpur
Mudaliyar RSRK, Assistant Professor Department of Physiology, College of Agriculture, Jabalpur
Nahatkar SB, Director IABM, JNKVV, Jabalpur
Nema RK. Professor and Head Department of Soil and Water Engineering, College of Agricultural Engineering, Jabalpur
Pahalwan DK, Director Farm, JNKVV, Jabalpur
Pandey SK, Professor and Head, Department of Horticulture, College of Agriculture, Jabalpur
Rai AK, Director, Directorate of Instrumentation, JNKVV, Jabalpur
Rai HK, Professor Department of Soil Science and Agricultural Chemistry, College of Agriculture, Jabalpur
Rawat Anay Assistant Professor, Department of Agronomy, College of Agriculture, Jabalpur
Sharma Amit, Dean Student Welfare, JNKVV, Jabalpur
Sharma PB, Professor and Head, Department of Agronomy, College of Agriculture, Jabalpur
Sharma Rajnee, Assistant Professor, Department of Horticulture, College of Agriculture, Jabalpur
Sharma Shailesh, Professor, College of Agricultural Engineering, Jabalpur
Shrivastava RN, Professor Department of Soil and Water Engineering, College of Agricultural Engineering, Jabalpur
Shukla RS, Professor and Head Department of Plant Breeding and Genetics, College of Agriculture, Jabalpur
Singh Makhan, PhD Scholar, Department of Forestry, College of Agriculture, Jabalpur
Singh Sanjay, Asstt. Professor and Incharge Farm, College of Agriculture, Jabalpur
Tagore GS, Assistant Professor, Department of Soil Science and Agricultural Chemistry, College of Agriculture, Jabalpur
Tiwari Akhilesh, Department of Horticulture, College of Agriculture, Jabalpur
Tiwari Gyanendra, Department of Plant Physiology, College of Agriculture, Jabalpur
Upadhaya SD, Ex Director Instruction, JNKVV, Jabalpur
Vani GK, Assistant Professor Department of Agricultural Economics and Farm management, College of Agriculture, Jabalpur
Verma Bhupesh, Department of Horticulture, College of Agriculture, Jabalpur
Yadav Vijay Kumar, Professor, Department of Plant Pathology, College of Agriculture, Jabalpur

Volunteers

Department of Agronomy	Mansi Joshi, Shivangi, Raghuwanshi, Anil Patel, Lakhan Bhalse, Shubham Choudhary, Hemat Kourav, Sangita Rawat, Priyambada Dwivedi, Kartik Mandoli, Ashi Thakur and Megha Singh
Department of Entomology	Jagrati Upadhyay, Kumari Pragya and Subrata Goswami
Department of Plant Pathology	Subhash SS Malempati and Gitanshu Dinkwar
Department of Plant Physiology	Supriya Debnath, Sachin Prakash Nagre, Salikram Mohare, Satyendra Thakur and Rohit Kumawat

Published by

Scientific Publishers, *Granthlok,* 1st Floor, 5-A, New Pali Road, Jodhpur -342001, Rajasthan, India

Correct Citation

Das SB, D Pahalwan, SB Nahatkar and D. Khare (2023) Agro-Biodiversity Exploration. Scientific Publisher, Jodhpur, Raj. **256p**
Cover page photograph: Courtesy MS Bhale

डॉ. प्रदीप कुमार बिसेन
पूर्व कुलपति
Dr. Pradeep Kumar Bisen
Ex. Vice Cancellor

जवाहरलाल नेहरू कृषि विश्वविद्यालय
कृषि नगर, आधारताल, जबलपुर ४८२ ००४ (म.प्र.)
Jawaharlal Nehru Krishi Vishwa Vidyalaya
Krishi Nagar, Adhartal, Jabalpur 482 004 (M.P.)

Foreword

Fast urbanization in India is driving biodiversity decline in cities. Therefore, to conserve biodiversity in urban ecosystems, urban green spaces have to be maintained. For this University campuses may play an important role. The campus of Jawaharlal Nehru Krishi Vishwa Vidyalaya, Jabalpur has 623.66 ha land distributed in six locations with three types of soils. 90.48% land is utilized for cultivation in both the season. It is home for great biodiversity and plays a role of lungs for the city of Jabalpur. However, a systematic review on the biodiversity of the university campus is still lacking to provide refugia for a diverse set of plant and animal species.

The team conducted a comprehensive review covering the diversity and distribution patterns of biodiversity and availability of natural resources on the university campus at Jabalpur. The campus harboured high biodiversity, with forest tree, fruit tree, shrubs, field crops, vegetable crops, ornamental plants, medicinal plants, aromatic plants and weeds; insects, mites, spiders and pathogens viz., fungi, bacteria, virus, nematodes including some endemic and endangered species. Hence, the university campus provides a unique opportunity for urban biodiversity research, conservation and education, as well as connecting the public with nature through citizen science.

There are more than 26,000 universities in the world (http://www.webometrics. info/en/node/54). The university campus is home to naturalists to provide a direct link to biodiversity research and education. At the same time, students at the university educated region-specific biodiversity to increase appreciation of regional biodiversity and natural experiences. As such, living biodiversity on campuses, such as plants, insects, pathogens etc have been recorded, protected and studied for educational and academic purposes by students and researchers.

The importance of biodiversity is being acknowledged in national policy. The team of the teachers and scientists worked meticulously to translate into action to address the biodiversity present and provide a baseline to work out the loss over the years in future. It will help to plan on strengthening in situ and ex situ conservation measures.

The Biodiversity audit performed at headquarter will make a valuable contribution and raise awareness on importance of biodiversity for agriculture and conserve nature. A remarkable exercise performed by the team and is highly appreciable. I hope this exercise will be a milestone in the history of JNKVV, Jabalpur and open many avenues for research.

(P.K. Bisen)

Ph.: 0761-2681706 (O) ; Fax: 0761-2681389; E-mail: bisenvcjnkvv@gmail.com ; web: www.jnkvv.org

प्रो. प्रमोद कुमार मिश्रा
कुलपति
Prof. Pramod Kumar Mishra
Vice Chancellor

जवाहरलाल नेहरू कृषि विश्वविद्यालय
कृषि नगर, आधारताल, जबलपुर ४८२ ००४ (म.प्र.)
Jawaharlal Nehru Krishi Vishwa Vidyalaya
Krishi Nagar, Adhartal, Jabalpur 482 004 (M.P.)
Ph.: 0761-2681706 (O), Fax: 0761-2681389
E-mail: mishravcjnkvv@gmail.com

Foreword

Incentive measures for biodiversity valuation and conservation needs to be start from institutional level. Institutions like educational institutes, rural institutes such as Gram Panchayat etc. In this direction efforts were made to compile and categorise scientifically the institutional level biodiversity by our scientists and editors of this book and bring this information to public domain. This information will definitely be useful for natural resources management from the point of view of its sustainable use at institutional level and replicating the same at macro level for the benefit of commons.

Institutional biodiversity encompasses plants, microbes, valuable germplasm of field crops, horticulture and medicinal plants etc. After compilation of this information the valuation of these precious biodiversity will be future task for our scientists. These biodiversity resources can be valued as "Value in Use" and "Value in Exchange" for generating internal resources for the benefit of the University.

I would like to congratulate all the contributors and editors for bringing this valuable informative on institutional level biodiversity at one platform in the form of this publication.

(P.K. Mishra)

Preface

The team conducted a comprehensive review covering the diversity and distribution patterns of biodiversity and availability of natural resources on the university campus at Jabalpur. The campus harboured high biodiversity, with 239 plant species including forest tree, fruit tree, shrubs, field crops, vegetable crops, ornamental plants, medicinal plants, aromatic plants and weeds; animal species that includes insects; earthworm; birds; dogs; cattle; snakes; and pathogens *viz.*, fungi, bacteria, virus, nematodes including some endemic and endangered species. Hence, the university campus provides a unique opportunity for urban biodiversity research, conservation and education, as well as connecting the public with nature through citizen science.

Fast urbanization in India is driving biodiversity decline in cities. Therefore, to conserve biodiversity in urban ecosystems, urban green spaces have to be maintained. For this, the University campuses may play an important role. The campus of Jawaharlal Nehru Krishi Vishwa Vidyalaya, Jabalpur has 623.66 ha land distributed in six locations with three types of soils. 90.48% land is utilized for cultivation in both the season. It is home for great biodiversity and plays a role of lungs for the city of Jabalpur. However, a systematic review on the biodiversity of the university campus is still lacking to provide refugia for a diverse set of plant and animal species. The University campus is an important component of green spaces in urban environment of the city of Jabalpur where private green spaces are lacking. Total **19422** trees of 239 species and 59 families represent 12% area of the campus. Out of that 52% species are exotic represented by 59.61% trees. Jabalpur campus hosts nearly 36 trees / hectare and 7.71 tree/head.

There are more than 26,000 universities in the world (http://www.webometrics.info/en/node/54). The university campus is home to naturalists to provide a direct link to biodiversity research and education. At the same time, students at the university educated region-specific biodiversity to increase appreciation of regional biodiversity and natural experiences. As such, living biodiversity on campuses, such as plants, insects, cattle *etc.* have been recorded, protected and studied for educational and academic purposes by students and researchers. Biodiversity on the university campus has been directly connected to the public.

Agriculture contributes about 16% to total greenhouse gas (GHG) emissions in India. The Indian agriculture emits 417.2 Mt CO_2 of total GHG emissions that comes from-enteric fermentation, 54%; agricultural soils, 19%; rice cultivation, 18%; manure management, 7%; and field burning and agricultural residue, 2%.

The University has reported a total of 2512 species from Jabalpur campus that includes 243 trees, 77 shrubs and annuals, 19 field crops, 1043 medicinal plants, 158 weeds, 497 insects, 371 fungi, 20 bacteria, 19 virus, 8 nematodes, 1 phytoplasma, 16 beneficial microorganism and 40 spiders.

Jabalpur
January 6, 2023

SB Das
Deep Pahalwan
SB Nahatkar
Dhirendra Khare

Acknowledgement

Salutation to the Scrupulous Concert of the Every Member of the Audit Team

A crew of scientists has shown a meticulous determination to perform the Biodiversity Audit of the University headquarters i.e., an audit of trees, shrubs, crops, weeds, insects, spiders, mites, fungi, bacteria, viruses, etc including an audit of soil, water, energy, and climate.

The University campus is an important component of green spaces in the urban environment of the city of Jabalpur where private green spaces are lacking. Total **19422** trees of 243 species and 59 families represent 12% area of the campus. Out of that 52% of species are exotic represented by 59.61% of trees. Jabalpur campus hosts nearly 36 trees/hectare and 7.71 trees/head. Biodiversity on the university campus has a direct connection to the public.

Agriculture contributes about 16% to the total greenhouse gas (GHG) emissions in India. Indian agriculture emits 417.2 Mt CO_2 of total GHG emissions that come from-enteric fermentation, 54%; agricultural soils, 19%; rice cultivation, 18%; manure management, 7%; and field burning and agricultural residue, 2%.

The University has reported a total of 2512 species from the Jabalpur campus that includes 243 trees, 77 shrubs and annuals, 19 field crops, 1043 medicinal plants, 158 weeds, 497 insects, 371 fungi, 20 bacteria, 19 viruses, 8 nematodes, 1 phytoplasma, 16 beneficial microorganism, and 40 spiders. This is the baseline prepared for the next generation to compare.

The Audit Team is thankful to the Hon'ble Vice Chancellor, JNKVV, Jabalpur, Director Research Services, Director Extension Services, and other authorities for unprecedented support to perform this exercise.

Jabalpur
January 6, 2023

SB Das
Deep Pahalwan
SB Nahatkar
Dhirendra Khare

Content

Medicinal & Aromatic crops

Introduction

Audit of the natural resources viz., trees, shrubs, crops, weeds, fungi, bacteria, virus, nematodes, soil microorganism, water, energy, climate have been performed of the Jabalpur campus of 463.06 hato provide a baseline.

Trees are the oldest living organisms on earth, and they never die because of old age. Russia has the largest number of trees in the world, followed by Canada, Brazil, then America, and then India where only 35 arab trees are left. India has only 28 tress per person. 32,000 ha of forests disappear from the earth every day. As per definition of forest 33% area shall covered under forest. Forest Survey report 2021 reported, the total forest and tree cover of the country is 80.9 million hectare which is 24.62 per cent of the geographical area of the country. Increase of 2,261 sq km in the total forest and tree cover of the country have been reported in last two years. As compared to the assessment of 2019, there is an increase of 2,261 sq km in the total forest and tree cover of the country. Out of this, the increase in the forest cover has been observed as 1,540 sq km and that in tree cover is 721 sq km. Area-wise 25.14% area of Madhya Pradesh ie., 77,493 sq Km has the largest forest cover in the country. Trees planted in one ha absorb as much carbon dioxide in a year as a car generates after driving 1,00,000 km. One ha of forest absorbs 15 tons of carbon dioxide and puts out 10 tons of oxygen. This is enough to meet the annual needs of 45 people. One tree gives enough oxygen in a day require by four people to survive. By planting 20 million trees, the earth and its people will be provided with 260 million more tons of oxygen. Those same 200 lakh trees will remove 100 lakh tons of CO_2. 22 LBS of carbon dioxide is sequestered every year by a mature tree. Total carbon stock in country's forest is estimated to be 7,204 million tonnes, an increase of 79.4 million.

Trees, shrubs and turf filter air by removing dust and absorbing other pollutants like carbon monoxide, sulfur dioxide and nitrogen dioxide. Row of trees reduces dust pollution up to 75% and noise pollution by 50%. Leaves absorb and filter the sun's radiant energy, keeping things cool in summer. Trees also lower the air temperature and reduce the heat intensity of the greenhouse effect by maintaining low levels of carbon dioxide. A tree produces as much cold as 1 A.C. does in 10 rooms for 20 hours. The area which is surrounded by trees remains 9 degrees colder than other areas. Three trees placed strategically around a single-family home can cut summer air conditioning needs by up to 50 %. By reducing the energy demand for cooling our houses, we reduce carbon dioxide and other pollution emissions from power plants. In one day, one large tree can lift up to 100 gallons of water out of the ground and discharge it into the air as oxygen and water vapour. Every year 5 arab trees are being planted but every year 10 arab trees are being cut. 50 lakh ha of land have been converted for cultivation of traditional crops from forest. 1crore 20 lakh ha to 1 crore 50 lakh ha of forests are lost every year i.e., equivalent to 48 football ground every minute and one football ground every 1.2 second. 28000 species are expected to become completely extinct in next 25 years due to deforestation. 1.6 arab People depend on trees for their livelihood. 80 % of the world's terrestrial plants and animals live in forest and are losing their habitants due to deforestation.

Audit on water one of the most important natural resource utilized to grow a crop is indispensable. Therefore, an inventory of water resources and its usages since inception have been prepared. Goals to improve efficient use of water with cost effective measures have been discussed.

The general energy audit is conducted to know the consumption, make a plan to reduce the amount of energy input into the system without negatively affecting the output and reduce carbon footprint.

The team has placed a great effort in auditing of weeds, Arthropods, Plant Pathogens, Bio-agents, soil and soil microorganisms. It is a baseline that may involve many dimensions in future to update and improve the audit on natural resources.

We associate
The roots of the peepal tree with Lord Brahma
The leaves with Lord Shiva
The Trunk with Lord Vishnu
The text of the sacred Bhagavad Gita states that
Lord Krishna identifies with the peepal tree and says
"Among trees, I am Ashvatha".

Ten wells equal a step well
Ten step wells equal a tank
Ten tanks equal a son
Ten sons equal a tree.

Matsya Purana 512

Those who plant trees in deserted and difficult to reach places
Give salvation to their previous and future generations.

Shiva Purana Uma Samhita 11/7

Those who want better future should plant good trees around tanks, and raise them like sons, as the trees are considered like sons in our religion.

Mahahbharat, Anu Parva 58/31

Trees are like gentlemen who help others.
They stand in the sun but give shade to others.
Their fruit is also for the use of others.

Vikram Charitam 65

The five biggest favours bestowed by trees are like five maha yagyas

They provide *Pranvayu* to life on earth
They offer fuel to the families
They give shade and resting place to the travellers
They arrange nests to the birds
They provide medicines from their leaves, roots and bark

Varaha Purana 162-41-42

Trees fulfil the desires of human beings by their leaves, flowers, shade, root, bark, wood, fragrance, gum, ash, charcoal, buds and new leaves.

Srimadbhagvat, Skandh, a-22 Shloka 34

By planting following, one can never go to hell therefore participate in plantation and care to establish a tree

One peepal (*Ficus religiosa*)
One neem (*Azadirachta indica*)

One bargad (*Ficus bengalensis*)
Two pomegranates (*Punica granatum)*
Two oranges *(Citrus sinensis)*
Three kaith (*Limoni aacidissima*)
Three vilva (*Aegle marmalos*)
Three aonla (*Phyllanthus embilica*)
Five mango (*Mangifera indica*)
Ten Imli (*Tamarindus indica*)
Ten flowering trees

Bhahvishya Purana, Rajdharma Kaustubh Khand, Varaha Purana 12-2-39

Those who donate a tree are given salvation by the same trees in the nether world.

Mahahbharat Anu Parva 58/30

Aonla (Indian gooseberry) is holiest of all fruits in the world.
Therefore it is important to plant aonla trees
By planting it, men and women get rid of the cycle of life, death and rebirth in this world.

Padma Purana, Srishti Khand, ch. 26, Shloka 2

O Tree!
You bear the burden of flowers, leaves and fruits, give relief to the people from sun, and give them succour during winter.
Thus you dedicate your entire life in helping others.
That is why you are greater than a gentleman.
Therefore
O Tree! Please accept my regards.

Bhamini Vilasa-86

Those who plant trees, for them they are like sons. There is no doubt that because of those trees, man attains heaven after his death.

Mahabharata Anuparva 58/27

CHAPTER 1

Site Map of the University Campus at Jabalpur

Shailesh Sharma
Department of Soil & Water Engineering
College of Agricultural Engineering, Jabalpur

S.S. Baghel
Department of Soil Science & Agricultural Chemistry
College of Agriculture, Jabalpur

The land at Jabalpur was transferred by the Government of Madhya Pradesh to College of Agriculture, Jabalpur on 22 July, 1955 and then to the University on October 1, 1964. The University has total 605.29 ha land at Jabalpur out of this 61.75 ha land was transferred to ICAR- Directorate of Weed Research, Jabalpur. From the remaining land 34.18 ha is utilized for establishment of both the colleges including hostels & administrative buildings and 25.43 ha for residential colony (Table 1.1). Total area under cultivation in the jurisdiction of the University at Jabalpur is 463.06ha. In all seven farms are available at Jabalpur under the University ear marked for specific purpose. Total 94.14% land allotted to the University is utilized for cultivation at Jabalpur. The whole land under cultivation remains green in both the seasons.

Krishi Nagar Farm of 69.85 ha, an establishment of year 1912 was transferred to the College of Agriculture, Jabalpur in the year 1955. An orchard of 13 ha of jackfruit and mango located at the other side of the railway line was transferred to Vehicle Factory, Jabalpur before establishment of the University (Plate 1.1).

Table 1.1 Distribution of the allotted land to the University at Jabalpur

Sr	Name of Unit	Total Area (ha)	Built up Area (ha)	Length of Farm Road (m)	Farm Road (ha)	Total cultivable land (ha)
1	KN Research Farm	56.85	6.12	2909.87	0.73	50.00
2	Seed Breeding Farm	119.58	1.63	6267.26	1.57	116.38
3	Fruit Research Farm, Imaliya	22.47	0.57	942.44	0.24	021.66
4	Vegetable Research Farm	84.64	1.43	5185.70	1.30	080.69
5	Forest Research Farm, Imaliya	6.23				6.23
6	Seed Farm Soybean	98.88	2.78	5065.59	1.27	094.83
7	Dusty Acre (Seed Farm Groundnut)	72.62	0.75	5055.41	1.26	070.61
9	Adhartal Tank including Fisheries Ponds	22.66				022.66
A	**Total**	**483.93**	**13.28**		**6.37**	**463.06**
10	College of Agriculture	25.58	8.87			
11	College of Agricultural Engineering	5.79	2.50			
12	New Hostel Suhagi	2.81	0.80			
B	**Total Colleges**	**34.18**	**12.17**			
C	**Krishi Nagar**	**25.43**				
	Grand total (A+B+C)	**543.54**	**25.45**	**25426.27**	**6.37**	**511.72**

Plate 1.1 Map showing distribution of the allotted land to the University at Jabalpur

Plate 1.2 The primogenital tree of the campus – The place where Acharya Rajneesh organized meditation camp

Administrative Building

The building hosted Vice Chancellor's secretariat, Office of the Registrar, Office of the Comptroller, Directorate of Instrumentation and Central Library (Plate 1.3).

Plate 1.3 University headquarters at Jabalpur

College of Agriculture, Jabalpur

Out of total 25.58 ha area allotted to the College 8.87 ha is under buildings. The College hosted 10 Departments, Auditorium 8 classrooms, examination halls and Office of the Dean in the main building apart from garden, botanical garden, medicinal garden, sports complex, central library, canteen, road, parking etc. (Plate 1.4).

Plate 1.4 College of Agriculture, Jabalpur

College of Agricultural Engineering, Jabalpur

Out of total 5.79 ha area allotted to the College 2.50 ha is under buildings. The College hosted 4 Departments, examination halls and Office of the Dean in the main building apart from workshop, garden, KVK, ATIC, new administrative building, play ground, canteen, road, parking etc. (Plate 1.5).

Plate 1.5 College of Agricultural Engineering, Jabalpur

The University Sports Complex

A well-equipped sports complex, with facilities for gymnasium, table tennis, badminton etc and all the outdoor facilities i.e., track and field events viz., races, jumps and throws; volleyball, kabaddi, kho-kho, cricket, hockey, football etc exists in the University with swimming pool (Plate 1.6).

Plate 1.6 The University Sports Complex

Residential Colony- Krishi Nagar

It is constructed in 25.43ha land hosting 433 residential quarters, guesthouse, play grounds, gardens dispensary, shopping complex, works section etc. It is a life time experience to live in isolated place close to nature (12% area covered with trees) for common cause with all the basic facilities (Plate 1.7).

Plate 1.7 Residential Colony Krishinagar

The University Guesthouse Complex

The University has well-furnished three guest houses to accommodate 46 persons apart from accommodation of 100 in Farmers hostel and 48 in Dormitory (Plate 1.8).

Plate 1.8 The University Guesthouses

Hostel

Jabalpur campus has well-furnished 16 hostels out of that 8 are for boys, 8 for girls and two international hostels to host total 779 students(Plate 1.9).

College	Boys	Girls	Total
College of Agriculture, Jabalpur	387	208	595
College of Agricultural Engineering, Jabalpur	84	72	156
International Hostel	12	16	28
Total	**483**	**296**	**779**

Plate 1.9 Hostels of the Colleges at Jabalpur

Farms of the University at Jabalpur

Fruit Research Farm

It is a farm of 22.47ha located in village Imaliya at the bank of perennial river Pariyat on west side of the road. The whole farm is dedicated for research on fruit crops. Orchards of mango, guava, pomegranate, papaya, citrus etc with forest of timber, poly houses and nursery are maintained for curricular research and research of All India Coordinated and other projects and Experiential Learning Programme on nursery management (Plate 1.10).

Plate 1.10 Map of Fruit Research Farm, Imaliya, Jabalpur

Vegetable Research Farm

It is a farm of 84.64ha located in village Maharajpur on east side of the road. The part of the farm is utilized for curricular and other All India Coordinated Research Projects on vegetables. Nearly 3/4 part of the farm is used for quality seed production of field crops (Plate 1.11).

Plate 1.11 Map of Vegetable Research Farm, Maharajpur, Jabalpur

Plate 1.12 Timeworn trees at Vegetable Research Farm, Maharajpur, Jabalpur

Forest Research Farm

It is a farm of 6.23ha located in village Imaliya at the bank of perennial river Pariyat on east side of the road. The whole farm is dedicated for curricular and other research on forest (Plate 1.13).

Plate 1.13 Forest Research Farm, Imaliya, Jabalpur

Seed Breeding Farm Soybean

A farm of 98.88ha located in the east of the University is ear marked as Seed Breeding Farm Soybean. The whole farm is dedicated for seed production and curricular research (Plate 1.14).

Plate 1.14 Seed Breeding Farm, Soybean

'Dusty Acre'- Seed Breeding Farm Groundnut

A farm of 72.62 ha is dedicated for seed production and curricular research (Plate 1.15).

Broadcast My Ashes in the Farm of the University

Farnham, Cecil H. "Dusty" (1907-1968)

Dr. CH Dusty Farnham from Department of Agronomy University of Illinois visited JNKVV in the year 1966.
He was very much attached with the University, city Jabalpur, river Narmada and Indian culture.
He expressed in his will that
on my death the cremation shall be as funeral and ashes shall be sprayed over the farms of the University. Unfortunately in the year 16 July 1968 Dr. Farnham left us to rest in peace.
As per his wishes, cremation of Dr. Farnham was organized by funeral and ashes were spread over the farms of the University. From that day the farm is named as ***Dusty area.***

Plate 1.15 'Dusty Acre'- Seed Breeding Farm Groundnut

Krishi Nagar Farm (Crop Production, Organic Farming and Natural Farming)

Krishi Nagar Farm of 69.85 ha, an establishment of year 1912 was transferred to the College of Agriculture, Jabalpur in the year 1955 with well net perennial canal for irrigation. An orchard of 13 ha of jackfruit and mango located at the other side of the railway line was transferred to Vehicle Factory, Jabalpur before establishment of the University.

Now 56.85ha, farm is utilized for cultivation of rice and wheat with canal irrigation. In the farm land a part is ear marked for research on organic and natural farming apart from research on crop production of field crops, Dairy and nursery (Plate 1.16).

Plate 1.16 Krishi Nagar Farm, Jabalpur

Seed Breeding Farm (Maintenance Breeding, Instructional Unit and Research Farm)

A well-equipped farm of 119.58 ha land located near Adhartal Tank is ear marked by the University for Maintenance Breeding, Instructional and experimentation of nearly 10 All Indian Coordinated Research Projects. This farm is the cradle of crescent for introduction of Soybean as oilseed crop for cultivation in Kharif season with exemplary yellow revolution in the state of Madhya Pradesh (Plate 1.17).

Plate 1.17 Seed Breeding Farm, Jabalpur

Adhartal Tank

An age old tank of 22.66ha built by Adhar Singh Minister of Rani Durgawat is under the administrative control of the University to utilize for agricultural purpose (Plate 1.18).

Plate 1.18 Adhartal tank, Jabalpur

Ambient Air Monitoring Report of Pollution Control Board of the University campus on 29.10.2020

Under National Air Quality Monitoring Programme four air pollutants viz., Sulphur Dioxide (SO2), Oxides of Nitrogen as NO2, Respirable Suspended Particulate Matter (RSPM / PM10) and Fine Particulate Matter (PM2.5) are identified for regular monitoring at all the locations. The monitoring of meteorological parameters such as wind speed and wind direction, relative humidity (RH) and temperature are also integrated with the monitoring of air quality.

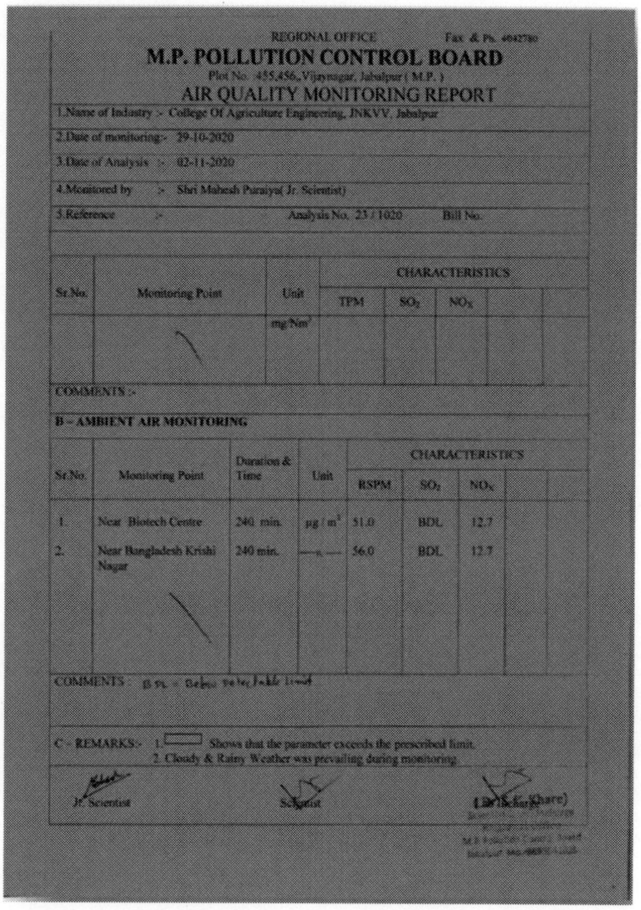

REGIONAL OFFICE Fax & Ph. 4042780

M.P. POLLUTION CONTROL BOARD

Plot No. 455,456,,Vijaynagar, Jabalpur (M.P.)

AIR QUALITY MONITORING REPORT

1.Name of Industry :- College Of Agriculture Engineering, JNKVV, Jabalpur

2.Date of monitoring:- 29-10-2020

3.Date of Analysis :- 02-11-2020

4.Monitored by :- Shri Mahesh Puraiya(Jr. Scientist)

5.Reference :- Analysis No. 23 / 1020 Bill No.

Sr.No.	Monitoring Point	Unit	CHARACTERISTICS				
			TPM	SO_2	NO_X		
		mg/Nm³					

COMMENTS :-

B – AMBIENT AIR MONITORING

Sr.No.	Monitoring Point	Duration & Time	Unit	CHARACTERISTICS				
				RSPM	SO_2	NO_X		
1.	Near Biotech Centre	240 min.	μg / m³	51.0	BDL	12.7		
2.	Near Bangladesh Krishi Nagar	240 min.	—"—	56.0	BDL	12.7		

COMMENTS : BDL = Below Detectable limit

C – REMARKS:- 1. ▭ Shows that the parameter exceeds the prescribed limit.
2. Cloudy & Rainy Weather was prevailing during monitoring.

Jr. Scientist Scientist I/C In-charge

Respirable Suspended Particulate Matter (RSPM / PM10) : PM with particle sizes less than 10 microns, which are known as respirable suspended particulates or PM10, can get deep into the lungs and cause a broad range of health effects, in particular, respiratory and cardiovascular illnesses. Technically, there isn't a safe level of PM10, as any amount of particulate matter in air isn't a good. Keeping exposure to PM10 concentrations below 54.0 μg/m³ is the best way to prevent any short or long-term health effects from developing. RSPM is 53.5 in the campus which is safe but the University shall take steps to keep it low.

Sulphur Dioxide (SO_2): Exposure to elevated level of sulphur dioxide can cause impairment of respiratory function, aggravation of existing respiratory disease (especially bronchitis) and cardiovascular disease. In the campus Sulphur Dioxide is below Detectable Limit, it shall not be more than 30 μg/m³, hence it is safe.

Oxides of Nitrogen (NO_x): Its low level can irritate the lungs and lower resistance to respiratory infection such as influenza. Individuals with respiratory problems, such as asthma, are more susceptible to the effects. In the campus Oxides of nitrogen is 12.7.

CHAPTER 2

Diversity Audit of Soil Types

HK Rai
Department of Soil Science and Agricultural Chemistry
College of Agriculture, Jabalpur

Soil, a dynamic living matrix, is an essential part of the terrestrial ecosystem. It is a critical resource not only to agricultural production and food security but also to the maintenance of most life processes. It is a source as well as sink to adsorbed, desorb, fix or release mineral elements and gases and grow plants and decompose crop residues incorporated in to it. Soils are classified into soil orders (12), sub-orders (63), great groups (>240), sub-groups (>1000), families, and series on the basis of distinguished characteristic features. The detail accounts on soil resources at J.N.K.V.V headquarter are given below (Table 2.1):

Table 2.1 Description of the soil type at Jabalpur campus

Sr.	ha	Soil type					
1.	**Krishi Nagar Research Farm**						
	56.85	Alfisol	Typic and vertic haplusterts	Fine loamy and mixed	Hyper-thermic	Sandy clay loam and sandy loam	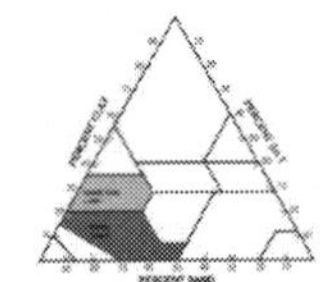
2.	**BSP- FC (Adhartal Tank)**						
	119.58	Vertisol	Typichaplusterts	Very fine mont-morrilonite	Hyper-thermic	Clayey	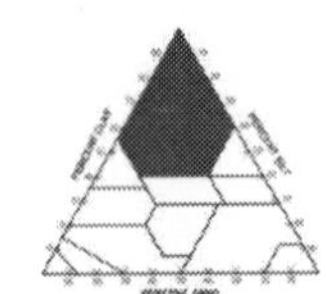
		Inceptisol	Verticustochrept	Fine mixed	Hyper-thermic	Clayey	
3.	**Fruit Research Farm, Imaliya**						
	22.47	Vertisol	Typichaplusterts	Very fine mont-morrilonite	Hyper-thermic	Clayey	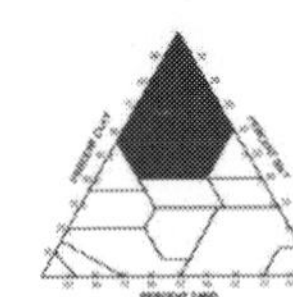
4.	**Vegetable Research Farm, Maharajpur**						
	84.64	Vertisol	Typichaplusterts	Very fine mont-morrilonite	Hyper-thermic	Clay loam, sandy clay and sandy clay loam	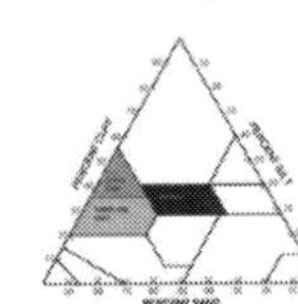
5.	**Forest Research Farm, Imaliya**						
	6.23	Vertisol	Typichaplusterts	Very fine mont-morrilonite	Hyper-thermic	Clayey	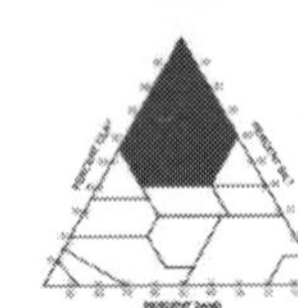
6.	**BSP Soybean**						
	98.88	Alfisol	Vertichaplusterts	Fine loamy and mixed	Hyper-thermic	Clayey	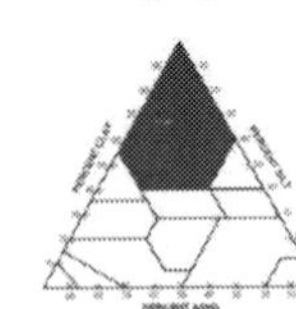
7.	**BSP Groundnut**						
	72.62	Inceptisol	Typicustochrept	Fine loamy and mixed	Hyper-thermic	Sandy loam to sandy clay loam	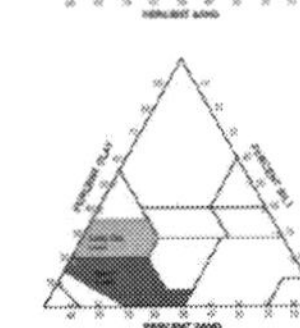
8.	**Adhartal Tank (Fisheries)**						
	22.66	Vertisol	Typichaplusterts	Very fine mont-morrilonite	Hyper-thermic	Clayey	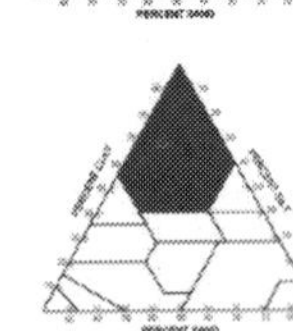

Representation of Soil order at Jabalpur campus in relation to the State

Soils at Jabalpur represents 52% area of the State of Madhya Pradesh based on soil order i.e., vertisols represents 37%, Alfisols 18% and Inceptisols 7% (Fig 2.1).

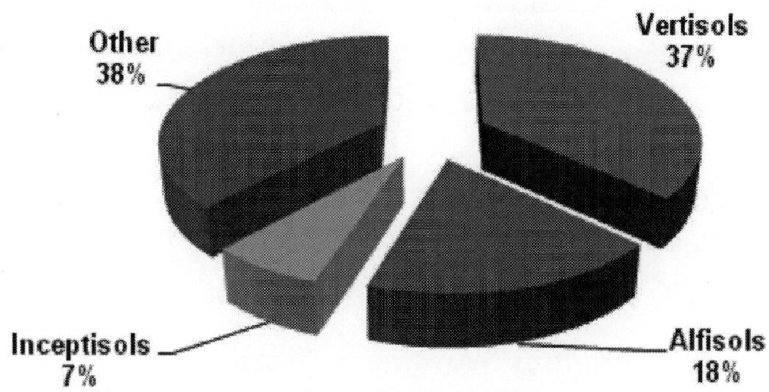

Fig 2.1 Representation of soil order at Jabalpur campus in relation to the State

Description of soil at Jabalpur campus

Vertisols

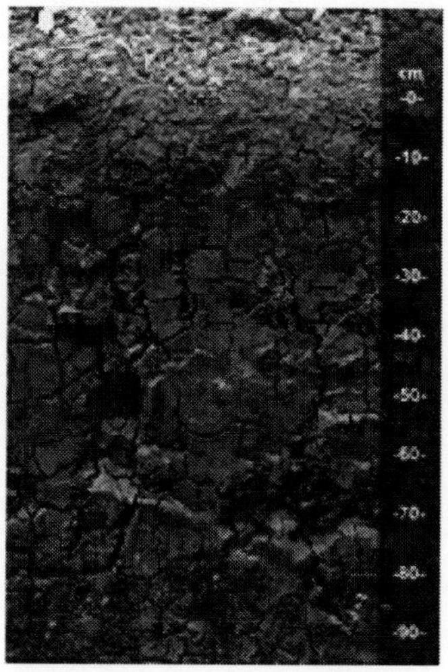

It is the most dominant soil order of Madhya Pradesh spread over an area of 16.21 m ha (36.53%) across Jabalpur, Narsinghpur, Hoshangabad, Damoh, Sagar, Raisen, Guna, Vidisha, Bhopal, Rajgarh, Shajapur, Sehore, Dewas, Khandwa, Khargone Indore, Ujjain, Dhar, Ratlam and Mandsaur districts. These soils have high clay content (40-60%) and the dominated by Montmorrilonite clay minerals of Smectite group. Most of the soils in this order are calcareous with neutral to slightly alkaline in reaction, high in cation exchange capacity [33-55 cmol (p^+) kg^{-1}], sticky and plastic when wet and shrink on drying. Black colour of these soils is due to presence of titeniferous magnetite minerals. These soils pose high water retention, low infiltration and narrow workability. Predominantly grown crops in these soils are wheat, cotton, groundnut, sorghum, rice, pulses and oilseeds including soybean.

Alfisols

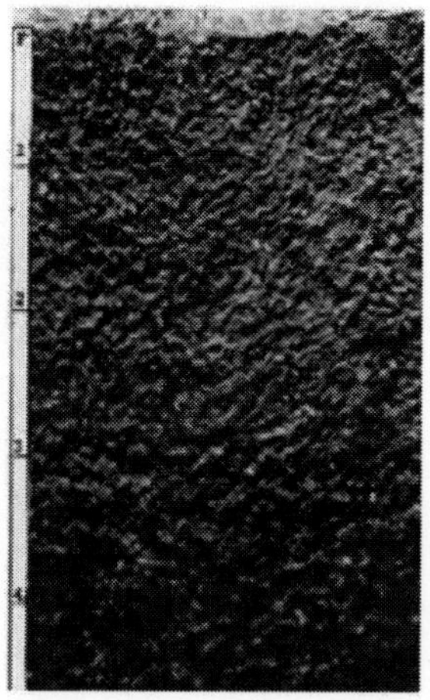

Alfisol forms in semi-arid to humid areas have a clay-enriched sub-soil and relatively high native fertility. "Alf" refers to aluminium and iron. In Madhya Pradesh Alfisol is spread over an area of about 8.11 m ha (18.30%) and mostly found in Satna, Rewa, Sidhi, Panna, Chhatarpur and Tikamgarh districts. These soils have also been found in patches under Bhind, Morena, Gwalior, Datia, Shivpuri and Balaghat districts. Major crops grown in these soils are rice, wheat, chickpea, millets, sesame, linseed, oilseeds etc

Inceptisols

In the state of Madhya Pradesh Inceptisol spread over an area of about 3.06 m ha (6.91%) mainly found in Chhindwara, Seoni and Betul districts. In patches these soils are also found in northern part of Jabalpur and southern parts of Sagar, Damoh, Khandwa and Khargone districts. Soil structure varies from granular to weak fine sub-angular blocky and lime is usually present in the form of fine grains or nodules in C horizon. Major crops grown in these soils are soybean, wheat, chickpea, sorghum, maize, lentil, linseed, fruits and vegetables etc.

Diversity of soil at JNKVV campus Jabalpur

In all 67% cultivated land of Jabalpur campus is Typichaplusterts i.e., 325.26 ha and 33% Vertichaplusterts (158.67ha); 40% Vertisols (195.79 ha), 32% Alfisol(155.73 ha) and 27% Inceptisol (132.41 ha); 47 % Fine loamy and mixed (228.35 ha); 41% land has very fine mont-morrilonite (195.79 ha), and 12% Fine mixed (59.79 ha); 88% soil is Hyper-thermic (424.14 ha) whereas 423% is clayey (210.03 ha) and 44% (214.11 ha) Sandy loam to sandy clay loam (Table 2.2, Fig 2.2 and 2.3).

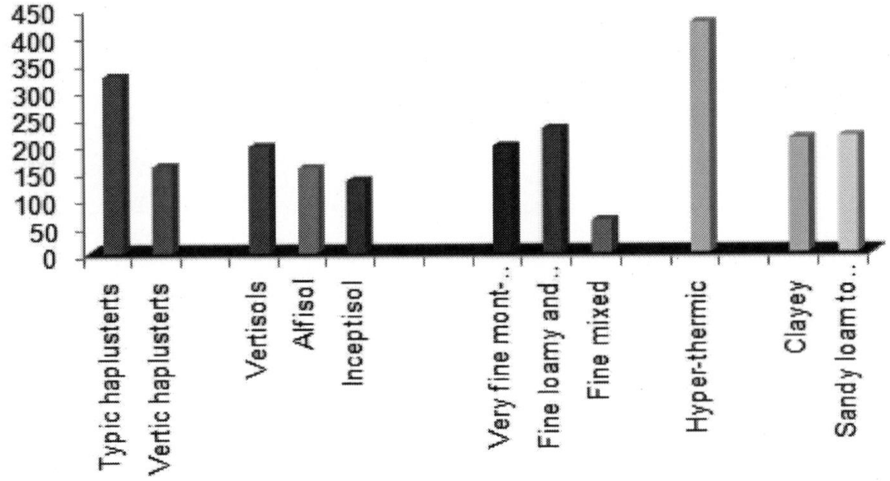

Fig 2.2 Classification of land in ha based on soil parameters at JNKVV campus

Table 2.2 Classification of land in ha based on soil parameters

Property	Ha
Typichaplusterts	325.26
Vertichaplusterts	158.67
Vertisols	195.79
Alfisol	155.73
Inceptisol	132.41
Fine loamy and mixed	228.35
Very fine mont-morrilonite	195.79
Fine mixed	59.79
Hyper-thermic	424.14
Clayey	210.03
Sandy loam to sandy clay loam	214.11

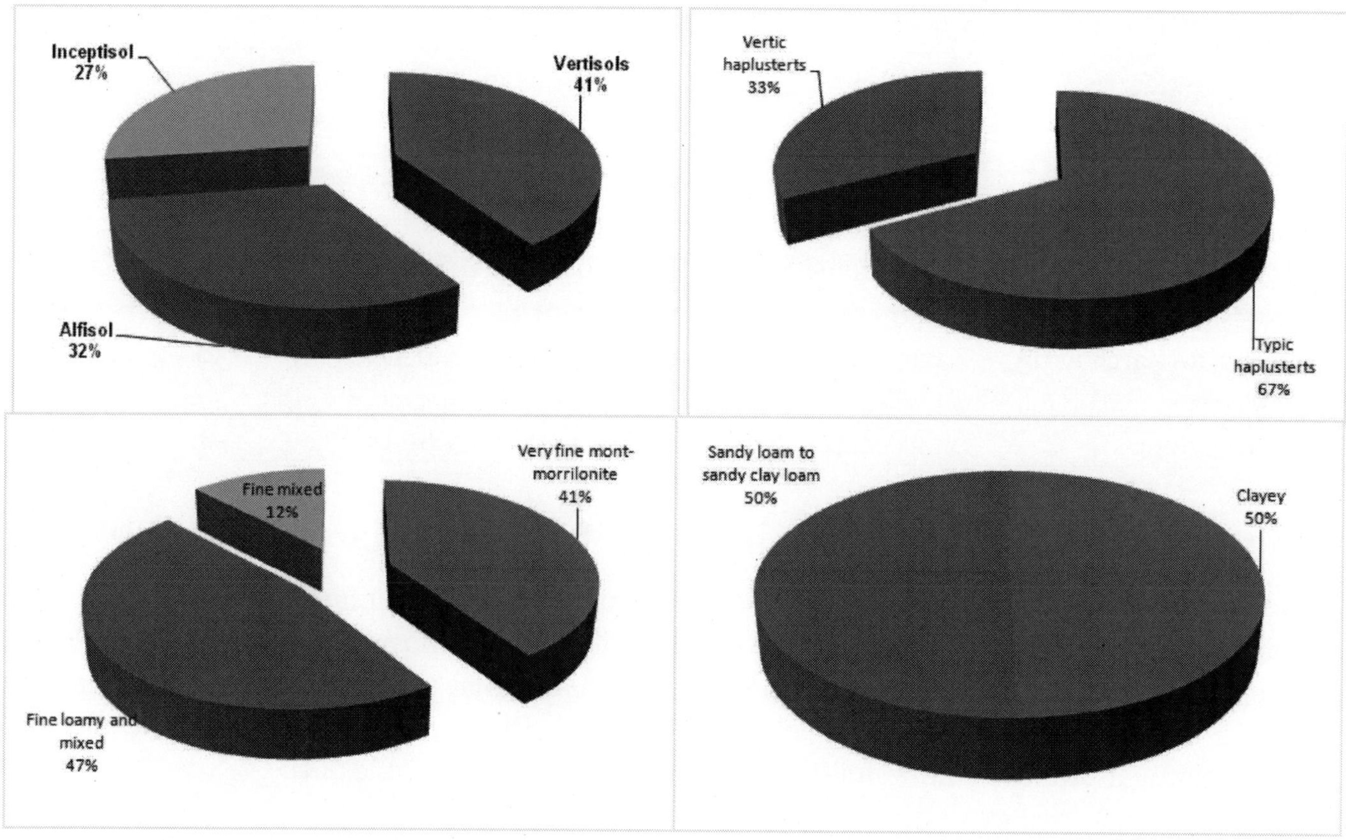

Fig 2.3 Percentage representation of land based on soil parameters

Properties and nutrients status in soils of Jabalpur campus

Properties and nutrients status in soils of Jabalpur campus was characterized by Dhakar (2017) and the findings are tabulated below:

Variability in soil properties of the University Farms atJabalpur

Bulk density of a soil indicates the degree of compactness and to some extent proportions of courser fractions and organic carbon content also. Variation in bulk density (BD) of soils at Krishi Nagar (1.33-1.40 Mg m^{-3}), BSP-FC (1.26-1.39 Mg m^{-3}) and Fruit Research (1.30-1.43 Mg m^{-3}) farms was less showing low compaction and good porosity, while at Vegetable Research (1.34-.1.53 Mg m^{-3}), BSP soybean (1.38-1.54 Mg m^{-3}) and BSP-Groundnut (1.35-1.49 Mg m^{-3}) farms were higher that revealed more compactness and presence of higher courser fractions and lower organic carbon content. Soils of Krishi Nagar, BSP-FC, Fruit Research and Vegetable Research farms exhibiting neutral reaction (pH), while the soils of BSP soybean and BSP-Groundnut farms showing slightly acidic reaction. It have been noted that presence of soluble salts (EC) in soils of different farms of Jabalpur campus was very less. Organic carbon content in soil across the farms was low (3.0 g kg-1) to high (11.9 g kg^{-1}). Whereas, cation exchange capacity of soil was maximum [56.6-68.4 {cmol (p^+) kg^{-1}}] at BSP-FC may be due to higher clay content. Soils at all the farms were calcareous in nature with higher content of calcium carbonate at BSP-FC and Fruit Research Farms (Table 2.3).

Table 2.3 Variability in soil properties of farms at Jabalpur

Farms	Bulk density (Mg m^{-3})	pH	EC (dSm^{-1})	OC (g kg^{-1})	CEC [cmol (p^+) kg^{-1}]	$CaCO_3$ (g kg^{-1})
Krishi Nagar Research Farm	1.33 - 1.40	6.01 - 7.70	0.07 - 0.27	5.0 - 11.9	38.1 - 49.2	07.5-17.5
BSP- FC	1.26 - 1.39	7.09 - 7.86	0.10 - 0.23	3.0 - 08.4	56.6 - 68.4	17.5-35.0
Fruit Research Farm	1.30 - 143	6.90 - 7.75	0.12 - 0.32	6.0 - 11.9	35.1 - 46.1	17.5-37.5
Vegetable Research Farm	1.34 - 1.53	6.29 - 7.23	0.04 - 0.12	3.0 - 11.1	48.1 - 56.1	05.0-37.5
BSP Soybean	1.38 - 1.54	5.70 - 6.65	0.04 - 0.17	4.4 - 11.1	45.2 - 59.2	12.5-27.5
BSP Groundnut	1.35 - 1.49	5.50 - 6.74	0.04 - 0.11	3.8 - 11.0	35.3 - 54.2	05.0-27.5

Variability in nutrients status in soils of Jabalpur farms

Major and secondary nutrients: It was noted that soils of different farms of Jabalpur campus contains low to medium available nitrogen, low to high available phosphorus, and medium to high in available potassium and sulphur. It have been also tested that available-K was highest at Fruit Research Farm, Imaliya followed by Vegetable Research Farm, Maharajpur. While, available-S was maximum at BSP Groundnut followed by Krishi Nagar Farm (Table 2.4).

Table 2.4 Variability for Major and secondary nutrientsin soils of Jabalpur farms

Farms	Variability in available nutrients (kg ha^{-1})			
	N	P	K	S
Krishi Nagar Research Farm	243-278	15.5-48.9	210-429	17.6-39.4
BSP- FC (Adhartal Tank)	208-298	6.5-33.33	227-421	13.2-28.6
Fruit Research Farm, Imaliya	230-285	8.7-21.5	484-745	10.4-35.3
Vegetable Research Farm, Maharajpur	208-298	5.2-36.9	297-650	10.4-47.5
BSP Soybean	208-298	12.5-46.9	219-392	15.7-38.7
BSP Groundnut	230-280	7.6-45.3	214-476	17.6-45.7

Micronutrients: It have been found that available-Zn at Fruit Research Farm, Imaliya and BSP Groundnut was medium to high, while at other farms it tested low to medium. Across the farms status of available-Cu was medium, whereas available-Fe and Mn contents found medium to high. However, available-B content in soils of different farms of Jabalpur campus was low to medium except at BSP Soybean farm where available-B was medium to high (Table 2.5).

Table 2.5 Variability for Micronutrients in soils of Jabalpur farms

Farms	Variability available micronutrients (mg kg^{-1})				
	Zn	Cu	Fe	Mn	B
Krishi Nagar Research Farm	0.58-2.52	0.82-3.25	26.2-43.7	2.2-18.3	0.41-2.28
BSP- FC (Adhartal Tank)	0.51-2.11	0.79-3.61	6.1-32.8	3.2-12.8	0.32-2.67
Fruit Research Farm, Imaliya	0.61-3.23	1.63-5.15	5.6-30.1	3.9-14.0	0.34-2.06
Vegetable Research Farm, Maharajpur	0.43-1.73	1.52-4.64	12.7-39.2	6.6-16.3	0.32-2.86
BSP Soybean	0.46-1.68	1.40-5.13	35.9-42.8	7.3-18.4	0.52-2.95
BSP Groundnut	0.63-2.34	1.08-4.17	21.2-43.9	6.2-18.3	0.34-2.83

CHAPTER 3

Irrigation Audit

MK Awasthi, RK Nema and RN Shrivastava
Department of Soil and Water Engineering
College of Agricultural Engineering, Jabalpur

Irrigation through Canal

The University had total 605.29 ha land at Jabalpur distributed in seven locations, out of this 511.72 ha is under cultivation and 61.75 ha have been transferred to ICAR- Directorate of Weed Research. The area was transferred to the College from Government of Madhya Pradesh for establishment of the College of Agriculture in the year 1955. The cultivated land is utilized for seed production, AICRP Research, curricular research, botanical garden, orchard, etc.

At the time of establishment in the year 1955, the source of irrigation was canal at Krishi Nagar Farm (56.85 ha), Urdana Nala for Dusty acre and Adhartal farm, Vehicle nala near Adhartal railway station for Vegetable Farm Maharajpur, Pariyat river for Horticulture Orchard at Imaliya apart from Adhartal tank of 22.66 ha for Adhartal farm and open well.

Canal Irrigation

Krishi Nagar Farm of 69.85 ha, an establishment of year 1912 was transferred to the College of Agriculture, Jabalpur in the year 1955 with well net perennial canal for irrigation. An orchard of 13 ha of jackfruit and mango located at the other side of the railway line was under the College establishment known as Richhai Farm. It was transferred to Vehicle Factory, Jabalpur before establishment of the University. From this farm a canal offshoot from Urdana nala and after under crossing railway line makes boundary of the KN farm at south-east direction. The offshoot point was about 900 m downstream from origin of Urdana nala to take water from Gokalpur tank though its waste weir. Urdana nala was being fed by another nala known as Bhadpuranala which joins it at 100 m upstream of canal offshoot point. The remains of canal offshoot headwork are still be noticed near Ankur Vidya Mandir, WWA, VFJ that suggests that a canal of size 60 cm bottom width, 120 cm depth and top width 300 cm with side slope 1:1 was exist. The fields of KN Farm were prepared for cultivation of rice with gradient to flow water from one field to other. KN Farm had gradient of 23 feet from canal to tail end i.e. Urdana nala. It is an example that the canal collect the water from Urdana nala and after irrigation in KN Farm, the excess water was again discharged in Urdana nala. The remains of canal headwork from which the canal used to fed from the Urdana nala, at Richhai farm, is shown in Plate 3.1. The main canal ends at pond in Katra village. The system of canal irrigation was stopped in the year 1982 mainly due to non function of the system. The location of the canal is presented in Plate 3.2 with one of the irrigation channel in Plate 3.3. The whole system is now defuncted and almost whole length of the canal is under settlement.

Plate 3.1 Remains of canal headworks

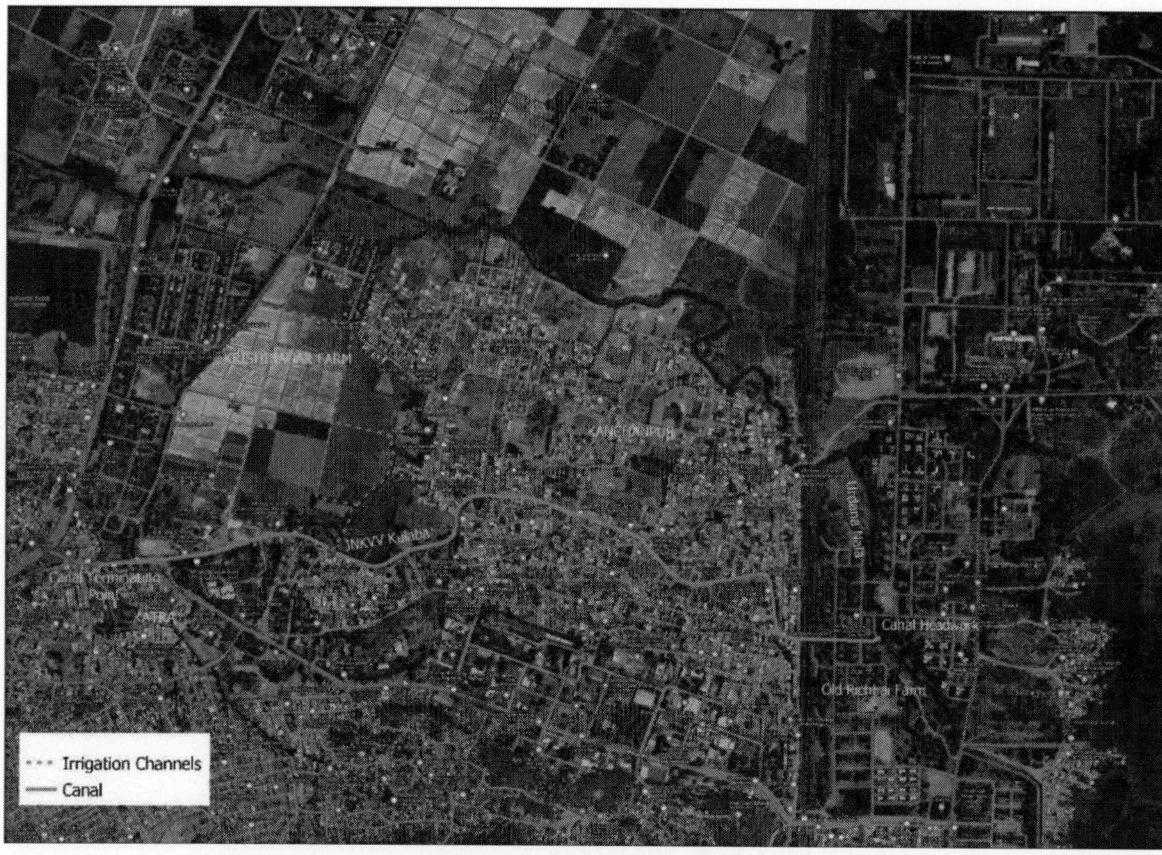

Plate 3.2 Canal layout at the side of Adhartal farm of the University

Plate 3.3 A view of the irrigation channel

Orchard of 13 ha at Richhai Farm was transferred to Vehicle Factory Jabalpur established in the year 1969.

Irrigation through river and nalas

Pariyat river– It is a perennial river originating from Kundam block and joins Narmada river. It is flowing in the north of Imalia Horticultural farm of the university. Two pumps of 10 HP each were installed on its bank for irrigation and irrigation was continued upto 1998. An area of 22.47 ha having plantation mainly of mangoes comes under this farm.

Vehicle nala 60 HP electric pump (now 30 HP) was installed at the head of Vegetable Farm, Maharajpur on perennial nala known as Vehicle Nala to fed water in Vegetable Farm Maharajpur, Soybean Farm, Dusty acres, old orchard of orange (currently converted in to Stadium), Adhartal Farm with the help of riser and pipeline crossed the national highway at culvert near guesthouse. It was a unique and successful system of irrigation established by Dr Basharat Khan remains functional upto 1992.

Dr Basharat Khan
Served
22.11.1958 to 3.3.1980
Developed successful riser based irrigation system in the University

Mainly 15 cm cast iron pipes buried below ground were used for carrying water. At risers the water was allowed to head- up by obstructing the flow and allowing the flow in only one desired direction. The risers were so designed and placed in each field that the irrigation was accomplished by gravity throughout that particular field. Approximately a network of 4.5 km pipes were laid on the sides of fields and are shown in Fig 3.1. Prof. Basharat Khan was the main force behind this irrigation project. Unfortunately, he untimely left for heavenly abode in 1980.

There is one perennial Urdana nala (Plate 3.4) flowing east-west in the south of Dusty acres. It supplies water to Dusty acres at five points spreading on its 1.6 km length. Total installed capacity of electric pumps is 54 HP.

Plate 3.4 Urdananala

A digitized and scaled map of Jabalpur campus showing all water resources is presented in Fig 3.1.

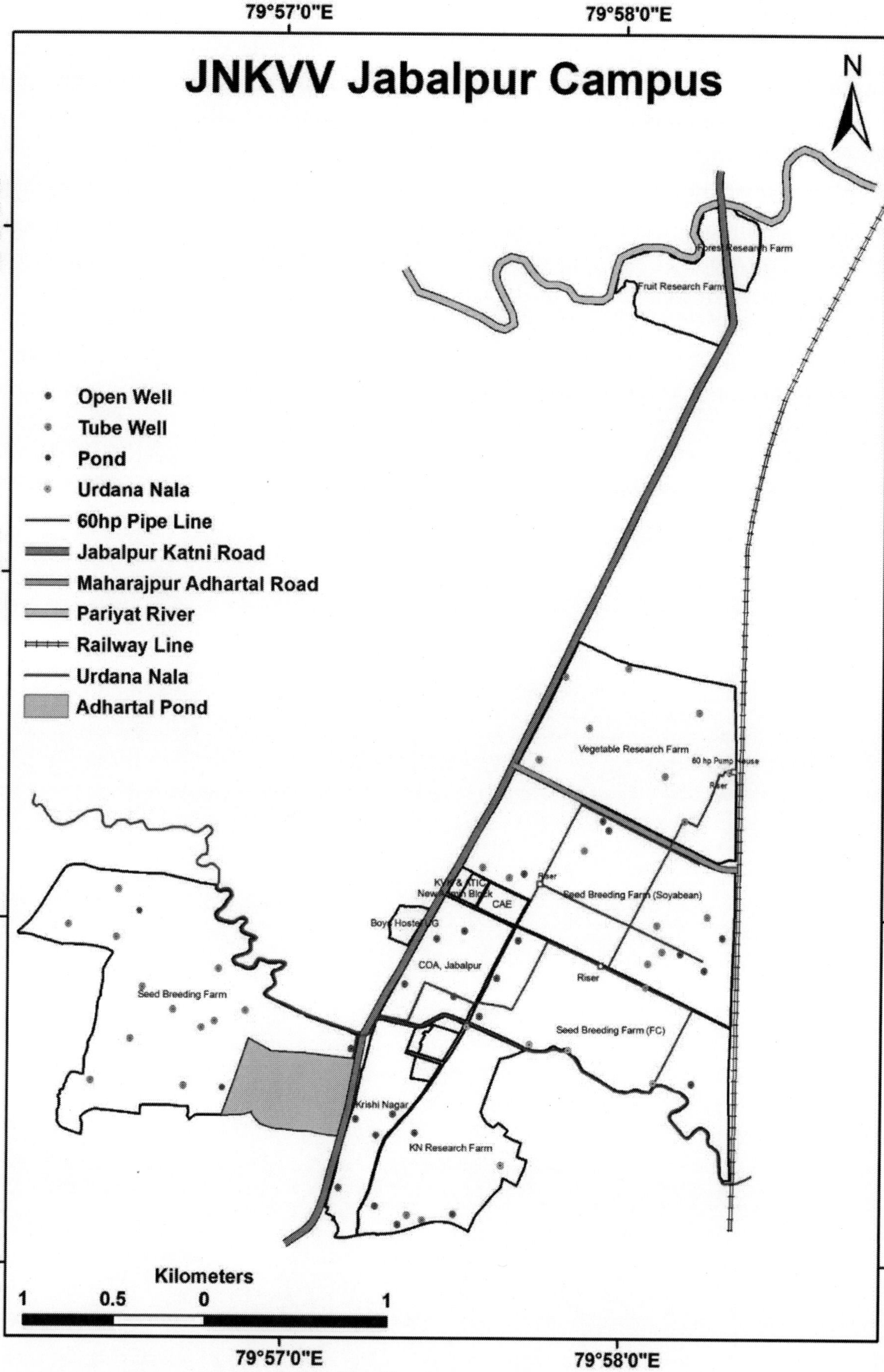

Fig 3.1 Water resources in Jabalpur campus

Irrigation through Open wells

Open wells were the only means of ground water extraction at the time of establishment of the college and university. It was the practice of irrigation in the whole area among the farmers due to high water table. There were 65 open wells exists in around Jabalpur campus. Out of that 30 were in the University campus and 11 in Maharajpur, 9 in Suhagi, 9 in Kanchanpur and 6 in village Adhartal.

The size of wells at campus ranges from 1.30 m to 6.40 m in diameter and depth from 3.66 m to 15.30 m. The locations of open wells are shown in Fig. 3.2. Later on most of the wells dried up due to lowering of water table. The dimensions and water storage of those wells are given in Table 3.1. The average diameter of wells found in the campus was 3.21 m with minimum 1.68 m (in Field No. 57) and the largest (6.18 m) and deepest well (14.63 m) was near present Communication Centre of the university. The average depth of wells of the campus was 9.25 m with the shallowest (3.66 m) in the field number 80. After monsoon nearly 4.70 m depth of water were available in the wells of the campus with maximum of 10.59 m water depth used to observe in well located near Communication Centre. The water storage used to reduce to 1.96 m on an average in pre monsoon season. Out of these 28 wells four wells used to dry up in summer.

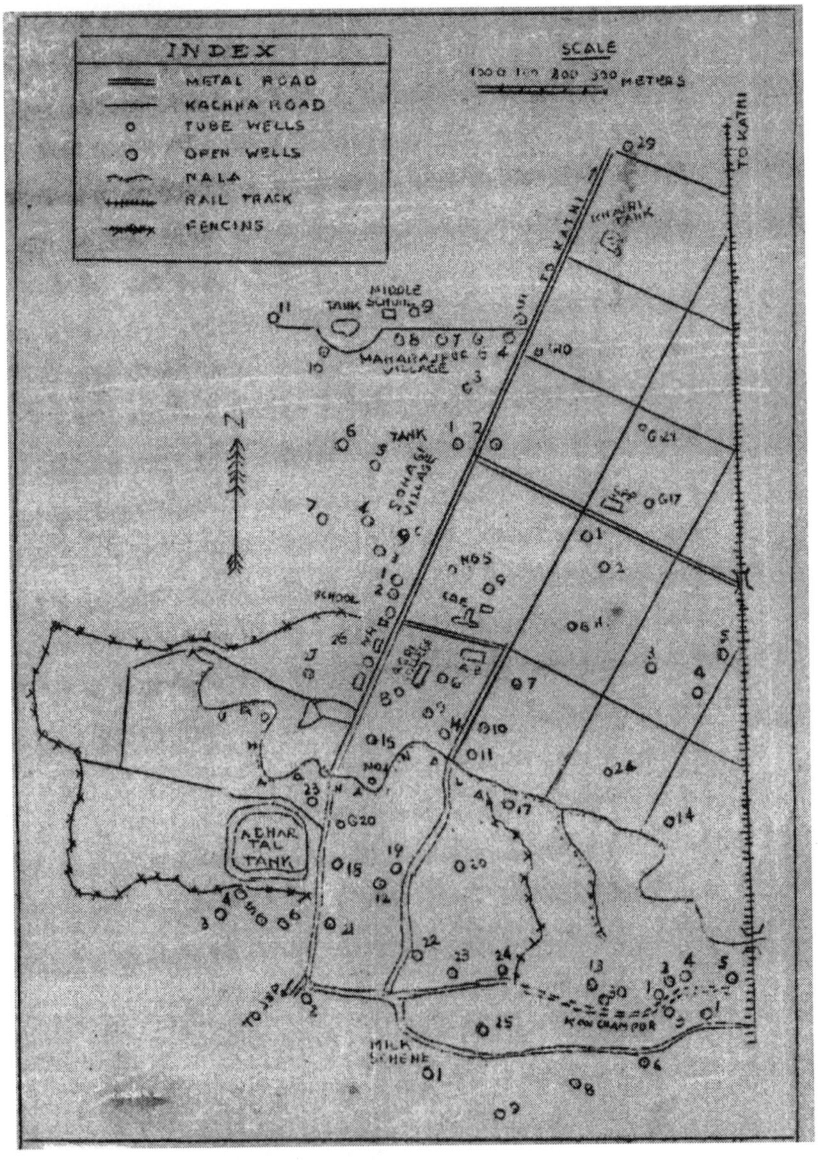

Fig 3.2 Location of open wells in Jabalpur campus

Table 3.1 Data of open well around Jabalpur campus of the University (1974-75)

No	Location	Diameter (m)	Depth from parapet(m)	Depth of storage (m)	
				Dec.74	May 75
1	Field No. 44	3.00	9.80	1.00	0.45
2	Field No. 52	2.90	9.60	3.17	0.45
3	Field No. 56	2.44	8.23	3.12	Nil
4	Field No. 57	1.68	8.84	3.74	0.94
5	Field No. 58	2.90	8.08	3.33	1.72
6	Near Communication centre	6.18/4.60	14.63	2.03	0.43
7	Field No. 67	3.96	9.28	3.92	0.02
8	S.W. to Agri. college	6.10	12.20	7.09	0.61
9	At Botany garden	3.79	10.31	3.45	Nil
10	Field No. 76	3.66	14.56	10.59	6.81
11	Field No. 76	3.05	8.84	4.86	1.29
12	Field No. 80	2.59	3.66	Nil	Nil
13	Field No. 81	2.44	3.66	Nil	Nil
14	Field No. 82	2.29	5.56	1.72	0.27
15	Field No. 85	3.05	10.06	5.06	0.74
16	Field No. 86	2.44	11.53	6.96	3.21
17	Field No. 89	2.13	7.47	3.00	2.39
18	Near F-45 Krishinagar	2.44	7.55	4.32	1.99
19	Near Circuit house	3.81	7.75	5.06	0.71
20	Near G-type quarter	2.44	9.15	7.22	5.87
21	Near Hostel No. 3	3.35	9.40	3.56	1.17
22	Behind Hostel No. 1	3.20	11.81	5.82	1.11
23	East of Hostel No. 1	3.05	8.54	4.32	2.25
24	Near Farm In-charge quarter	2.90	6.88	4.50	1.01
25	B.D.O. centre	3.05	9.91	3.88	2.00
26	Near faculty office	2.74	8.84	4.98	3.99
27	Near Dairy servant quarters	4.65	10.31	6.44	4.04
28	Near Eastern Bank of Adhartal tank	3.43	12.63	9.20	3.60

Irrigation through Tube wells

In the end of seventh decade of the nineteenth century the tube wells came in to existence in Jabalpur campus. At present, 31 tube wells with 111.5 HP powers spread over five farms are established at the headquarters. Out of these six are in Horticulture farm with depth ranges from 200 to 300 m, motor size 5 to 7 HP and usual operating hours of 4 to 6 hrs per day; nine in Groundnut unit with depth ranges from 100 to 150 m installed mostly with 5 HP pumps; two in Dusty Acres with 6 hours per day operation; two in Krishinagar farm with 5 HP motors and 8 hours per day operating; and twelve in BSP Tank area with 100 to 150 m depth, 5 to 7.5 HP pumps and operating hours of 10 to 12 hrs. The Details of existing tube wells are given in Table 3.2. Out of 35 tubewells four wells are not functioning.

A study conducted in nineties by the than Groundwater project identified different potential zones of ground water which is given in Fig 3.3. A triangular area with its one side as railway track in the east and one side as Urdana nala in the south is found as unconfined zone where limited groundwater can be drawn. Just adjoining to this triangular area lying at west, the area is found suitable for tapping confined aquifer and comparatively better well yield can be taken from deep aquifers. This area extends upto university boundary in the north to either side of NH7. The Krishinagar farm area and BSP unit open wells could be taken in the past, but now it is not suitable for open wells due to lowering of water table in this area.

Table 3.2 Details of Tube well at Jabalpur campus

Vegetable Farm, Maharajpur							
Well No.	24	25	28	31	32	40	35
Well depth (m)	250-300	250-300	250-300	150-200	150-200	200-250	-
Current status	Not Operating	Operating	Operating	Operating	Operating	Operating	Not Operating
Operation Hrs.	-	-	6 hrs/day	5-6 Hrs/ day	4Hrs/day	4-6 Hrs/ day	-
Motors HP	5	5	7	5	5	5	-
Latitude	23.2288	23.2284	23.2259	23.2266	23.2236	23.2214	23.2244
Longitude	79.9670	79.9638	79.9650	79.9705	79.9688	79.9698	79.9625
Breeder Seed Production Unit Soybean							
Well No.	57	44	55	63a	63b	60	50
Well depth (m)	150-200	100-150	100-110	100-120	100-120	-	150-200
Current status	Not Operating	Operating	Operating	Operating	Operating	Operating	Not Operating
Operation Hrs.	-	4Hrs/day	4-6 Hrs/ day	2-4 Hrs/ day	2-4 Hrs/ day	2-4 Hrs/ day	-
Motors HP	-	7.5	7	5	5	5	5
Latitude	23.2168	23.2200	23.2164	23.2151	23.2145	23.2187	23.2191
Longitude	79.7910	79.9648	79.9685	79.9687	79.9680	79.9611	79.9598
BSP Unit, Adhartal Farm							
Well No.	9	3	25	30	33	44	17
Well depth (m)	100-110						
Current status	Operating	Operating	Operating	Operating	Operating	Operating	Operating
Operation Hrs.	04-Jun	4-6Hrs	10-12Hrs	12-14Hrs	10-12Hrs	10-12Hrs	10-12Hrs
Motors HP	5	5	5	7.5	5	5	5
Latitude	23.2108	23.2133	23.2180	23.2163	23.2157	23.2113	23.2142
Longitude	79.9425	79.9431	79.9419	79.9394	79.9418	79.9460	79.9468
Breeding Seed Production Unit, Adhartal Farm							
Well No.	10b	12	22	19	6	2	
Well depth (m)							
Current status	Not operating	Operating	Not operating	Operating	Not operating	Operating	
Operation Hrs.		12-14Hrs		10-12Hrs		10-12Hrs	
Motors HP		7.5	-	5	-	5	
Latitude	23.2088	23.2086	23.2084	23.2122	23.2117	23.2122	
Longitude	79.9406	79.9452	79.9471	79.9482	79.9467	79.9446	
Dusty Acre							
Well No. / Nala	1	2	3	4	5	6	
Well depth (m)							
Current status	Operating	Operating	Operating	Operating	Operating	Operating	
Operation Hrs.	6Hrs/day						
Motors HP	5	7	7	15	15	10	

Latitude	23.2134	23.2088	23.2103	23.2106	23.2114	23.2048	
Longitude	79.9679	79.9683	79.9641	79.9621	79.9591	79.9608	
Dairy Farm							
Well No.	1	2					
Well depth (m)							
Current status	Operating	Operating					
Operation Hrs.	6Hrs/day	6Hrs/day					
Motors HP	5	5					
Latitude	23.2024	23.2021					
Longitude	79.9562	79.9569					

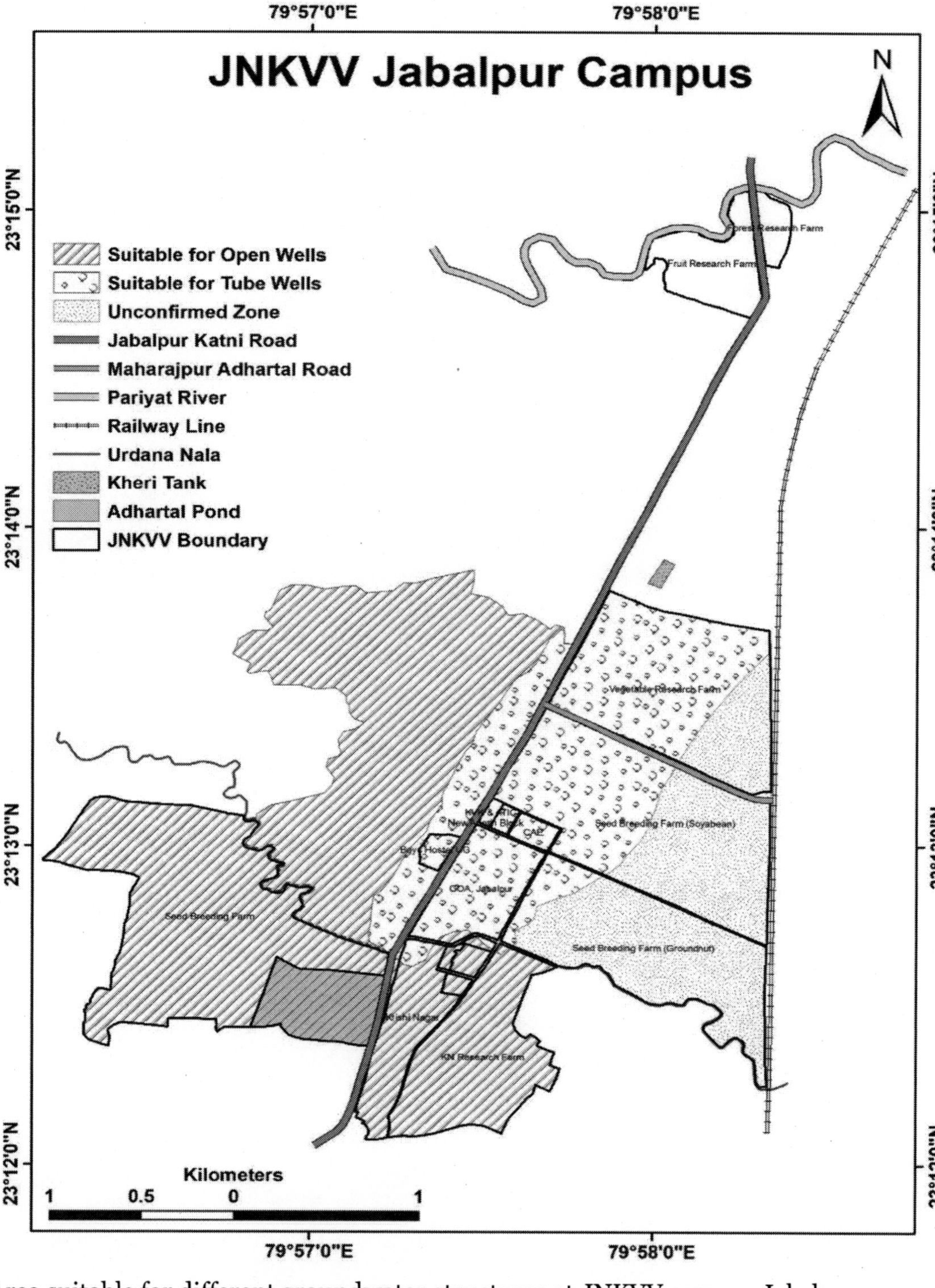

Fig 3.3. Area suitable for different groundwater structures at JNKVV campus Jabalpur

CHAPTER 4

Biodiversity Audit of Beneficial Microorganism in Soil

NG Mittra
Department of Soil Science and Agricultural Chemistry
College of Agriculture, Jabalpur

Soil-related ecosystem services Soil formation and maintenance are inextricably linked to biodiversity. Micro-organisms are vital to soil health. Reduction in soil biodiversity can impair various soil processes, including decomposition, nutrient retention and nutrient cycling. Microorganisms give the soil disease-suppressive qualities to protect plants from pathogens. In some agro-ecosystems, shade from trees provides protection to earthworm populations and thus promotes improvements to soil structure.

Out of 16 benign microorganisms reported from Jabalpur campus of the University, 87 % are bacteria and 13% fungus (Fig 4.1). Out of 14 benign Bacteria 10 are for fixation of Nitrogen, 1 each for solubilisation of Phosphorus and Potash whereas, one is bio-decomposer and one plant growth promoter (Fig 4.2) (Table 4.1).

Table 4.1 Diversity for Beneficial Microorganism in Soil

Sr.	Scientific name	Nature of Benefits	Associated crop
Bacteria			
1	*Azospirillum* sp.	Associative symbiotic nitrogen fixer	All cereals
2	*Azotobacter* sp.	Free living nitrogen fixer	All cereals, vegetable
3	*Bacillus* sp.	Phosphorous solubilizer	All crops
4	*Bacillus* sp.	Bio-decomposer	Agricultural waste
5	*Bradyrhizobium japonicum*	Symbiotic nitrogen fixer	Soybean
6	*Bradyrhizobium* sp.	Symbiotic nitrogen fixer	Pigeonpea, Cowpea, Peanut, Mungbean
7	*Cyanobacteriahetero* sp (Blue green algae)	Photosynthetic nitrogen fixer	Rice
8	*Fraturia aurentia*	Potash solubilizer	All crops
9	*Pseudomonas fluorescence*	Plant growth promoter	All crops
10	*Rhizobium cicerum*	Symbiotic nitrogen fixer	Chickpea
11	*Rhizobium leguminosarum*	Symbiotic nitrogen fixer	Field pea, Lentil Fababean
12	*Rhizobium leucaena*	Symbiotic nitrogen fixer	Sesbania
13	*Rhizobium meliloti*	Symbiotic nitrogen fixer	Fenugreek
14	*Rhizobium phaseoli*	Symbiotic nitrogen fixer	Cluster bean, French bean
Fungus			
1	*Trichoderma viride*	Act as Bio-fertilizer and Bio-control agent	For all crops
2	*Glomus* sp.	P, Zn, Cu mobilizer	For all crops

Presently, Bio-fertilizerProduction Unit is producing 18 types of bio-fertilizers viz., Rhizobium, Azotobacter, Azospirillum, Acetobacter, BGA, Phosphate Solubilizer (PSB), Mycorrhiza, Potash Solubilizer (KSB), Zinc Solubilizer (ZSB), Pseudomonas, Biofertisol, EM culture, Trichoderma, Bio-digester I (Fungal) and Bio-digester II (Bacterial), Bacto- booster, Myco-booster and Myco-safe.

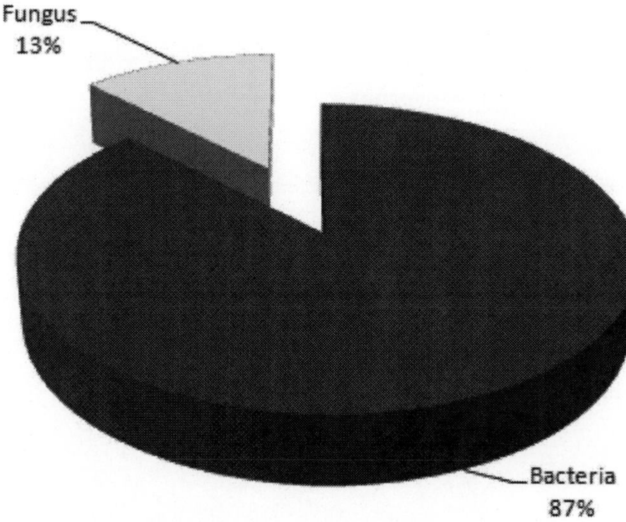

Fig 4.1 Diversity for Benign Soil Microorganism

Dr Jay Narayan Dubey

Served

(29.11. 1962 to 30.6.1994)

Awarded life time annual royalty by Sussex University, London on identification of ***Lignite as a carrier of Azotobacter***

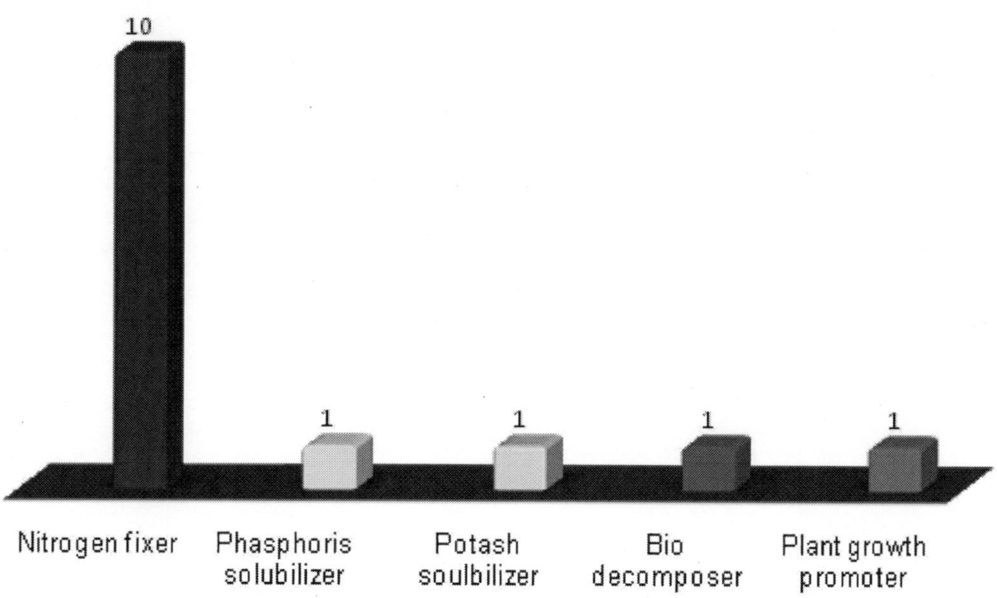

Fig 4.2 Diversity for Benign Soil Bacteria

CHAPTER 5

Biodiversity Audit of Trees at BSP Soybean, College of Agricultural Engineering and KVK

Sanjay Singh
Department of Plant Breeding and Genetics
College of Agriculture, Jabalpur

Breeder Seed Production Unit on Soybean consists of 98.88 ha irrigated land of Alfisols, Vertic haplusterts, fine loamy and mixed, Hyper-thermic and clayey soil. The farm contains 251 trees of 21 species (Table 5.1 Fig 5.1). The farm is dedicated for Breeder Seed Production (Plate 5.1).

Area of Krishi Vigyan Kendra, Jabalpur (Plate 5.2) and College of Agricultural Engineering, Jabalpur (Plate 5.3) are adjacent to the Unit.

Table 5.1 Biodiversity at BSP Soybean

Sr	Name			Family	Number	Filed Number
	Botanical	English	Hindi			
1	*Acacia nilotica*	Gum arabic	Desi babul	Fabaceae	01	46
2	*Aegle marmelos*	Stone apple	Bael	Rutaceae	03	54,55,57
3	*Azadirachta indica*	Indian lilac	Neem	Meliaceae	11	45,46,47,55
4	*Bambusa vulgaris*	Bamboo	Bans	Gramineae	25	44,45,46,47
5	*Cordia dichotoma*	Sabestan plum	Lasora	Boraginaceae	01	65
6	*Dalbergia sissoo*	Rose wood	Shisham	Faboideae	10	44,45,47,49,50
7	*Eucalyptus spp.*	Eucalyptus	Nilgiri	Myrtaceae	10	42,47,57
8	*Ficus bengalensis*	Banyan	Bargad	Moraceae	01	57
9	*Ficus racemosa*	Cluster fig	Gular	Moraceae	03	57,60, 64
10	*Ficus religiosa*	Sacred fig	Peeipal	Moraceae	08	42,49,50,55,57,58, 62,63
11	*Gmelina arborea*	Beechwood	Khamer	Lamiaceae	93	46
12	*Madhuca indica*	Buttercup	Mahua	Sapotaceae	06	47,58
13	*Mangifera indica*	Mango	Aam	Anacardiaceae	45	43,44,46,48,49,54,55, 56,57, 59,60,61,63
14	*Millettia pinnata*	Indian beech	Kanji	Fabaceae	03	47,48
15	*Moringa oleifera*	Drumstick	Munga	Moringaceae	01	58
16	*Saraca asoca*		Ashok	Detarioideae	12	58
17	*Syzygium cumini*	Black plum	Jamun	Myrtaceae	09	46,56,58,59,60
18	*Tectona grandis*	Teak	Sagone	Lamiaceae	05	46,49,63,64
19	*Terminalia arjuna*		Arjun	Combretaceae	02	58
20	*Terminalia bellirica*	Beach almond	Bahera	Combretaceae	01	64
21	*Zizyphus jujuba*	Indian jujube	Ber	Rhamnaceae	01	46
	Total				**251**	

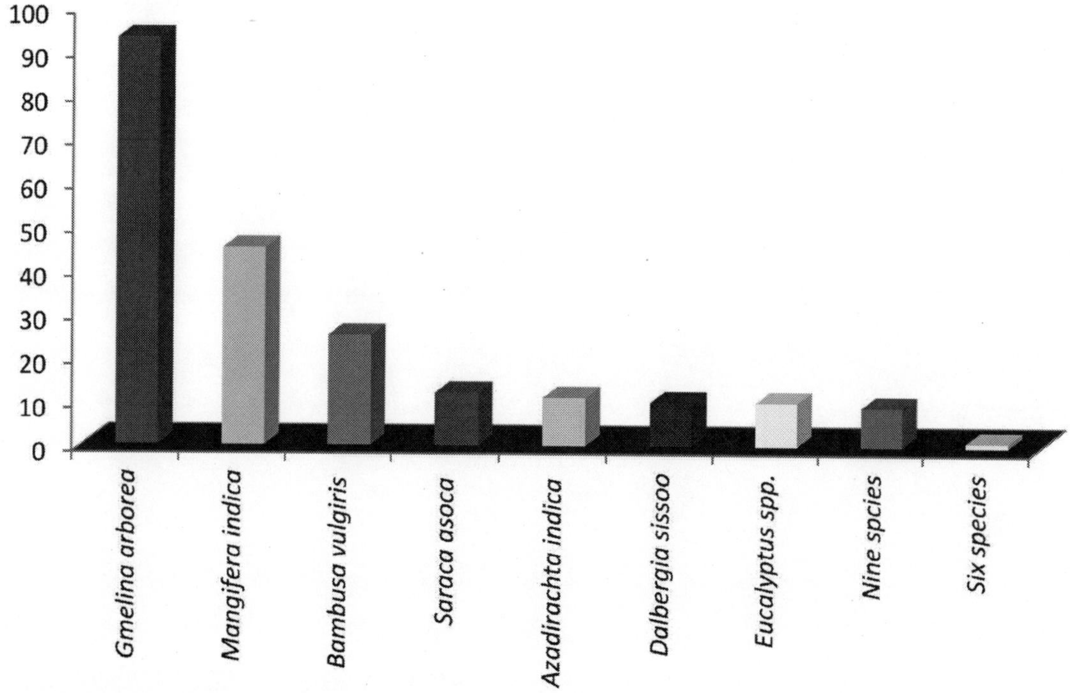

Fig 5.1 Representation of tree species at BSP Soybean

Plate 5.1 BSP Soybean, Jabalpur

Plate 5.2 KVK, Jabalpur

Plate 5.3 KVK, Jabalpur

CHAPTER 6

Biodiversity Audit of Tree Species at Dusty Acre (BSP Groundnut)

R K Bajpai
Department of Forestry
College of Agriculture, Jabalpur

A farm of 6.23 ha contains total 4129 trees of 31 species. The highest representation is of *Guava* followed by *Karanja, Teak, Gamhar, Shisham and Arjun.* Out of these, *Arjun* is established naturally whereas others were planted; 99 clumps of bamboo were planted in two locations; 15 species have 11-50; whereas 10 species have < 10 number of trees each (Table and Fig. 6.1).

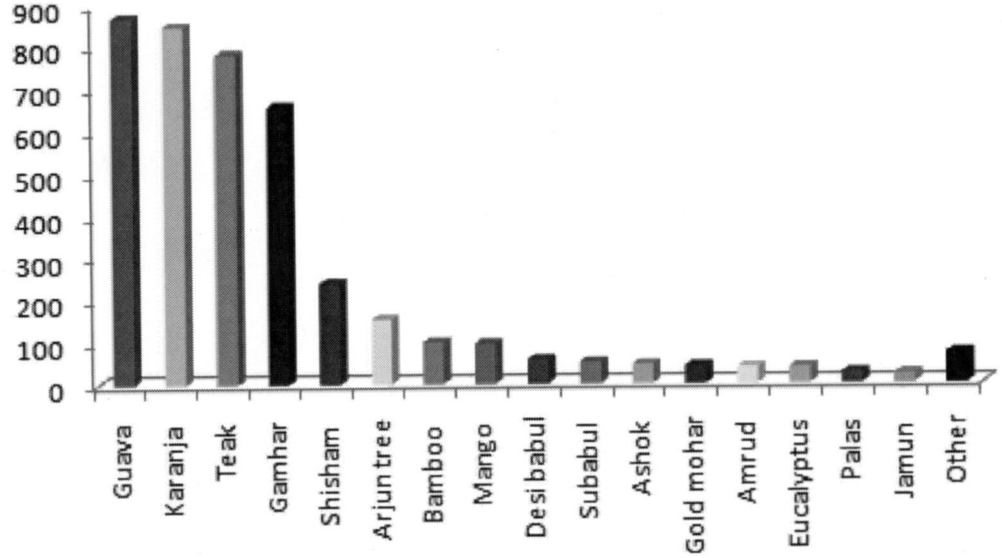

Fig 6.1 Representation of tree species at Dusty Acre

Plate 6.1

Table 6.1 Diversity for trees at Dusty Acre

	Botanical	English	Hindi	Family	Number	Location
1.	*Acacia auriculiformis*	Australian babul		Mimosaceae	4	Dusty Acre
2.	*Acacia nilotica*		Desi babul	Mimosaceae	57	Dusty Acre
3.	*Albizia alebbeck*	Black siris		Mimosaceae	3	Dusty Acre
4.	*Artocarpus heterophyllus*	Jackfruit		Moraceae	1	Dusty Acre
5.	*Azadirachta indica*		Neem	Meliaceae	17	Dusty Acre
6.	*Bambusa vulgaris*	Bamboo		Poaceae	99clump	Dusty Acre
7.	*Bombax ceiba*	Silk cotton		Malvaceae	6	Dusty Acre
8.	*Butea monosperma*		Palas	Fabaceae	25	Dusty Acre
9.	*Cassia fistula*		Amaltas	Caesalpiniaceae	2	Dusty Acre
10.	*Cassia siamea*	Kassod		Caesalpiniaceae	3	Dusty Acre
11.	*Ceiba pentandra*	Pakok		Malvaceae	4	Dusty Acre
12.	*Dalbergia sissoo*		Shisham	Caesalpiniaceae	239	Dusty Acre
13.	*Delonix regia*	Gold mohar		Caesalpiniaceae	42	Dusty Acre
14.	*Eucalyptus spp.*	Eucalyptus		Myrtaceae	39	Dusty Acre
15.	*Ficus bengalensis*	Banyan		Moraceae	4	Dusty Acre
16.	*Ficus racemosa*	Cluster figs		Moraceae	8	Dusty Acre
17.	*Ficus religiosa*		Pipal	Moraceae	5	Dusty Acre
18.	*Gmelina arborea*		Gamhar	Lamiaceae	656	Dusty Acre
19.	*Leucaen aleucocephela*		Subabul	Mimosaceae	51	Dusty Acre
20.	*Mangifera indica*	Mango		Anacardiaceae	96	Dusty Acre
21.	*Morus alba*	Mulberry		Moraceae	2	Dusty Acre
22.	*Pithecellobium dulce*		Vilayati imli	Mimosaceae	4	Dusty Acre
23.	*Pongamia pinnata*		Karanja	Fabaceae	846	Dusty Acre
24.	*Psidium guajava*	Guava		Myrtaceae	867	Dusty Acre
25.	*Psidium guajava*		Amrud	Myrtaceae	39	Dusty Acre
26.	*Saraca asoca*		Ashok	Detarioideae	46	Dusty Acre
27.	*Syzygium cumini*		Jamun	Myrtaceae	21	Dusty Acre
28.	*Tectona grandis*	Teak		Verbenaceae	780	Dusty Acre
29.	*Terminalia arjuna*		Arjun tree	Combretaceae	154	Dusty Acre
30.	*Terminalia bellirica*		Bahera	Combretaceae	4	Dusty Acre
31.	*Zizyphus jujuba*			Rhamnaceae	5	Dusty Acre
32.	*Artocarpus heterophyllus*	Jackfruit		Moraceae	1	B.Tech Hostel
33.	*Azadirachta indica*		Neem	Meliaceae	7	B.Tech Hostel
34.	*Cassia siamea*	Kassod		Caesalpiniaceae	3	B.Tech Hostel
35.	*Delonix regia*	Gold mohar		Caesalpiniaceae	42	B.Tech Hostel
36.	*Ficus bengalensis*	Banyan		Moraceae	2	B.Tech Hostel
37.	*Ficus racemosa*	Cluster figs		Moraceae	8	B.Tech Hostel
38.	*Ficus religiosa*		Pipal	Moraceae	3	B.Tech Hostel
39.	*Gmelina arborea*	White teak	Gamhar	Lamiaceae	513	Near culvert
40.	*Leucaen aleucocephela*		Subabul	Mimosaceae	8	B.Tech Hostel

	Botanical	English	Hindi	Family	Number	Location
41.	*Mangifera indica*	Mango		Anacardiaceae	18	B.Tech Hostel
42.	*Pithecellobium dulce*		Vilayati imli	Mimosaceae	4	B.Tech Hostel
43.	*Psidium guajava*	Guava		Myrtaceae	867	Near culvert
44.	*Saraca asoca*		Ashok	Detarioideae	46	B.Tech Hostel
45.	*Syzygium cumini*		Jamun	Myrtaceae	9	B.Tech Hostel
46.	*Tectona grandis*	Teak		Verbenaceae	738	Near culvert
47.	*Terminalia arjuna*		Arjun	Combretaceae	2	B.Tech Hostel
48.	*Terminalia bellirica*		Bahera	Combretaceae	4	B.Tech Hostel
	Total Tree				**2275**	

Plate 6.2 Agro forestry experiment on Karanj

Plate 6.3 Agro-forestry experiment on Gamhar

Plate 6.4 Plantation of Bamboo

Plate 6.5 Agro-forestry experiment on *Gmelina arborea*

Plate 6.6 Agro-forestry experiment with Shisham

Plate 6.7 Agro-forestry experiment on *Pongamia pinnata*

CHAPTER 7

Biodiversity Audit of Tree Species at Breeder Seed Production FC Unit

RS Shukla
Department of Plant Breeding and Genetics
College of Agriculture, Jabalpur

It is a well-equipped farm of 119.58 ha land located near Adhartal Tank ear marked for Maintenance Breeding, Instructional and experimentation of nearly 10 All Indian Coordinated Research Projects. This farm is the cradle of crescent for introduction of Soybean as oilseed crop for cultivation in Kharif season with exemplary yellow revolution in the state of Madhya Pradesh. The soil is Vertisol, Typichaplusterts, Very fine mont-morrilonite, Hyper-thermic and Clayey.

No tree is present with in the field in this farm, however (Table 7.1) total 337 trees of 17 species are present at the boundary of the field.

Table 7.1 Diversity for Tree Species at Breeder Seed Production (FC) Unit

Sr	Name			Family	Number
	Botanical	**English**	**Hindi**		
1.	*Acacia cassia*		Gurar	Mimosaceae	01
2.	*Acacia senegal L.*	Gum Arabic		Mimosaceae	66
3.	*Azadirachta indica*		Neem	Meliaceae	22
4.	*Bauhinia variegata*		Kachanar	Caesalpiniaceae	01
5.	*Butea monosperma*	Sacred tree	Palash	Fabaceae	01
6.	*Dalbergia sissoo*	Shisham		Caesalpiniaceae	01
7.	*Erythrina suberosa*		Hadhua	Fabaceae	01
8.	*Ficus religiosa*	Sacred fig		Moraceae	01
9.	*Hyophor belagenicaulis*	Bottle palm		Arecaceae	01
10.	*Mangifera indica*	Mango	Aam	Anacardiaceae	03
11.	*Melia azadirachata*		Mahaneem	Meliaceae	02
12.	*Phoenix dactylifera*	Date palm		Arecaceae	02
13.	*Phyllanthus embilica*		Amla	Euphorbiaceae	02
14.	*Santalum album*		Chandan	Sanatalaceae	04
15.	*Saraca asoca*		Ashoka	Detarioideae	01
16.	*Tamarindus indica*	Tamarind		Caesalpiniaceae	175
17.	*Tectona grandis*		Sagon	Verbenaceae	53
18.	**Total**				**337**
19.	**Shrubs**				
20.	*Leucaen aleucocephela*		Soobabul	Mimosaceae	07
21.	*Zizyphus mauritiana*	Indian Jujube	Ber	Rhamnaceae	88
22.	*Vachellia nilotica*		Gandbabul	Fabaceae	13
23.			Rasutula		03
24.	*Epipremnum aureum*	Spring Queen			01
25.	*Bambusa vulgaris*	Bamboo		Poaceae	02
26.	**Total**				**114**

CHAPTER 8

Biodiversity Audit of Tree Species at Krishi Nagar Farm Adhartal

P. B. Sharma
Department of Agronomy
Participation of students

Mansi Joshi, Shivangi Raghuwanshi, Anil Patel, Lakhan Bhalse, Shubham Choudhary, Hemat Kourav, Sangita Rawat, Priyambada Dwivedi, Kartik Mandoli, Ashi Thakur and Megha Singh
Department of Agronomy
College of Agriculture, Jabalpur

Krishinagar farm Adhartal consist of 56.85 ha irrigated land of Alfisols, Vertic Typic and vertichaplusterts, fine loamy and mixed, Hyper-thermic and Sandy clay loam and sandy loam soil. The farm contains 251 trees of 21 species. The farm an establishment of 1912 is utilized for seed production of rice and wheat apart from research on organic and natural farming, curricular research, nursery and dairy. The farm has 702 trees of 22 species present on the boundary of the farm (Table 8.1, Plate 8.1 and 8.2). Detail of the part of nursery is presented separately in Chapter 9.

Plate 8.1 Plantation at Krishi Nagar Farm

Table 8.1 Bio-Diversity for Tree Species at Krishi Nagar Farm Adharta

	Botanical	English	Hindi	Family	Number	Location
	Tree					
1.	*Aegle marmelos*		Bel	Rutaceae	04	Krishi Nagar Farm
2.	*Annona squamosa*	Custard apple		Annonaceae	16	Krishi Nagar Farm
	Annona squamosa	Custard apple		Annonaceae	11	Bangladesh colony
3.	*Azadirachta indica*		Neem	Meliaceae	05	Krishi Nagar Farm
	Azadirachta indica		Neem	Meliaceae	05	IFS
	Azadirachta indica		Neem	Meliaceae	05	Dairy
	Azadirachta indica		Neem	Meliaceae	03	Girls hostel
	Azadirachta indica		Neem	Meliaceae	11	Bangladesh colony
4.	*Bambusa vulgaris*	Bamboo		Poaceae	03	Dairy
5.	*Carissa carandas*		Karonda	Apocynaceae	60	IFS

	Botanical	English	Hindi	Family	Number	Location
6.	*Cascabela thevetia*		Kaner	Apocynaceae	03	Bangladesh colony
7.	*Eucalyptus globulus*	Eucalyptus		Myrtaceae	09	Krishi Nagar Farm
	Eucalyptus globulus	Eucalyptus		Myrtaceae	38	KVK Orchard
	Eucalyptus globulus	Eucalyptus		Myrtaceae	127	Near culvert
8.	*Ficus bengalensis*	Banyan		Moraceae	02	Krishi Nagar Farm
	Ficus benghalensis	Banyan		Moraceae	02	Girls hostel
9.	*Ficus religiosa*		Peepal	Moraceae	04	Krishi Nagar Farm
	Ficus religiosa		Peepal	Moraceae	02	Dairy
	Ficus religiosa		Peepal	Moraceae	03	Near culvert
10.	*Leucaen aleucocephela*		Subabul	Mimosaceae	04	Krishi Nagar Farm
	Leucaen aleucocephela		Subabul	Mimosaceae	01	IFS
	Leucaen aleucocephela		Subabul	Mimosaceae	02	Dairy
	Leucaen aleucocephala		Subabul	Mimosaceae	01	KVK Orchard
	Leucaen aleucocephala		Subabul	Mimosaceae	06	Bangladesh colony
11.	*Mangifera indica*	Mango		Anacardeaceae	11	Krishi Nagar Farm
	Mangifera indica	Mango		Anacardeaceae	04	Dairy
	Mangifera indica	Mango		Anacardeaceae	06	Girls hostel
	Mangifera indica	Mango		Anacardeaceae	13	Bangladesh colony
12.	*Moringa oleifera*	Drumstick		Moringaceae	02	Krishi Nagar Farm
	Moringa oleifera	Drumstick		Moringaceae	05	IFS
	Moringa oleifera	Drumstick		Moringaceae	02	Dairy
	Moringa oleifera	Drumstick		Moringaceae	03	Girls hostel
13.	*Neolamarckia cadamba*		Kadam	Rubiaceae	34	KVK Orchard
14.	*Phoenix dactylifera*		Khajur	Arecaceae	02	Krishi Nagar Farm
15.	*Phyllanthus emblica*		Aonla	Euphorbiaceae	06	Bangladesh colony
16.	*Pithecellobium dulce*		Jungle Jalebi	Mimosaceae	03	Girls hostel
	Pithecellobium dulce		Jungle Jalebi	Mimosaceae	03	Bangladesh colony
17.	*Psidium guajava*	Guava		Myrtaceae	05	Krishi Nagar Farm
	Psidium guajava	Guava		Myrtaceae	06	IFS
	Psidium guajava	Guava		Myrtaceae	02	Dairy
	Psidium guajava	Guava		Myrtaceae	139	KVK Orchard
	Psidium guajava	Guava		Myrtaceae	04	Girls hostel
	Psidium guajava	Guava		Myrtaceae	13	Bangladesh colony
18.	*Saraca asoca*		Ashoka	Detarioideae	12	Krishi Nagar Farm
	Saraca asoca		Ashoka	Detarioideae	06	Dairy
	Saraca asoca		Ashoka	Detarioideae	07	Bangladesh colony
19.	*Syzygium cumini*		Jamun	Myrtaceae	02	Krishi Nagar Farm
	Syzygium cumini		Jamun	Myrtaceae	02	Dairy
20.	*Tamarindus indica*	Tamarind		Caesalpiniaceae	68	KVK Orchard
21.	*Terminalia arjuna*		Arjun	Combretaceae	12	Near culvert
22.	*Zizyphus mauritiana*		Ber	Rhamnaceae	03	IFS
	Total				**702**	

	Botanical	English	Hindi	Family	Number	Location
	Shrubs					
1	*Cascabela thevetia*		Kaner	Apocynaceae	05	Krishi Nagar Farm
	Cascabela thevetia		Kaner	Apocynaceae	03	IFS
2	*Delonix regia*		Gulmohar	Caesalpiniaceae	37	Bangladesh colony
	Delonix regia		Gulmohar	Caesalpiniaceae	06	Near culvert
3	*Hibiscus rosasinensis*		Gudhal	Malvaceae	14	Bangladesh colony
	Hibiscus rosa –sinensis		Gudhal	Malvaceae	04	IFS
	Hibiscus rosa-sinensis		Gudhal	Malvaceae	03	Krishi Nagar Farm
4	*Lantana camara*	Lantana		Verbenaceae	03	Krishi Nagar Farm
	Lantana camara	Red Sage		Verbenaceae	13	IFS
5	*Pithecellobium dulce*		VilayatiImli	Mimosaceae	02	Krishi Nagar Farm
6	*Ricinus communis*	Castor		Euphorbiaceae	13	IFS
	Ricinus communis	Castor		Euphorbiaceae	13	Near culvert
7	*Tabermaeontana divaricata*	Crepe Jasmine		Apocynaceae	03	Bangladesh colony
8	*Zizyphus mauritiana*		Ber	Rhamnaceae	11	IFS
	Zizyphus mauritiana		Ber	Rhamnaceae	07	Bangladesh colony

Plate 8.2 Plantation at Krishi Nagar Farm

CHAPTER 9

Biodiversity Audit of Trees at Residential Area, Krishi Nagar

Makhan Singh* and RK Bajpai
Department of Forestry
College of Agriculture, Jabalpur
* Ph D Scholar

It is the residential area of the University at the headquarters constructed in 25.43ha land hosting 433 residential quarters, guesthouse, play grounds, gardens dispensary, shopping complex, works section etc (Plate 9.1 to 9.4). It is a life time experience to live in isolated place close to nature (18% area covered with trees) for common cause with all the basic facilities. It contains 2665 trees of 43 species (Table 9.1).

Table 9.1 Diversity for Trees at Krishi Nagar

Sr	Scientific Name	Name	Number	Location
1.	*Acarus sapota*	Chiku	1	Near Ven Ganga GH
2.	*Aegle marmelos*	Beal	5	Residential area
3.	*Aegle marmelos*	Beal tree	16	Near Ven Ganga GH
4.	*Albizia lebbeck*	Balcksiris	3	Kaushal Bhawan
5.	*Albizia procera*	White siris	7	Residential area
6.	*Alstonia scholaris*	Chitwan	38	Residential area
7.	*Alstonia scholaris*	Chitwan	4	Narmada Tapti GH
8.	*Alstonia scholaris*	Chitwan	3	Dispensary
9.	*Anthocephalus cadamba*	Kadam	3	Near Ven Ganga GH
10.	*Artocarpus heterophyllus*	Jackfruit	12	Residential area
11.	*Artocarpus heterophyllus*	Jackfruit	13	Near Ven Ganga GH
12.	*Azadirachta indica*	Neem	41	Residential area
13.	*Azadirachta indica*	Neem	5	Kaushal Bhawan
14.	*Azadirachta indica*	Neem	49	NTVGG
15.	*Azadirachta indica*	Neem	5	Shastri Bhawan
16.	*Azadirachta indica*	Kachanar	7	Dispensary
17.	*Bauhinia variegate*		2	Narmada Tapti GH
18.	*Calliandra surinamensis*	Dixie pink	27	Residential area
19.	*Calliandra surinamensis*	Dixie pink	14	Guest House Road
20.	*Calliandra surinamensis*	Dixie pink	12	NTVGG
21.	*Callistemon rigidus*	Bottle brushes	2	Dispensary
22.	*Cascabela thevetia*	Kaner	5	Shastri Bhawan
23.	*Cassia siamea*	Kassod tree	1	Residential area
24.	*Dalbergia sissoo*	Shisham	2	Kaushal Bhawan
25.	*Delonix regia*	Gold Mohar	2	Residential area
26.	*Delonix regia*	Gold Mohar	4	NTVGG
27.	*Delonix regia*	Gold Mohar	3	Dispensary
28.	*Eucalyptus* spp.	Eucalyptus	39	Residential area
29.	*Eucalyptus* spp.	Eucalyptus	828	NTVGG
30.	*Eucalyptus* spp.	Eucalyptus	5	Dispensary
31.	*Ficus banjamiana*	Weeping Fig	69	Residential area
32.	*Ficus banjamiana*	Weeping Fig	21	Kaushal Bhawan
33.	*Ficus banjamiana*	Weeping Fig	16	Guest House Road
34.	*Ficus benghalensis*	Banyan tree	7	Residential area
35.	*Ficus benghalensis*	Banyan tree	8	Near Ven Ganga GH
36.	*Ficus racemosa*	Cluster fig	2	Kaushal Bhawan
37.	*Ficus racemosa*	Cluster fig	2	Dispensary
38.	*Ficus religiosa*	Pipal	11	Residential area
39.	*Ficus religiosa*	Pipal	9	NTVGG
40.	*Hyophor belagenicaulis*	Bottle Palm tree	21	Residential area

Sr	Scientific Name	Name	Number	Location
41.	*Hyophor belagenicaulis*	Bottle Palm tree	62	Kaushal Bhawan
42.	*Hyophor belagenicaulis*	Bottle Palm tree	15	Near Ven Ganga GH
43.	*Hyophor belagenicaulis*	Bottle Palm tree	5	Dispensary
44.	*Kigelia pinnata*	Cucumber tree	5	Kaushal Bhawan
45.	*Lagerstroemia speciosa*	Pride of India	57	NTVGG
46.	*Leucaen aleucocephela*	Subabul	88	Residential area
47.	*Leucaen aleucocephela*	Subabul	3	Kaushal Bhawan
48.	*Leucaen aleucocephela*	Subabul	150	NTVGG
49.	*Leucaen aleucocephela*	Subabul	25	Dispensary
50.	*Mangifera indica*	Mango	83	Residential area
51.	*Mangifera indica*	Mango	7	Kaushal Bhawan
52.	*Mangifera indica*	Mango	60	Near Ven Ganga GH
53.	*Mangifera indica*	Mango	12	Dispensary
54.	*Manilkara hexandra*	Khirni	1	Narmada Tapti GH
55.	*Morus alba*	Mulberry	7	Kaushal Bhawan
56.	*Nerium oleander*	Nerium	15	Near Ven Ganga GH
57.	*Peltophorum pterocarpum*	Yellow flame Tree	5	Near Ven Ganga GH
58.	*Phyllanthus emblica*	Aonla	17	Residential area
59.	*Phyllanthus emblica*	Aonla	17	Near Ven Ganga GH
60.	*Pithecellobium dulce*	Vilayati imli	16	Residential area
61.	*Pithecellobium dulce*	Vilayati imli	5	Near Ven Ganga GH
62.	*Pithecellobium dulce*	Vilayati imli	5	Shastri Bhawan
63.	*Pithecellobium dulce*	Vilayati imli	3	Dispensary
64.	*Polyalthia longifolia*	Ashok	73	Residential area
65.	*Polyalthia longifolia*	Ashok	90	Kaushal Bhawan
66.	*Polyalthia longifolia*	Ashok	223	NTVGG
67.	*Polyalthia longifolia*	Ashok	55	Near Ven Ganga GH
68.	*Polyalthia longifolia*	Ashok	17	Guest House Road
69.	*Pongamia pinnata*	Karanj	17	Near Ven Ganga GH
70.	*Pongamia pinnata*	Karanja	3	NTVGG
71.	*Prosopis cineraria*	Khejari	2	Near Ven Ganga GH
72.	*Prunus dulcis*	Badam	2	Near Ven Ganga GH
73.	*Psidium guajava*	Guava	27	Residential area
74.	*Psidium guajava*	Guava	45	Near Ven Ganga GH
75.	*Syzygium cumini*	Jamun	19	Residential area
76.	*Syzygium cumini*	Jamun	4	Kaushal Bhawan
77.	*Syzygium cumini*	Jamun	7	Near Ven Ganga GH
78.	*Syzygium cumini*	Jamun	2	Narmada Tapti GH
79.	*Tabebuia argentea*	Trumpet tree	8	Dispensary
80.	*Tectona grandis*	Teak	3	Residential area
81.	*Terminalia arjuna*	Arjun	3	Shastri Bhawan
82.	*Terminalia arjuna*	Arjun	5	Dispensary
83.	*Thespesia populnea*	Paras Pipal	22	Residential area
84.	*Zizyphus jujuba*	Ber	39	Residential area
85.	*Ziziphus jujuba*	Ber	12	NTVGG
	Total Tree		**2665**	

Plate 9.1 Trees near girls hostel and Kaushal Bhawan

Plate 9.2 Ashok Avenue Krishi Nagar

Plate 9.3 Lush green guesthouse of the University

Plate 9.4 Plantation an annual event of the oldest ladies club of Jabalpur

CHAPTER 10

Biodiversity Audit of Trees and Shrubs at Herbal Garden, College of Agriculture, Administrative Block, and Jawahar Stadium

Gyanendra Tiwari, Sachin Prakash Nagre, Salikram Mohare, Satyendra Thakur and Rohit Kumawat
Department of Plant Physiology,
College of Agriculture, Jabalpur

A systematic survey was conducted during March-April, 2022 to record trees and shrub diversity with number of individual species covering Herbal Garden (Block-A to E), around JNKVV Administrative Building, College of Agriculture to Jawahar Stadium up to Uldana Nala. Species of trees and shrubs are found were presented in different tables (Table 10.1 to 10.11) and depicted in Plates 10.1 to 10.14.

Table 10.1. Diversity for Trees at Herbal Garden- Block A, College of Agriculture, Jabalpur

Sr	Scientific Name	Common name	Family	Number
1.	*Acacia auriculiformis*	Australian babool	Fabaceae	2
2.	*Adansonia Digitata*	Kalpvriksha	Bombacaceae	2
3.	*Aegle marmelos*	Bael	Rutaceae	4
4.	*Albizia lebbeck*	Kala siris	Mimosaceae	2
5.	*Alsonia scholaris*	Saptaparni	Apocynaceae	2
6.	*Annona squamosa*	Sitaphal	Annonaceae	10
7.	*Anthocephalus cadamba*	Kadam	Rubiaceae	3
8.	*Artocarpus lakoocha*	Badhal	Moraceae	5
9.	*Azadirachta indica*	Neem	Meliaceae	12
10.	*Bixa orellana*	Sinduri	Bixaceae	2
11.	*Bombax cieba*	Semal	Bombacaceae	3
12.	*Butea monosperma*	Flame-of-the-forest	Fabaceae	4
13.	*Caesalpinia crista*	Gataran	Caesalpiniaceae	3
14.	*Calendrina* sp.	Kalendra	Mimosaceae	5
15.	*Carissa carandas*	Karonda	Apocyanaceae	13
16.	*Cassia fistula*	Amaltas	Fabaceae	3
17.	*Cassia siamea*	Rajvriksha	Caesalpiniaceae	2
18.	*Casuarina equisetifolia*	Jangli saru	Casuarinaceae	1
19.	*Ceiba pentandra*	Kapok	Malvaceae	4
20.	*Celastrus paniculatus*	Malkangni	Celastraceae	5
21.	*Cicca acida*	Harfarewri	Phyllanthaceae	4
22.	*Dalbergia sissoo*	Sissoo	Fabaceae	2
23.	*Dendrocalamus strictus*	Katanga bans	Poaceae	100
24.	*Dillenia indica*	Chalta champa	Dilleniaceae	1
25.	*Diospyros montana*	Jungle tendu	Ebenaceae	5
26.	*Emblica officinalis*	Amla	Euphorbiaceae	50
27.	*Eucalyptus globulus*	Blue gum	Myrtaceae	1
28.	*Ficus benghalensis*	Bargad	Moraceae	2
29.	*Ficus glomerata*	Cluster fig	Moraceae	10
30.	*Ficus religiosa*	Peepal	Moraceae	1
31.	*Glycosmis pentaphylla*	Ban nimbu	Rutaceae	5
32.	*Grevillea robusta*	Silver oak	Proteaceae	1
33.	*Justicia adhatoda*	Adulsa	Acanthaceae	300
34.	*Kigelia africana*	Balam khira	*Bignoniaceae*	6
35.	*Lagerstroemia flosregini*	Landia	Lythraceae	4
36.	*Lagerstroemia parviflora*	Dhaura	Lythraceae	7
37.	*Lawsonia inermis*	Mehndi	Lythraceae	15
38.	*Livistona chinensis*	Fan palm	Arecaceae	8
39.	*Mangifera indica*	Mango	Anaacardiaceae	1
40.	*Melaleuca citrina*	Bottle brush	Myrtaceae	2

Sr	Scientific Name	Common name	Family	Number
41.	*Melia azadirachata*	Bakain	Meliaceae	1
42.	*Mimusops elengi*	Maulshree	Sapotaceae	5
43.	*Mitragyna parviflora*	Mundi	Rubiaceae	5
44.	*Moris alba*	Mulberry	Moraceae	1
45.	*Murraya panniculata*	Madhu kamini	Rutaceae	500
46.	*Parkia biglandulosa*	Parkia	Mimosaceae	1
47.	*Peltophorum pterocarpum*	Peeli gulmohar	Mimosaceae	2
48.	*Pithecellobium dulce*	Jungli jalebi	Fabaceae	10
a.	*Plumeria indica*	Champa	Apocynaceae	29
49.	*Polyalthia longifolia*	Ashok tree	Annonaceae	15
50.	*Prosopis cineraria*	Shami	Fabaceae	4
51.	*Pterocarpus marsupium*	Bijasaar	Fabaceae	2
52.	*Pterospermum acerifolium*	Kanak chamka	Malvaceae	6
53.	*Putranjiva roxburghii*	*Putranjiva*	Putranjivaceae	1
54.	*Saraca asoca*	Sita ashok	Fabaceae	1
55.	*Schleichera oleosa*	Kusum	Sapindaceae	3
56.	*Sterculia alata*	Sterculia	Sterculiaceae	2
57.	*Swietenia mahagoni*	Mahagoni	Meliaceae	1
58.	*Syzygium cumini*	Jamun	Myrtaceae	2
59.	*Terminalia arjuna*	Arjun, koha	Combretaceae	68
60.	*Terminalia catappa*	Jangli badam	Combretaceae	1
61.	*Thevetia peruviana*	Peeli kanera	Apocynaceae	10

Plate 10.1

Table 10.2 Diversity for Trees at Herbal Garden- Block B including behind glasshouse and Office of the ATARI

Sr.	Scientific Name	Common Name	Family	Number
1.	*Aegle marmelos*	Bel	Rutaceae	13
2.	*Albizzia lebbeck*	Kala Siris	Leguminosae	1
3.	*Artocarpus lacucha*	Badhal	Moraceae	2
4.	*Azadirachta indica*	Neem	Meliaceae	16
5.	*Bauhinia variegata*	Kachnar	Fabaceae	2
6.	*Borassus flabellifer*	Toddypalm	Arecaceae	3
7.	*Buchanania lanzan*	Chironji	Anacardiaceae	4
8.	*Butea frondosa*	Palas	Leguminosae	2
9.	*Cassia fistula*	Amaltas	Caesalpiniaceae	5
10.	*Cordia dichotoma*	Lasuda	Ehretiaceae	3
11.	*Dendrocalamus strictus*	Bas	Poaceae	50
12.	*Eucalyptus globules* Lobit.	Eucalyptus	Myrtaceae	22
13.	*Ficus benghalensis*	Bargad	Moraceae	2
14.	*Ficus glomerata*	Umar	Moraceae	1
15.	*Gmelina arborea*	Khamer	Lamiaceae	32
16.	*Hevea brasiliensis*	Rubber	Euphorbiaceae	1
17.	*Hyophorbe lagenicaulis*	Bottle palm	Arecaceae	22
18.	*Juniperus communis*	Juniper	Cupressaceae	5
19.	*Lagerstroemia indica*	Landia	Lythraceae	1
20.	*Leucaena leucocephala*	Subabul	Mimosaceae	3
21.	*Mangifera indica*	Mango	Anacardiaceae	7
22.	*Media azedarach*	Mahaneem	Meliaceae	3
23.	*Mimusops elengi*	Malshree	Sapotaceae	1
24.	*Moringa oleifera*	Munga	Moringaceae	1
25.	*Morus alba*	Sehtuth	Moraceae	3
26.	*Murraya exotica*	Madhu Kamini	Rutaceae	6
27.	*Neolamarckia cadamba*	Kadam	Rubiaceae	1
28.	*Phyllanthus emblica*	Aonla	Euphorbiaceae	99
29.	*Phyllantun acidus*	Harparevdi	Phyllanthaceae	5
30.	*Pithecellobium dulce*	Jungle jalebi	Mimosaceae	4
31.	*Polyalthea longifolia var.pendula*	Ashok	Annonaceae	16
32.	*Pongamia pinnata*	Karanj	Fabaceae	3
33.	*Putranjiva roxburghii*	Putranjiva	Putaranjivaceae	8
34.	*Roystonea regia*	Royal palm	Arecaceae	1
35.	*Semecarpus anacardium*	Bhilwa	Anacardiaceae	4
36.	*Spathodea companulata*	Fountain tree	Bignoniaceae	1
37.	*Syzygium cumini*	Jamun	Myrtaceae	1
38.	*Tabebuia rosea*	Basant ki Rani	Bignoniaceae	3
39.	*Terminalia arjuna*	Arjun	Combretaceae	15
40.	*Terminalia bellirica*	Baheda	Combretaceae	5

Plate 10.2 Herbal garden near glass house

Plate 10.3 Herbal garden Block B

Table 10.3 Diversity for Trees at Block C Herbal Garden (Opposite building of the College towards Department of Plant Pathology)

Sr.	Scientific Name	Common name	Family	Number
1.	*Acacia catechu*	Katha	Fabaceace	2
2.	*Acacia nilotica*	Babul	Mimosaceae	1
3.	*Acacia auriculiformis*	Australian babul	Mimosaceae	1
4.	*Acacia catechu*	Khair	Mimosaceae	5
5.	*Achras sapota*	Sapota	Sapotaceae	4
6.	*Aegel marmelos*	Bel	Rutaceae	1
7.	*Albizia procera*	Safed seris	Mimosaceae	3
8.	*Albizzia lebbek*	Kala seris	Leguminaceae	4
9.	*Alstonia scholaris*	Saptparni	Apocynaceae	4
10.	*Annona muricata*	Mamfal	Annonaceae	4
11.	*Annona reticulata*	Ramfal	Annonaceae	2
12.	*Artabotrys hexapetalus*	Hari champa	Annonaceae	5
13.	*Artocarpus lakoocha*	Badhel	Moraceae	6
14.	*Azadirachta indica*	Neem	Meliaceae	5
15.	*Bambusa vulgaris*	Bas (bunch)	Poaceae	4
16.	*Bambusa vulgaris*	Pila bas	Poaceae	70
17.	*Bauhinia variegata*	Sizeria	Caesalpiniaceae	17
18.	*Bixa orellana*	Sinduri	Bixaceae	2
19.	*Bombax malabaricum*	Semal	Bombacaceae	3
20.	*Borassus flabellifer*	Tad	Arecaceae	1
21.	*Butea frondosa*	Palas	Leguminosae	4
22.	*Callistemon lanceolatus*	Bottle brush	Myrtaceae	1
23.	*Careya arbarea*	Kumbhi	Lecythidaceae	32
24.	*Carissa carandas*	Karonda	Apocynaceae	2
25.	*Cascabela thevetia*	Kanner	Apocynaceae	10
26.	*Cassia siamea*	Rajvriksha	Caesalpiniaceae	2
27.	*Ceiba pentandra*	Kapok	Malvaceae	3
28.	*Cicca acida*	Harparevdi	Phyllanthaceae	11
29.	*Cinnamomum tamala*	Tejpatta	Lauraceae	3
30.	*Copernicia alba*	Royal palm	Arecaceae	2
31.	*Cordia macleoidi*	Dahiman	Boraginaceae	8
32.	*Crataeva religia*	Varun	Capparaceae	1
33.	*Dalbergia latifolia*	Shisham	Fabaceae	2
34.	*Dalbergia sissoo*	Sisham	Fabaceae	10
35.	*Elaeis guineensis*	Oil palm	Arecaceae	20
36.	*Emblica officinalis*	Aonla	Euphorbiaceae	14
37.	*Erythria variegata*	Pangara	Fabaceae	2
38.	*Erythrina orientalis*	Pangara	Fabaceae	3
39.	*Eucalyptus globulus*	Eucalyptus	Myrtaceae	73
40.	*Ficus glomerata*	Gular	Moraceae	1
41.	*Ficus religiosa*	Pipal	Moraceae	2
42.	*Gardenia gummifera*	Dikamali	Rubiaceae	4
43.	*Gossypium arboreum*	Kapas	Malvaceae	2
44.	*Grevilleae robusta*	Silver oak	Proteaceae	1
45.	*Hibiscus rosa-sinensis*	Gudhal	Malvaceae	25

Sr.	Scientific Name	Common name	Family	Number
46.	*Hyophorbe lagenicaulis*	Bargad	Moraceae	3
47.	*Juniperus communis*	Juniper	Cupressaceae	1
48.	*Kigelia pinnata*	Jharfanus	Bignoniaceae	8
49.	*Lagerstroemia indica*	Legerus stomia	Lythraceae	10
50.	*Leucaena leucocephala*	Subabul	Fabaceae	9
51.	*Leucas aspera*	Tumbi	Fabaceae	1
52.	*Litoria umarensis*	Umar	Pelodryadiaceae	1
53.	*Livistona chinensis*	Fan palm	Arecaceae	13
54.	*Magnolia champa*	Chalta champa	Magnoliaceae	1
55.	*Mallotus Philipinensis*	Kamala tree	Euphorbiaceae	10
56.	*Mangifera Indica*	Mango	Anacardiaceae	10
57.	*Manilkara hexandra*	Kherni	Sapotaceae	4
58.	*Melia azedarach*	Bakain	Meliaceae	26
59.	*Mimusops elengi*	Malshree	Sapotaceae	2
60.	*Mitragyna parviflora*	Mundi	Rubiaceae	1
61.	*Morus alba*	Sehtuth	Moraceae	4
62.	*Neolamarckia cadamba*	Kadam	Rubiaceae	2
63.	*Oroxylum indicum*	Shivnag	Bingoniaceae	8
64.	*Parkia biglandulosa*	Parkia	Mimosaceae	6
65.	*Parkinsonia aculeata*	Palas	Fabaceae	2
66.	*Peltophorum pterocarpum*	Peltaphorum	Mimosaceae	10
67.	*Pinus roxburghii*	Chir pine	Pinaceae	1
68.	*Pinus syivestris L.*	Pynus	Pinaceae	2
69.	*Pithecellobium dulce*	Jangal jalebhi	Fabaceae	13
70.	*Polyalthea longifolia*	Ashok	Annonaceae	21
71.	*Pongamia pinnata*	Karanj	Fabaceae	4
72.	*Prosopis cineraria*	Shami	Fabaceae	5
73.	*Psidium cattleianum*	Chinise amrud	Myrtaceae	1
74.	*Pterospermum acerifolium*	Kanak champa	Malvaceae	5
75.	*Putranjiva roxburghii*	Putaranj jiva	Putaranjivaceace	9
76.	*Santalum album*	Chandan	Sanatalaceae	2
77.	*Saraca asoca*	Sita ashok	Fabaceae	4
78.	*Schleichera oleosa*	Kusum	Sapindaceae	4
79.	*Semecarpus anacardium*	Bhilwa	Anacrdiaceae	2
80.	*Sterculia foetida*	Jangli badam	Sterculiaceae	1
81.	*Strychnos nux-vomica*	Kuchla	Strychnaceae	4
82.	*Syzygium cumini*	Jamun	Myrtaceae	6
83.	*Syzygium fr uiticosum*	Kath jamun	Myrtaceae	1
84.	*Tabebuia Rosea*	Basant ki rani	Bignoniaceae	25
85.	*Tamarindus indica*	Tamarind	Fabaceae	1
86.	*Terminalia bellirica*	Baheda	Combretaceae	8
87.	*Thespesia populnea*	Paras pipal	Malvaceae	5
88.	*Thevetia peruviana*	Kaner	Apocynaceae	8
89.	*Thuja occidentalis*	Thuja	Cupressaceae	8
90.	*Vachellia nilotica*	Babul	Fabaceae	4
91.	*Ziziphus mauritiana*	Ber	Rhamnaceae	1

Plate 10.4 Herbal garden block C

Plate 10.5 Plantation towards Department of Plant Pathology

Table 10.4 Diversity for Trees at Medicinal Garden- Block-D Nursery & Conservatory of the Department of Plant Physiology

Sr	Scientific Name	Common name	Family	Number
1.	*Acacia nilotica*	Babul	Mimosaceae	1
2.	*Acacia conncinna*	Shikakai	Mimosaceae	10
3.	*Acacia mangium*	Australlion teak	Mimosaceae	2
4.	*Achras sapota*	Chiku	Sapotaceae	1
5.	*Adansonia Digitata*	Kalpvraksh	Malvaceae	1
6.	*Adenanthera parvania*	Lal chandan	Fabaceae	2
7.	*Aegle marmelos*	Bel	Rutaceae	13
8.	*Albizia procera*	Safed siris	Fabaceae	1
9.	*Albizzia lebbeck*	Kala seris	Leguminosae	2
10.	*Annona muricata*	Mamfal	Annonaceae	21
11.	*Annona reticulata*	Ramfal	Annonaceae	2
12.	*Anogeissus latifolia*	Dacher	Combretaceae	1
13.	*Areca catechu*	Supari	Arecaceae	1
14.	*Artocarpus lacucha*	Badhal	Moraceae	2
15.	*Azadirachta indica*	Neem	Meliaceae	11
16.	*Bambusa tulda*	Gathia bas	Poaceae	2
17.	*Bambusa vulgaris*	Yellow bas	Poaceae	2
18.	*Bauhinia vahli*	Mahul	Leguminosae	1
19.	*Bauhinia variegata*	Kachnar	Fabaceae	7
20.	*Bersera spp.*	Pink pipal	Burseraceae	1
21.	*Bixa orellana*	Sinduri	Bixaceae	2
22.	*Boswellia serrata*	Salai	Burseraceae	2
23.	*Butea monosperma*	Palas	Fabaceae	1
24.	*Callistemon lanceolatus*	Bottle brush	Myrtaceae	1
25.	*Cassia fistula*	Amaltas	Fabaceae	3
26.	*Cinnamomum camphora*	Kapoor	Lauraceae	2
27.	*Cinnamomum tamala*	Tejpatta	Lauraceae	2
28.	*Cinnamomum zeylanicum*	Dalchini	Lauraceae	1
29.	*Citrus limetta*	Mosambi	Rutaceae	1
30.	*Cleistanthus collinus*	Garari	Phyllanthaceae	1
31.	*Cocos nucifera*	Coconut	Arecaceae	4
32.	*Crataeva nurvala*	Varun	Cappraceae	2
33.	*Delonix regia*	Gulmohar	Fabaceae	1
34.	*Dendrocalamus strictus*	Hara bas	Poaceae	3
35.	*Dillenia indica*	Chalta champa	Magnoliaceae	1
36.	*Elaeis guineensis*	Oil palm	Arecaceae	24
37.	*Elaeocarpus angustifolius*	Rudraksh	Elaeocarpaceae	4
38.	*Emblica officinalis*	Aonla	Euphorbiaceae	68
39.	*Erythria variegata*	Pangara	Fabaceae	5
40.	*Eucalyptus citriodora*	Odomos tree	Myrtaceae	116
41.	*Eucalyptus globules*	Nilgiri	Myrtaceae	123
42.	*Euginea uniflora*	Tropical cherry	Myrtaceae	1
43.	*Ficus benghalensis*	Bargad	Moraceae	1
44.	*Ficus benjamina*	Ficus	Moraceae	2
45.	*Ficus elastica*	Rubber plant	Moraceae	1
46.	*Ficus glomerata*	Gular	Moraceae	1
47.	*Ficus religiosa*	Pipal	Moraceae	2
48.	*Ficus virens*	Pakhar	Moraceae	1

Sr	Scientific Name	Common name	Family	Number
49.	*Gliricidia sepium*	Glericedia	Fabaceae	5
50.	*Gmelina arborea*	Khamer	Lamiaceae	13
51.	*Grevillea robusta*	Silver oak	Proteaceae	5
52.	*Helicteres isora*	Maror fali	Malvaceae	1
53.	*Hyophorbe lagenicaulis*	Bottle palm	Arecaceae	3
54.	*Jacaranda mimosifolia*	Jakaranda	Bignoniaceae	2
55.	*Juniperus communis*	Juniper	Cupressaceae	3
56.	*Leucaena leucocephala*	Subabul	Fabaceae	6
57.	*Litsea glutinosa*	Maida tree	Lauraceae	3
58.	*Mallotus philippensis*	Melotus	Euphorbiaceae	5
59.	*Mangifera indica*	Mango	Anacardiaceae	3
60.	*Melia azedarach*	Bakain	Meliaceae	3
61.	*Michelia champaka*	Swarna champa	Magnoliaceae	4
62.	*Moringa oleifera*	Munga	Moringaceae	1
63.	*Murraya koenigii*	Mithi neem	Rutaceae	1
64.	*Neolamarckia cadamba*	Kadam	Rubiaceae	6
65.	*Nerium alba*	Kanner (safed)	Apocynaceae	2
66.	*Nerium indicum*	Kanner (lal)	Apocynaceae	4
67.	*Oroxylum indicum*	Shivnag	Bignomiaceae	68
68.	*Peltaphorum pterocarpum*	Peltaphorum	Fabaceae	4
69.	*Pimenta dioica*	Allspice	Myrtaceae	3
70.	*Pithecellobium dulce*	Manila tamarind	Fabaceae	6
71.	*Plumeria alba*	Champa (safed)	Apocynaceae	4
72.	*Plumeria rubra*	Champa (red)	Apocynaceae	4
73.	*Polyalthia longifolia var. pendula*	Pendula ashok	Annonaceae	2
74.	*Polyalthia longifolia*	Ashok (Spreading type)	Annonaceae	7
75.	*Prosopis cineraria*	Shami	Mimosaceae	1
76.	*Pterocarpus marsupium*	Bijasar	Fabaceae	11
77.	*Pterospermum acerifolium*	Kanak champa	Malvaceae	7
78.	*Salmalia malabarica*	Semal	Bombacaceae	1
79.	*Santalum album*	Safed chandan	Sanatalaceae	6
80.	*Saraca asoca*	Sita ashok	Fabaceae	12
81.	*Semecarpus anacardium*	Bhilma (marking nut)	Anacardiaceae	1
82.	*Sesbania augsata*	August	Arecaceae	10
83.	*Sterculia guttata*	Sterculia	Sterculiaceae	4
84.	Sterculia *urens*	Gullugum	Malvaceae	6
85.	*Stereospermum suavealens*	Garud	Bignoniaceae	2
86.	*Syzygium aromaticum*	Laung	Myrataceae	1
87.	*Syzygium cumini*	Jamun	Myrtaceae	2
88.	*Tabebuia pentaphylla*	Basant ki rani (Yellow)	Bignoniaceae	40
89.	*Tabebuia rosea*	Basant ki Rani (Pink)	Bignoniaceae	5
90.	*Terminalia catappa*	Badam	Combretaceae	4
91.	*Theobroma cacao*	Cocao	Sterculiaceae	2
92.	*Thevetia alba*	Safed kanner	Apocynaceae	1
93.	*Thevetia peruviana*	Kanner (pili)	Apocynaceae	3
94.	*Thuja occidentalis*	Thuja	Cupressaceae	21
95.	*Triadica sebifera*	Wax tree	Euphorbiaceae	2
96.	*Vitex negundo*	Nirgundi	Verbenaceae	6
97.	*Ziziphus mauritiana*	Ber	Rhamnaceae	4

Plate 10.6 Herbal garden Block D

Plate 10.7 Conservatory of the Department of Plant Physiology

Table 10.5 Diversity for Trees at Jawahar Stadium to B Tech Hostel Square towards Uldana Nalla including Herbal Garden –Block-E

S.No.	Scientific Name	Common Name	Family	Number
1.	*Aegle marmelos*	Bel	Rutaceae	3
2.	*Azadirachta indica*	Neem	Meliaceae	34
3.	*Bauhinia purpurea*	Pink Kachnar	Caesalpiniaceae	6
4.	*Bauhinia variegata*	Kachnar	Caesalpiniaceae	5
5.	*Bombax ceiba*	Semal	Malvaceae	2
6.	*Careya arborea*	Kumbhi	Barringtoniaceae	14
7.	*Cassia fistula*	Amaltas	Caesalpiniaceae	4
8.	*Cicca acida*	Harfarevri	Euphorbiaceae	5
9.	*Dalbergia sissoo*	Sissoo	Fabaceae	10
10.	*Ficus benghalensis*	Bargad	Moraceae	2
11.	*Ficus religiosa*	Pipal	Moraceae	3
12.	*Gmelina arborea*	Khamer	Vebenaceae	110
13.	*Haloptelea integrifolia*	Chilbil	Ulmaceae	2
14.	*Madhuca latifolia*	Mahua	Sapotaceae	1
15.	*Mangifera indica*	Mango	Anacardiaceae	15
16.	*Oroxylum indicum*	Shivnag	Bignoniaceae	68
17.	*Peltaphorum pterocarpum*	Copper pod tree	Caesalpiniaceae	3
18.	*Pithecellobium dulce*	Jangal Jalebi	Mimosaceae	5
19.	*Polyalthia longifolia*	Ashok	Annonaceae	30
20.	*Pongamia pinnata*	Kanji	Fabaceae	3
21.	*Saraca indica*	Sita Ashok	Fabaceae	4
22.	*Syzygium cumini*	Jamun	Myrtaceae	2
23.	*Tamarindus indica*	Imli	Fabaceae	3
24.	*Terminalia arjuna*	Arjun	Cobretaceae	12
25.	*Thespesia populnea*	Paras Pipal	Malvaceae	18
26.	*Zizyphus mauritiana*	Ber	Rhamnaceae	2
27.	*Zizyphus xylopyra*	Ghont Ber	Rhamnaceae	2

Plate 10.8 Diversity for Trees at Jawahar Stadium

Table 10.6 Diversity for Trees around the University Administrative Block

Sr.	Scientific Name	Common Name	Family	Number
1.	*Bombax ceiba*	Semal	Malvaceae	6
2.	*Borassus flabelliferl*	Tadi Palm	Arecaceae/ palmae	5
3.	*Ceiba pentandra*	Kapok	Malvaceae	10
4.	*Delonix regia*	Gulmohar	Fabaceae	2
5.	*Dillenia indica*	Chalta champa	Magnoliaceae	1
6.	*Eucalyptus globules Lobit.*	Eucalyptus	Myrtaceae	8
7.	*Ficus benghalensis*	Bargad	Moraceae	1
8.	*Ficus religiosa*	Pipal	Moraceae	4
9.	*Grevillea robusta*	Silver Oak	Proteaceae	1
10.	*Hyophorbe lagenicaulis*	Bottle palm	Arecaceae	4
11.	*Livistona chinensis*	Fen palm	Arecaceae	2
12.	*Livistona rotundifolia*	Table palm	Arecaceae	1
13.	*Mangifera indica*	Mango	Anacardiaceae	10
14.	*Manilkara hexandra*	Khirni	Sapotaceae	2
15.	*Mimusops elengi*	Maulshree	Sapotaceae	1
16.	*Morus alba*	Shehtoot	Moraceae	1
17.	*Peltophorum pterocarpum*	Peltaforum	Fabaceae	4
18.	*Pithecellobium dulei*	Jungle jalebi	Fabaceae	5
19.	*Polyalthia longifolia*	Ashok	Annonaceae	143
20.	*Putranjiva roxburghii*	Putranj jiva	Annonaceae	4
21.	*Saraca asoca*	Sita ashok	Fbaceae	4
22.	*Spathodea*	Fountain tree	Bignoniaceae	3
23.	*Sphaeranthus indicus*	Mundi	Rubiaceae	1
24.	*Syzygium cumini*	Jamun	Myrtaceae	2
25.	*Tabebuia pentaphylla*	Basant ki rani	Bignoniaceae	5

Plate 10.9 Diversity for Trees at the University headquarters

Table 10.7 Diversity for trees at the Garden of College of Agriculture, Jabalpur

Sr.	Scientific Name	Common Name	Family	Number
1.	*Callistemon lanceolatus*	Bottle brush	Myrtaceae	4
2.	*Hyophorbe lagenicaulis*	Bottle Palm	Arecaceae	15
3.	*Mangifera indica*	Mango	Anacardiaceae	2
4.	*Mimusops elengi*	Maulshree	Sapotaceae	1
5.	*Polyalthia longifolia.*	Ashok	Annonaceae	61
6.	*Syzygium cumini*	Jamun	Myrtaceae	1

Plate 10.10 Garden of College of Agriculture, Jabalpur

Table 10.8 Diversity for trees at behind College of Agriculture, Jabalpur (Office of the DRS, DES, Seed Technology etc)

Sr.	Scientific Name	Common Name	Family	Number
1.	*Acacia auriculiformis*	Australian Babul	Mimosaceae	2
2.	*Albizia lebbeck*	Kala sirus	Fabaceae	2
3.	*Albizia procera*	Safed sirus	Mimosaceae	3
4.	*Alstonia scholaris*	Saptparni	Apocynaceae	2
5.	*Annona reticulata*	Ramfal	Annonaceae	1
6.	*Careya arborea roxb.*	Kumbhi	Lecythidaceae	1
7.	*Ceiba pentandra*	Kapok	Malvaceae	1
8.	*Dalbergia sissoo*	Sisu	Fabaceae	10
9.	*Delonix regia*	Gulmohar	Fabaceae	9
10.	*Eucalyptus globules Lobit.*	Eucalyptus	Myrtaceae	11
11.	*Ficus benghalensis*	Bargad	Moraceae	1
12.	*Ficus religiosa*	Pipal	Moraceae	1
13.	*Hyophorbe lagenicaulis*	Bottle Palm	Arecaceae	17
14.	*Kigelia africana*	Kigelia	Bignoniaceae	1
15.	*Kigelia pinnata*	Sausage tree	Bignoniaceae	2
16.	*Lagerstroemia parviflora.*	Landia	Lythraceae	2
17.	*Lagerstroemia indica*	Legerus stomia	Lythraceae	2
18.	*Mangifera indica*	Mango	Anacardiaceae	3
19.	*Millettia pinnata*	Karanj	Fbaceae	1
20.	*Peltophorum pterocarpum*	Peltaforum	Fbaceae	11
21.	*Polyalthia longifolia*	Sita ashok	Annonaceae	12
22.	*Saraca asoca*	Ashok	Fabaceae	21
23.	*Senna siamea*	Kesia semia	Fabaceae	2
24.	*Syzygium cumini*	Jamun	Myrtaceae	1
25.	*Tabebuia entaphuylla*	Basant ki rani	Bignoniaceae	5
26.	*Thespesia popunea*	Paras Pipal	Malvaceae	1

Plate 10.11 Plantation behind College of Agriculture, Jabalpur

Table 10.9 Diversity for trees at the Cycle Stand near Plant Pathology to MS Swaminathan PG Hostel, College of Agriculture, Jabalpur

Sr.	Scientific Name	Common Name	Family	Number
1.	*Azadirachta indica*	Neem	Meliaceae	9
2.	*Careya arbarea*	Kumbhi	Lecythidaceae	5
3.	*Emblica officinalis*	Aonla	Euphorbiaceae	2
4.	*Ficus glomerata*	Umar	Moraceae	3
5.	*Ficus religiosa*	Pipal	Moraceae	1
6.	*Lagerstroemia indica*	Landia	Lythraceae	3
7.	*Leucaena leucocephala*	Subabul	Annonaceae	1
8.	*Melia azedarach*	Bakain	Meliaceae	1
9.	*Pithecellobium dulce*	Jangal jalebi	Fabaceae	6
10.	*Plumeria alba*	Champa	Apocynaceae	3
11.	*Polyalthia longifolia.*	Ashok	Annonaceae	30
12.	*Pongamia pinnata*	Karanj	Fabaceae	1
13.	*Putranjiva roxburghii*	Putranjiva	Putranjivaceae	4
14.	*Salmalis malaberica*	Semal	Malvaceae	6
15.	*Syzygium cumini*	Jamun	Myrtaceae	1
16.	*Syzygium saliafelium*	Kath Jamun	Myrtaceae	1
17.	*Tabebuia rosea*	Basant ki rani	Bignoniaceae	4
18.	*Ziziphus mauritiana*	Ber	Rhamnaceae	4

Plate 10.12 Plantation near Dr MS Swaminathan Postgraduate Hostel

Table 10.10 Diversity for trees at Jawahar Stadium Road Jawahar stadium to the Hostel of the College of Agricultural Engineering Hostel Square (West to East)

Sr.	Scientific Name	Common Name	Family	Number
1.	*Albizia procura*	Safed siris	Mimosaceae	17
2.	*Albizzia lebbeck*	Kali siris	Mimosaceae	2
3.	*Artocarpus lacucha*	Badhal	Moraceae	4
4.	*Azadirachta indica*	Neem	Meliaceae	4
5.	*Bauhinia variegata*	Kachnar	Fabaceae	40
6.	*Cassia firtula* Linn.	Amaltas	Fabaceae	6
7.	*Cicca acida*	Herparewri	Phyllanthaceae	8
8.	*Eucalyptus globules Lobit.*	Eucalyptus	Myrtaceae	2
9.	*Feronia elephandum*	Kaith	Rutaceae	1
10.	*Haloptelea integrifolia*	Chilbil	Ulmaceae	2
11.	*Lagerstroemia indica*	Legerus stomia	Lythraceae	9
12.	*Leucaena leucocephala*	Subabul	Fabaceae	2
13.	*Mangifera indica*	Mango	Anacardiaceae	16
14.	*Peltophorum pterocarpum*	Peltaforum	Caesalpiniaceae	3
15.	*Pithecellobium dulce*	Jungle jalebi	Fabaceae	8
16.	*Polyalthia longifolia.*	Nakli Ashok	Annonaceae	6
17.	*Sapindus mukorossi*	Ritha	Sapindaceae	3
18.	*Syzygium cumini*	Jamun	Myrtaceae	3
19.	*Tabebuia Rosea*	Basath kirani	Bignoniaceae	1
20.	*Terminalia arjuna*	Arjun	Combretaceae	3
21.	*Thespesia populnea*	Paras pipal	Malvaceae	10
22.	*Ziziphus mauritiana*	Ber	Rhamnaceae	5

Plate 10.13 Plantation near Jawahar Stadium

Table 10.11 Diversity for trees at College of Agricultural Engineering and Krishi Vigyan Kendra

Sr.	Scientific Name	Common Name	Family	Number
1.	*Azadirachta indica*	Neem	Meliaceae	4
2.	*Eucalyptus globulus Lobit.*	Eucalyptus	Myrtaceae	7
3.	*Mangifera indica*	Mango	Anacardiaceae	54
4.	*Thespesia populnea*	Paras pipal	Malvaceae	1
5.	*Dalbergia sissoo*	Sisham	Fabaceae	10
6.	*Emblica officinalis*	Aonla	Euphorbiaceae	3
7.	*Madhuca latifolia*	Mahua	Sapotaceae	1
8.	*Ficus benghalensis*	Bargad	Moraceae	2
9.	*Aegle marmelos*	Bel	Rutaceae	1
10.	*Polyalthia longifolia.*	Ashok	Annonaceae	88
11.	*Delonix regia*	Gulmohar	Fabaceae	3
	Total			**174**

Plate 10.14 Plantation near College of Agricultural Engineering, Jabalpur

CHAPTER 11

Biodiversity Audit of Tree Species at BSP Unit of Vegetable Crops

Akhilesh Tiwari, Rahul Dongre, Bhupesh Verma
Department of Horticulture
College of Agriculture, Jabalpur

Breeder Seed Production Unit of Vegetables at Maharajpur consists of 84.64ha irrigated land, Vertisols, Typichaplusterts, Very fine mont- morrilonite, Hyper-thermic Clay loam, sandy clay and sandy clay loam. The farm contains 605 trees of 28 species representing 19 families. The farm has nearly 7 trees per ha (Fig. 11.1).

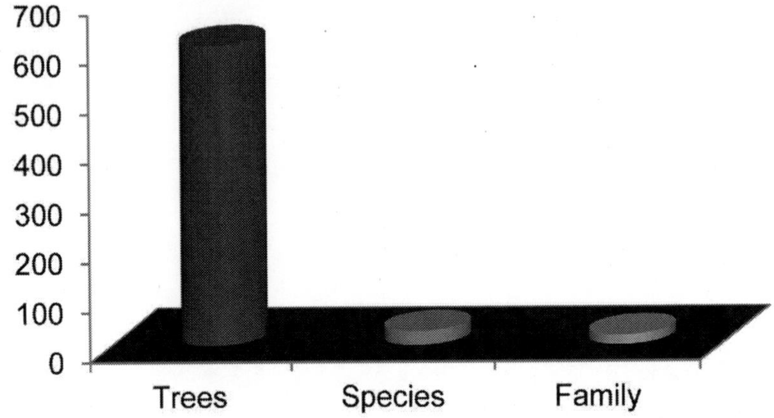

Fig 11.1 Tree Diversity at Breeder Seed Production Unit of Vegetables, Maharajpur

The diversity at this farm is about 7 trees per ha which is very low because this farm is mainly used for breeder seed production and thus shades of the tree restrict the proper growth of the field crops and thus affect seed production. The highest tree density is of neem (175), followed by bael (91), and custard apple (81) (Table 11.1 and Plates 11.1).

Plate 11.1 Plantation of neem as avenue tree

Table 11.1 Bio-Diversity for Tree species at Breeder Seed Production Unit of Vegetables, Maharajpur

Sr.	Name		Family	Number	Field Number
	Botanical	Common			
		Samera		02	24
1.	*Aegle mormelos* LCorrea	Bael	Rutaceae	91	
2.	*Annona squamosa* L.	*Custard apple*	Annonaceae	81	
3.	*Azadirachta indica* A.Juss.	*Neem*	Meliaceae	175	30,31,36,37,24,25, Main road
4.	*Bambusa vulgaris*	Bamboo	Poaceae	8	Fencing weed
5.	*Butea monosperma*	*Palash*	Fabaceae	10	37,24
6.	*Carissa carandas*	Karonda	Apocynaceae	5	37, Main road
7.	*Citrus aurantiifolia*	*Lime*	Rutaceae	02	29
8.	*Dalbergia sissoo*	*Shisham*	Caesalpiniaceae	38	24,37,40, Fencing weed
9.	*Delonix regia* (Bojer) Raf.	*Gulmohar*	Caesalpiniaceae	02	30
10.	*Eucalyptus globulus* Labill.	*Eucalyptus*	Myrtaceae	33	25,37,40, Main road, Fencing weed
11.	*Ficus religiosa* L.	*Peepal*	Moraceae	08	25,30,31,35
12.	*Leucaena leucocephala* Lam.	Subabul	Mimosaceae	03	
13.	*Madhuca longifolia* Konig	*Mahua*	Sapotaceae	01	30
14.	*Mangifera indica* L.	*Mango*	Anacardeaceae	31	30,36,37,25,35,40, main road
15.	*Millettia pinnata*	Kanjee	Fabaceae	4	Fencing weed
16.	*Moringa oleifera* Lam.	Drumstick	Moringaceae	04	30,35
17.	*Morus alba*	*Shahtoot*	Moraceae	3	29
18.	*Musa paradisiaca* Linn	*Banana*	Musaceae	01	29
19.	*Neolamarckia adamba*	*Kadamb*	Rubiaceae	02	25
20.	*Pithecellobium dulce* (Roxb.) Benth		Mimosaceae	08	
21.	*Psidium guajava* L.	*Guava*	Myrtaceae	20	24,25,30,35
22.	*Saraca asoca* (Roxb.) Willd	*Ashok*	Detarioideae	02	30
23.	*Syzygium cumini* Skeels		Myrtaceae	16	
24.	*Tamarindus indica* L.	*Tamarind*	Caesalpiniaceae	36	24,25,30,37, 40
25.	*Terminalia arjuna*	Arjun	Combretaceae	2	Fencing weed
26.	*Terminalia bellirica* (Gaertn.) Roxb.	*Baheda*	Combretaceae	02	31,40
27.	*Thespesia populnea*	*Paras peepal*	Malvaceae	01	29
28.	*Ziziphus mauritiana* Lam.	*Ber*	Rhamnaceae	14	25,30.37

Plate 11.2 BSP Unit of Vegetable Crops, Maharajpur Farm

CHAPTER 12

Biodiversity Audit of Forestry Farm, Imiliya

R K Bajpai
Department of Forestry
College of Agriculture, Jabalpur

A Farm of 6.23 ha dedicated to the Department of Forestry contains total 702 trees of 28 species representing 16 families (Plate 12.1). The highest representation (17%) is of *Subabul;* two species have 50-100; 15 species have 11-50; whereas 10 species have < 10 number of trees each. The tree density per ha is 113, species density is 5 and density of family is 3 per ha (Table 12.1 and Fig 12.1).

Table 12.1 Representation of tree species at Forestry Farm, Imiliya

S. N.	Name		Number	Family
	Scientific Name	**Common Name**		
1.	*Acacia auriculiformis*	Australian babul	15	Mimosaceae
2.	*Acacia nilotica*	Desi babul	63	Mimosaceae
3.	*Aegle mormelos*	Beal	20	Rutaceae
4.	*Albizia lebbeck*	Kala siris	7	Mimosaceae
5.	*Albizia procera*	White siris	8	Mimosaceae
6.	*Annona squamosa*	Sarifa	5	Annonaceae
7.	*Azadirachta indica*	Neem	23	Meliaceae
8.	*Bambusa vulgaris*	Bamboo	19	Poaceae
9.	*Butea monosperma*	Palas	26	Fabaceae
10.	*Cassia siamia*	Cassia	5	Caesalpiniaceae
11.	*Dalbergia sissoo*	Shisham	24	Caesalpiniaceae
12.	*Eucalyptus* spp.	Eucalyptus	77	Myrtaceae
13.	*Ficus religiosa*	Pipal	7	Moraceae
14.	*Leucaen aleucocephela*	Subabul	116	Mimosaceae
15.	*Limonia acidissimal*	Kaitha	14	Rutaceae
16.	*Madhuca indica*	Mahua	8	Sapotaceae
17.	*Mangifera indica*	Mango	2	Anacardiaceae
18.	*Manilkara hexandra*	Khirni	8	Sapotaceae
19.	*Moringa oleifera*	Moringa	10	Moringaceae
20.	*Phyllanthus emblica*	Aonla	42	Euphorbiaceae
21.	*Pithecellobium dulce*	Vilayati imli	5	Mimosaceae
22.	*Pongamia pinnata*	Karanja	17	Fabaceae
23.	*Syzygium cumini*	Jamun	33	Myrtaceae
24.	*Tamarindus indica*	Imli	23	Fabaceae
25.	*Tectona grandis*	Teak	42	Verbenaceae
26.	*Terminalia arjuna*	Arjun	9	Combretaceae
27.	*Terminalia bellirica*	Bahera	36	Combretaceae
28.	*Ziziphus jujuba*	Ber	38	Rhamnaceae
	Total Trees		**702**	

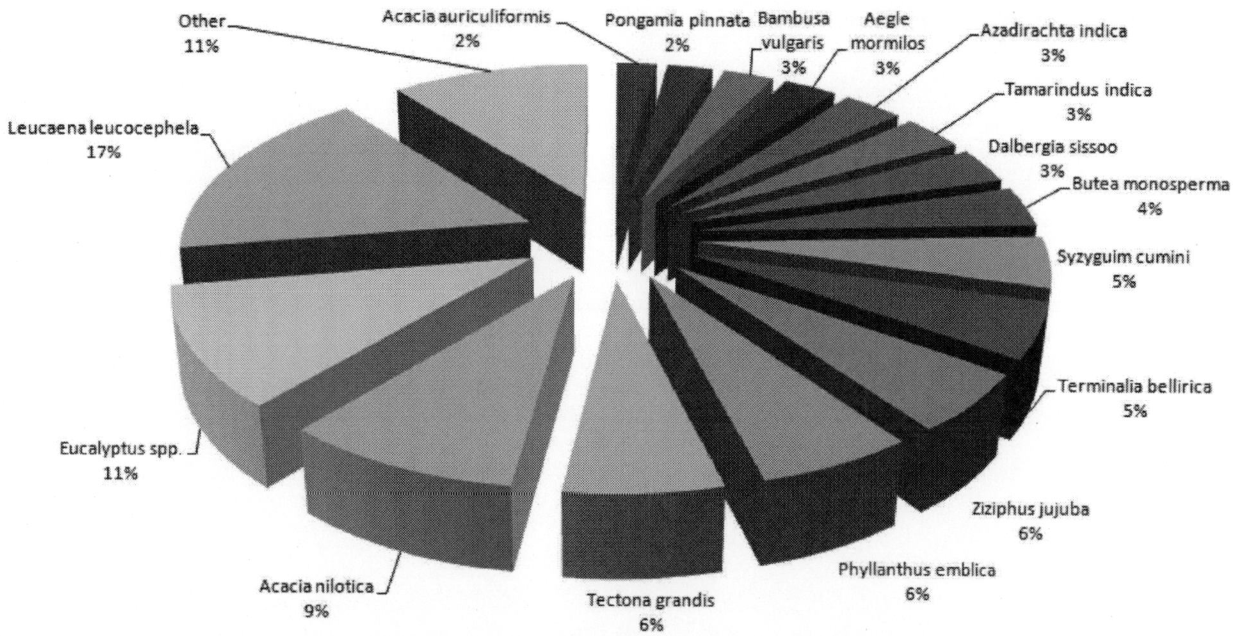

Fig 12.1 Representation of Tree Species at Farm of Department of Forestry

Plate 12.1 Forestry Farm, Imaliya Department of Forestry

CHAPTER 13

Biodiversity Audit of Horticulture Nursery, Adhartal

BP Bisen
Department of Horticulture
College of Agriculture, Jabalpur

Horticulture Nursery located at Adhartal farm on 3.5 ha irrigated land contains 227 trees of 41 species; 132 ornamental shrubs of 23 species; 388 plants of four rose species apart from 27 species of ornamental annuals (Table 13.1 and Fig 13.1). In all, 95 plant species exists in the nursery.

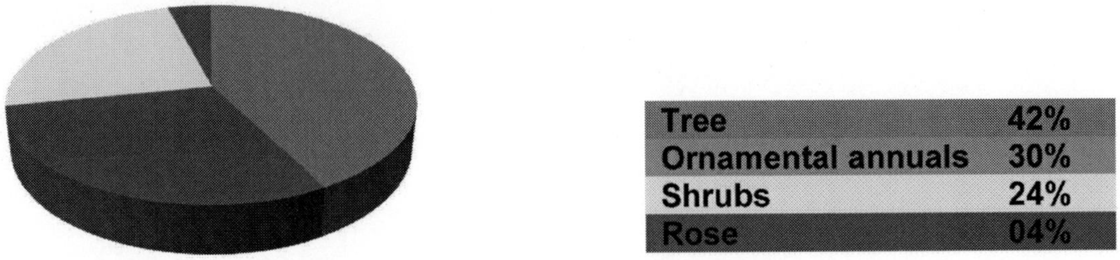

Fig 13.1 Percent representation of plant species at Adhartal Nursery

Table 13.1 Diversity for plant species at Adhartal Nursery

Sr	Name			Family	Number	Location
	Botanical	English	Hindi			
			Tree			
1.	*Achras zapota*	Sapota	Chiku	Sapotaceae	01	Nursery
2.	*Aegle marmelos*	Bengal Quince	Bael	Rutaceae	01	Nursery
3.	*Alstonia scholaris*	Devil's Tree	Saptparni	Aprocynaceae	01	Nursery
4.	*Artocarpus heterophyllus*	Jackfruit	Kathal	Moraceae	02	Nursery
5.	*Azadirachta indica*		Neem	Meliaceae	02	Nursery
6.	*Bassia latifoliasyn.Maduka indica*		Mahua	Sapotaceae	19	Nursery
7.	*Bauhinia purpurea*		Kachnar	Caesalpiniaceae	01	Nursery
8.	*Bombax malabaricum*	Red cotton silk	Semal	Bombaceae	02	Nursery
9.	*Callistemon lanceolatus*			Myrtaceae	02	Nursery
10.	*Cassia fistula*	Bottle Brush		Caesalpiniaceae/	01	Nursery
11.	*Citrus aurantiifolia*	Acid lime	/Niboo	Rutaceae	13	Nursery
12.	*Dalbergia sissoo*	Sissoo	Shisham	Caesalpiniaceae	01	Nursery
13.	*Delonix regia Syn. Poinciana regia*	Red Flamboyant	Amaltas	Caesalpiniaceae	01	Nursery
14.	*Emblica officinalis Gaertn*	Goose berry	Aonla	Euphorbiaceae	16	Nursery
15.	*Eucalyptus sp.*	Blue Gum	Safeda	Myrtaceae	18	Nursery
16.	*Ficus benjamina*	Weeping Fig	Pukar	Moraceae	01	Nursery
17.	*Ficus carica*	Fig	Anjeer	Moraceae	01	Nursery
18.	*Ficus elastica*	Rubber		Moraceae	02	Nursery
19.	*Ficus religiosa*		Pipal	Moraceae	01	Nursery
20.	*Ficus retusa*		Chilkan	Moraceae	01	Nursery
21.	*Grevillea robusta*	Silver oak		Proteaceae	01	Nursery
22.	*Inga dulcis*		VilaytiiImali	Caesalpiniaceae	02	Nursery
23.	*Mangifera indica* L	Mango	Aam	Anacardiaceae	27	Nursery
24.	*Michelia champaka*		SwarnaCahmpa	Magnoliaceae	01	Nursery
25.	*Morus alba*	Mulberry	Sehtoot	Moraceae	01	Nursery
26.	*Musa paradisiaca*	Banana	/ Kela	Musaceae	01	Nursery
27.	*Pinus* sp.	Pine Tree		Pinaceae	01	Nursery
28.	*Plumeria alba*	Temple tree		Aprocynaceae	01	Nursery
29.	*Plumeria rubra*	Red frangipani	Champa	Aprocynaceae	01	Nursery
30.	*Polyalthia longifolia*		Ashok	Annonaceae	21	Nursery
31.	*Polyalthia longifolia var. pendula*		Ashok	Annonaceae	09	Nursery
32.	*Psidium guajava* L	Guava	Aamrood	Myrtaceae	45	Nursery
33.	*Pterospermum acerifolium*		KanakChampa	Sterculiaceae	01	Nursery

Sr	Name			Family	Number	Location
	Botanical	English	Hindi			
34.	*Punica granatum* L	Pomegranate	Anar	Punicaceae	09	Nursery
35.	*Putranjivar oxburghii*		PutranJiva	Euphorbiaceae	02	Nursery
36.	*Roystonea regia*	Royal palm		Palmaceae	01	Nursery
37.	*Santalum album*	Sandal Wood	Chandan	Santalaceae	01	Nursery
38.	*Spondias cythera*	Indian Hog Plum	Amra	Anacardiaceae	01	Nursery
39.	*Tamarindus indica*	Tamarind	Imli/	Caesalpiniaceae	12	Nursery
40.	*Tectona grandis*	Teak	Sagwan	Verbenaceae	01	Nursery
41.	*Terminalia catapa*	Indian almond	Badam	Combretaceae	01	Nursery
		Ornamental Shrub				
1.	*Acalypha hispida*	Acalypha		Euphorbiaceae	04	
2.	*Bixa orellana* L		Sindoor	Bixaceae	01	
3.	*Bougainvillea species*	Bougainvillea		Nyctaginaceae	13	
4.	*Carissa carndus* L		Karonda	Apocynaceae	15	
5.	*Cestrum nocturnam*		Din Ka Raja	Solanaceae	02	
6.	*Cestrum norcturnum*		Rat Kee Rani	Solanaceae	12	
7.	*Codiaeum variegatum*	Croton		Euphorbiaceae	23	
8.	*Coffea arabica*	Coffee		Rubiaceae	01	
9.	*Euphorbia pulcherrima*	Poinsettia		Euphorbiaceae		
10.	*Hibiscus rosa-sinensis*	China rose		Malvaceae	06	
11.	*Ixora coccinea*		Rukmani	Euphorbiaceae	04	
12.	*Jasminum sambac*		Mogra/Bela	Oleaceae	28	
13.	*Lagerstroemia indica*		Shravani	Lythraceae	02	
14.	*Lantana camara*		Lantana	Verbenaceae	02	
15.	*Murraya panniculata*		Kamini	Rutaceae	02	
16.	*Mussaenda frondosa*	Mussaenda		Rubiaceae	02	
17.	*Mussaenda rosa*	Mussaenda		Rubiaceae	01	
18.	*Nerium indicum*	Red kaner		Apocynaceae	01	
19.	*Nyctanthes arbor-trisitis*	Tree of Sadness	HarShringar	Oleaceae	03	
20.	*Russelia juncea*	Coral bush		Scrophulariaceae	01	
21.	*Tabernaemontana divaricata*		Double chandani	Apocynaceae	03	
22.	*Tecoma stans*	Yellow bells		Bignoniaceae	05	
23.	*Thevetia peruviana*		PiliKaner	Apocynaceae	01	
24.		Xeroa				
		Winter Season Annuals				
1.	*Ageratum houstonianum*	Floss Flower		Compositae		
2.	*Althea rosea*	Holly Hock		Malvaceae		
3.	*Alyssum maritimum*	Sweet Alyssum		Cruiciferae		
4.	*Antirrhinum majus*	Snap Dragon		Scrophulariaceae		
5.	*Calendula officinalis*	Pot Marigold		Compositae		
6.	*Chrysanthemum coronarium*	Margaret		Compositae		
7.	*Cosmos bipinnatus*	Cosmoea		Compositae		
8.	*Dahlia variabilis*	Dahlia		Compositae		
9.	*Delphinium hybridum*	Larkspur		Ranunculaceae		
10.	*Eschecholtzia californica*	California poppy		Papaveraceae		
11.	*Helichryum bracteatum*	Straw Flower		Compositae		
12.	*Lathyrus odoratus*	Sweet pea		Leguminosae		
13.	*Mesembryanthemum criniflorum*	Ice Plant		Aizoaceae		
14.	*Nemasia strumosa*	Nargis		Scrophuliaraceae		

Sr	Name			Family	Number	Location
	Botanical	English	Hindi			
15.	*Papaver rhoeas*	Shirley poppy		Papaveraceae		
16.	*Petunia hybrida*	Petunia		Solanaceae		
17.	*Salvia splendens*	Salvia		Labiatae		
18.	*Tagetes erecta*	African Marigold		Asteraceae		
19.	*Tagetes patula*	French Marigold		Asteraceae		
20.	*Verbena hybrida*	Verbena		Verbanaceae		
21.	*Viola tricolor var.hortensis*	Pansy		Violaceae		
22.	*Aster amellus*		Aster	Asteraceae		
23.	*Iberis sempervirens*		Candytuft	Brassicaceae		
24.	*Dianthus caryophyllus*		Carnation	Caryophyllaceae		
25.	*Chrysanthemum coronarium*		Chrysanthemum	Asteraceae		
26.	*Centaurea cineraria*		Cineraria	Asteraceae		
27.	*Celosia cristata*		Cockscomb	Amaranthaceae		
28.	*Centaurea cyanus*		Cornflower	Asteraceae		
29.	*Dahlia pinnata*		Dahelia	Asteraceae		
30.	*Dianthus barbatus*		Dianthus	Caryophyllaceae		
31.	*Linum grandiflorum*		Flax	Linaceae		
32.	*Myosotis scorpioides*		Forget me not	Boraginaceae		
33.	*Gladiolus palustris*		Galdulus	Iridaceae		
34.	*Gazania linearis*		Gazania	Asteraceae		
35.	*Pelargonium graveolens*		Geranium	Geraniaceae		
36.	*Gerbera jamesonii*		Gerbera	Asteraceae		
37.	*Osteospermum fruticosum*		Osteospermum	Asteraceae		
38.	*Phlox panniculata*		Phlox	Polemoniaceae		
39.	*Dianthus plumarius*		Pink	Caryophyllaceae		
40.	*Primula vulgaris*		Primerose	Primulaceae		
41.	*Helianthus annus*		Sunflower	Asteraceae		
42.	*Dianthus barbatus*		Sweet william	Caryophyllaceae		
43.	*Polianthes tuberosa*		Tuberose	Asparagaceae		
44.	*Zinnia elegans*		Zinnia	Asteraceae		
		Summer/Rainy Season Annuals				
1	*Gaillardia pulchella*	Blanket Flower		Compositeae		
2	*Gomphrena globosa*	Globe Amaranth		Amaranthaceae		
3	*Impatiens balsamina*	Touch-Me-Not	Gulmundi	Balsaminaceae		
4	*Kochia scoparia*	Kochia		Amaranthaceae		
5	*Portulaca grandiflora*	Rose Moss		Portulaceae		
6	*Vinca rosea*		SadaBahar	Apocynaceae		
		Lily				
		Rose				
1	*Rosa chininsisvar.minima*	Miniature Rose	Rosaceae	02		
2	*Rosa damascena*	Noorjahan	Rosaceae	361		
3	*Rosa hybrida*	Gulab/Rose	Rosaceae	23		
4	*Rosa wichuriana*	Climbing Rose	Rosaceae	02		

CHAPTER 14

Biodiversity Audit of Fruit Research Farm, Imaliya

Rajani Sharma and SK Pandey
Department of Horticulture
College of Agriculture, Jabalpur

Biodiversity at Fruit Research Farm Imaliya (Plate 14.1) is presented in Table 14.1 shows that among the fruit crops the highest number is of guava (821) followed by ber (530) and mango (363). These are the dominating fruits crops of Kymore Plateau and Satpura hill. Other fruits crops with more than 100 fruit trees are custard apple (103), kagzi lime (166), Indian gooseberry (209), and pomegranate (106). Fruit plants of other regions viz., grapes, almond, mulberry, litchi, coconut etc are also planted at this farm. In total, 3194 trees are present on 22.47 ha land with average 142 trees/ ha.

Plate 14.1 Plantation of Guava, Fruit Research Farm, Imaliya

Table 14.1 Biodiversity at Fruit Research Farm Imaliya

Sr	Name			Family	Number
	Botanical	English	Hindi		
	Fruit trees				
1.	*Acarus sapota L*	Sapota	Cheeku	Sapotaceae	58
2.	*Aegle marmelos (L) Correa*		Bael	Rutaceae	12
3.	*Annona reticulata L*		Ramphal	Annonaceae	03
4.	*Annona squamosa L*	Custard apple	Seetaphal	Annonaceae	103
5.	*Arthocarpus heterophyllus Lamarck*	Jackfruit	Kathal	Moraceae	49
6.	*Carissa carandas L*	Bengal currant	Karonda	Apocynaceae	54
7.	*Citrus aurantiifolia* Swingle (L)	Kagzi Lime		Rutaceae	166
8.	*Citrus grandis / Citrus maxima* BurmMerr.	Pummelo	Chakotra	Rutaceae	01
9.	*Citrus jambhiri* Lush	Rough Lemon		Rutaceae	01
10.	*Citrus limonia* Osbeck	Rangpur Lime		Rutaceae	12
11.	*Citrus paradise* Macfad.	Grapefruit	Angur	Rutaceae	01
12.	*Citrus sinensis* L	Sweet Orange	Mosambi	Rutaceae	49
13.	*Cocos nucifera* L	Coconut	Nariyal	Arecaceae	05
14.	*Cordia myxa* Linnaeus		Lasoda	Boraginaceae	07
15.	*Diospyros melanoxylon* Roxburgh	persimmon	Tendu	Ebenaceae	02
16.	*Emblica officinalis* Gartner	Ggooseberry	Aonla	Euphorbiaceae	209
17.	*Ficus carica* L	Fig	Anjeer	Moraceae	-
18.	*Grewia asiatica* L	-	Phalsa	Proteaceae	38
19.	*Limonia acidissima* L	Wood Apple	Kaitha	Rutaceae	01
20.	*Litchi chinensis* Sonnerat		Litchi	Sapindaceae	21
21.	*Madhuca longifolia* (J.Konig)		Mahua	Sapotaceae	03
22.	*Mangifera indica* L	Mango	Aam	Anacardeaceae	363
23.	*Manilkara hexandra* (Roxburgh) Dubard	Rayan	Khirni	Sapotaceae	09
24.	*Moringa oleifera* Lamarck	Drumstick	Munga	Moringaceae	45
25.	*Morus alba* Linnaeus	Mulberry	Shahtoot	Moraceae	03
26.	*Musa paradisiaca* L	Banana	Kela	Musaceae	13
27.	*Phoenix dactylifera* L	Date palm	Khajoor	Arecaceae	42
28.	*Prunus amygdalus* Batsch	Almond	Badam	Rosaceae	06
29.	*Psidium guajava* L	Guava	Amrud	Myrtaceae	821
30.	*Punica granatum* L	Pomegranate	Anar	Punicaceae	106
31.	*Syzygium cumini* Skeels		Jamun	Myrtaceae	71
32.	*Tamarindus indica* L	Tamarind	Imli	Caesalpiniaceae	28
33.	*Vitis vinifera* Linnaeus	Grapes	Angoor	Vitaceae	08
34.	*Ziziphus mauritiana* Lamarck		Ber	Rhamnaceae	530
	Forest trees				
35.	*Azadirachta indica* A. Juss		Neem	Meliaceae	11
36.	*Dalbergia sissoo* Roxburgh		Shisham	Caesalpiniaceae	06
37.	*Eucalyptus gigantea* Dehnh	Eucalyptus		Myrtaceae	09
38.	*Ficus religiosa* L	Sacred fig	Peepal	Moraceae	13
39.	*Leucaena leucocephela* (Lamarck) de Wit		Subabul	Mimosaceae	23
40.	*Pithecellobium dulce* (Roxburgh) Bentham		VilaytiImlie	Mimosaceae	33
41.	*Syzygium jambos* L	Rose Apple		Myrtaceae	02
42.	*Tectona grandis* Linnaeus	Teak	Sagoun	Verbenaceae	219
43.	*Vachellia nilotica* L		Desi Babul	Fabaceae	37

Tree Diversity at Fruit Research Farm Imaliya

Among the forest trees (Fig 14.1) the high density is reported of teak (219) planted in the year 1978 followed by desi babul (37), VilayatiImli (33) and subabul (23). The other forest trees are having very less density. There are 34 fruit trees and 10 forest trees on this farm showing wide plant biodiversity with 44 fruits and forest tree (Plate 14.2 to 14.5). The farm has 142.00/ha tree density withvery high plant biodiversity index.

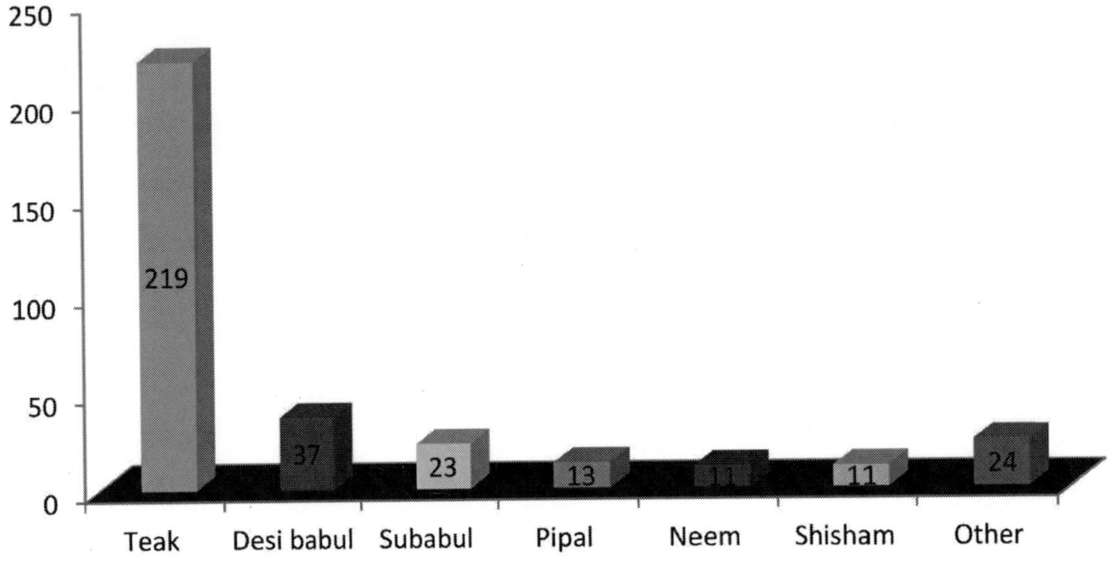

Fig 14.1 Representation of tree species at Fruit Research Farm, Imaliya

Plate 14.2 Mango plantation at Fruit Research Farm, Imaliya

Plate 14.3 Jamun plantation at Fruit Research Farm, Imaliya

Plate 14.4 Bael plantation at Fruit Research Farm, Imaliya

Plate 14.5 Sapota plantation at Fruit Research Farm, Imaliya

CHAPTER 15

Crop Diversity Audit of Breeder Seed Production

DK Pahalwan and Ajay Jaiswal
Directorate of Farms
JNKVV, Jabalpur
E-mail: seeds.jnkvv@gmail.com

Out of total cultivable area 435.97 ha available in the University at Jabalpur, Breeder seed production of 20 varieties of five field cropsis organized on 217.7 ha during Kharif; 33 varieties of seven field cropson 206.62 ha during rabi and 10 varieties of three field cropson 45.2 ha during zayad. Out of 11 crops two represents cereal, six pulse, three oilseed and one fodder crop. The maximum area (311.17ha) of Breeder seed production is under cereal i.e., 163.7 ha under wheat and 148.2 ha under rice; total 109.1 ha under legumes i.e., maximum 55.5 ha under soybean followed by 24.51 ha under chickpea. Total 45.2 ha land under breeder seed production programme has 300% and 206.62ha land has 200% cropping intensity (Table and Fig. 15.1).

Table 15.1 Crop diversity in Breeder seed production

Sr	Botanical Name	English name	Hindi name	Variety	Unit	Area (ha)	Total
A	**Kharif season 2020**						
1	*Oryza sativa*	Rice	Chawal	Kranti	BSP (G. Nut)	30	
2				IR 64		6.5	
3				MTU 1010		7.0	
4				JR 81		6.0	
5				JR 201		0.3	
6				PS 4		1.1	
7				JR 206		4.8	**55.7**
1	*Cajanus cajan*	Pigeonpea	Arhar	TJT 501	BSP (G. Nut)	6.5	
2				BDN 711		2	**8.5**
1	*Glycine max*	Soybean	Soybean	JS 20-69	BSP (G. Nut)	1.5	
2				JS 20-98		6.5	**8.00**
1	*Oryza sativa*	Rice	Dhan	Kranti	BSP (FC)	12	
2				MTU 1010		0.4	
3				PB 1		0.2	
4				IR 64		0.5	
5				P 1460		0.2	
6				PS 4		1.7	
7				JR 81		1.8	
8				PS 5		0.2	
9				IR 36		0.5	
10				JR 206		7	**24.5**
1	*Cajanus cajan*	Pigeonpea	Arhar	BDN 711	BSP (FC)	0.3	**0.3**
1	*Glycine max*	Soybean	Soybean	JS 20-98	BSP (FC)	20	
2				JS 20-116		6	**26.0**
1	*Oryza sativa*	Rice	Dhan	Kranti	BSP (Soy)	19	
2				JR 206		13	
3				JR 81		1.5	
4				PS 4		1.6	
5				PB 1		2.4	
6				IR 64		1.5	
7				MTU 1010		3.2	
8				Imp. Chhinor		0.4	
9				Imp. Jeera Shankar		0.4	**43.0**
1	*Vigna mungo*	Black gram	Urd	IPU 2-43	BSP (Soy)	1.5	
2				T 9		1.5	**3.00**
1	*Cajanus cajan*	Pigeonpea	Arhar	Rajeshwari	BSP (Soy)	1.2	
1	*Glycine max*	Soybean	Soybean	JS 20-98	BSP (Soy)	13	
2				JS 20-116		0.4	**13.4**

Sr	Botanical Name	English name	Hindi name	Variety	Unit	Area (ha)	Total
1	*Oryza sativa*	Rice	Dhan	Kranti	BSP (Veg)	18	
2				JR 206		7	**25.0**
1	*Vigna radiata*	Greengram	Moong	Sikha	BSP (Soy)	1	**1.00**
1	*Glycine max*	Soybean	Soybean	JS 20-98	BSP (Soy)	7.5	
2				JS 20-94		0.6	**8.1**
B	**Rabi season 2020-21**						
1	*Triticum aestivum*	Wheat	Gehu	GW 322	BSP (G.Nut)	30.00	
2				GW 3288		12.00	
3				GW 3382		10.00	**52.0**
1	*Cicer arietinum*	Chickpea	Chana	JG 36		2.40	
2				JG 63		1.50	**3.90**
1	*Pisum sativum*	Pea	Matar	KashiNandini		2.50	
2				PSM 3		1.80	**4.30**
1	*L. usitatissimum*	Linseed	Alsi	JLS 79		0.20	**0.20**
1	*Triticum aestivum*	Wheat	Gehu	JW 3336	BSP (FC)	1.50	
2				JW 3382		10.50	
3				JW 322		10.00	
4				HI 8759		1.24	
5				GW 366		1.00	
6				DDW 47		1.20	
7				JW 3288		1.50	
8				GW 273		1.00	
9				JW 3211		1.00	
10				LOK 1		3.50	
11				C 306		0.83	
12				GW 451		0.40	**33.67**
1	*Cicer arietinum*	Chickpea	Chana	JG 14		0.71	
2				JG 12		3.75	
3				JG 36		4.21	
4				JG 24		1.31	
5				JKG 5		0.93	**10.91**
				Total			
1	*Pisum sativum*	Pea	Matar	Arkel		0.48	
2				PSM 3		0.75	
3				KashiNandani		0.93	
4				IPFD 11-05		0.80	**2.96**
1	*Brassica nigra*	Mustard	Sarso	Push Tarak		0.55	**0.55**
1	*Lens culinaris*	Lentil	Masoor	RVL 31		0.83	**0.83**
1	*L. usitatissimum*	Linseed	Alsi	JLS 73		1.10	**1.10**
1	*Triticum aestivum*	Wheat	Gehu	GW 322	BSP (Soy)	9.00	
2				JW 3288		0.20	
3				GW 273		13.30	
4				JW 3382		1.80	
5				LOK 1		7.30	
6				JW 3211		0.60	
7				C 306		1.20	
8				HI 8759		7.20	
9				DDW 47		1.20	
10				GW 451		3.00	**44.80**

Sr	Botanical Name	English name	Hindi name	Variety	Unit	Area (ha)	Total
1	*Cicer arietinum*	Chickpea	Chana	JG 36		6.50	**6.50**
1	*Pisum sativum*	Pea	Matar	KashiNandini		0.80	
2				PSM 3		0.40	
3				IPFD 12-02		1.20	
4	*Pisum sativum*	Pea	Matar	JP 885		0.50	
1	*Lens culinaris*	Lentil	Masoor	IPL 316		0.80	**0.80**
1	*L. usitatissimum*	Linseed	Alsi	JLS 73		0.40	**0.40**
1	*Triticum aestivum*	Wheat	Gehu	GW 322	BSP (Veg)	28.00	
2		Wheat		HI 8759		3.30	
3				DBW 47		1.40	**32.70**
1	*Cicer arietinum*	Chickpea	Chana	JG 36		3.00	
2				JG 14		0.20	**3.20**
1	*Pisum sativum*	Pea	Matar	KashiNandini		1.00	
2				JP 885		0.10	
3				IPFD 10-12		0.30	**1.40**
1	*Lens culinaris*	Lentil	Masoor	RVL 11-6		0.80	
1	*L. usitatissimum*	Linseed	Alsi	JLS 27		1.50	
2			Alsi	JLS 79		0.20	**1.70**
	T. alexandruinum	Egyptian clover	Berseem	JB 05-09		1.00	**1.00**
C	**Zayad 2021**						
1	*Glycine max*	Soybean	Soybean	JS 20-98	BSP (G.Nut)	3.00	
1	*Glycine max*	Soybean	Soybean	JS 20-98	BSP (FC)	6.00	
2				JS 20-116		5.50	**11.50**
1	*Glycine max*	Soybean	Soybean	JS 20-98	BSP (Soy)	5.50	
2				JS 20-34		1.50	
3				JS 20-116		0.50	**7.50**
1	*Glycine max*	Soybean	Soybean	JS 20-98	BSP (Veg)	11.00	
2				JS 20-34		2.50	**13.50**
1	*Vigna mungo*	Urid bean	Urad	IPU 2-43	BSP (FC)	1.00	
2				Indira Urad 1		1.00	**2.00**
1	*Vigna mungo*	Urid bean	Urad	Indira Urad 1	BSP (Soy)	1.00	
1	*Vigna radiata*	Mung bean	Moong	PDM-139	BSP (FC)	1.00	
2				Sikha		1.00	
3				TMB 37		0.20	
4				Virat		1.00	**3.20**
1	*Vigna radiata*	Mung bean	Moong	Sikha	BSP (Soy)	1.00	
2				Virat		1.00	
3				Kanika		0.50	**2.50**
1	*Vigna radiata*	Mung bean	Moong	Sikha	BSP (Veg)	1.00	**1.00**

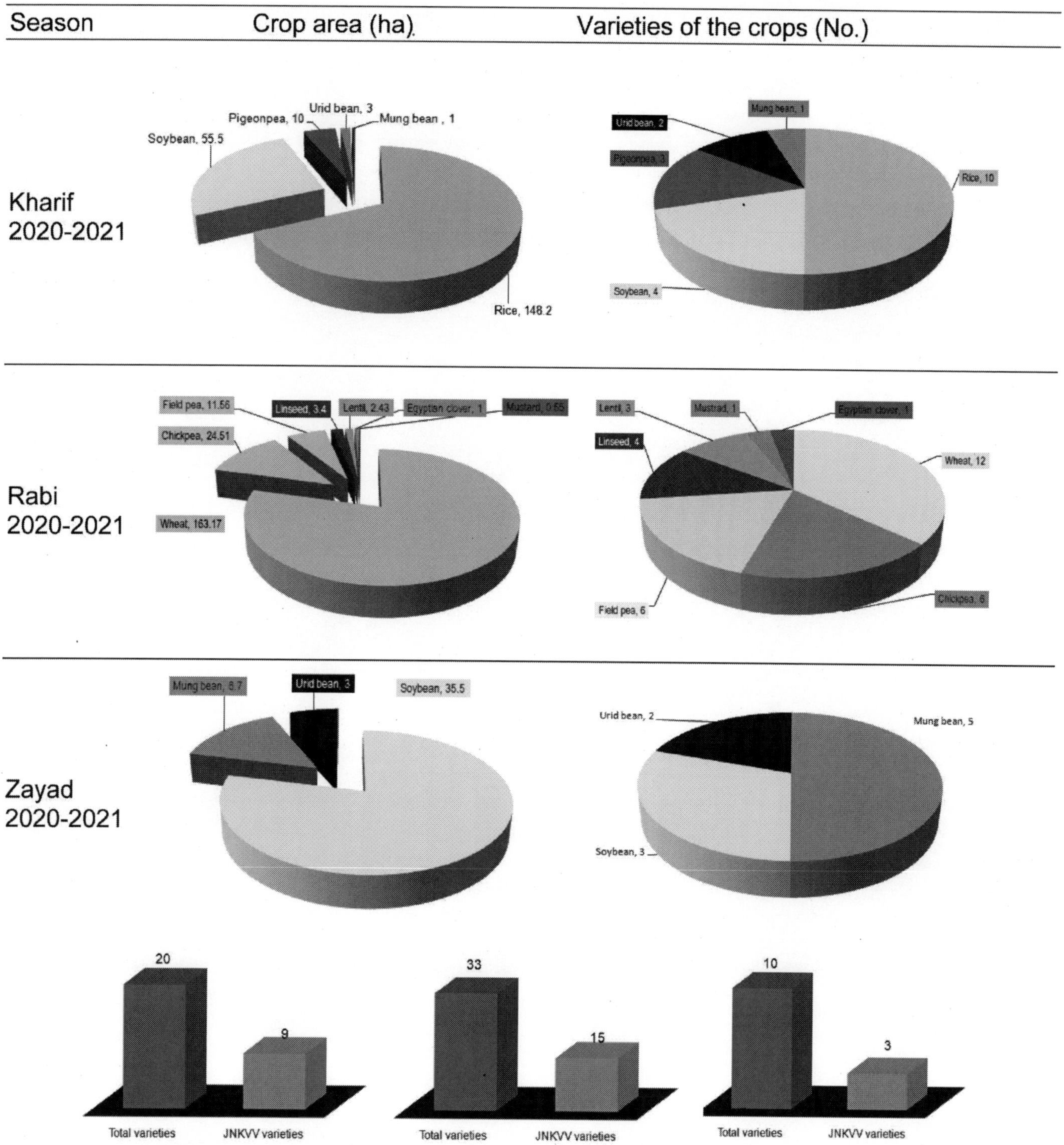

Crop	*Kharif* Varieties Total	JNKVV
Rice	10	4
Soybean	4	4
Pigeonpea	3	1
Urid bean	2	0
Mung bean	1	0

Crop	*Rabi* Varieties Total	JNKVV
Wheat	12	4
Chickpea	6	6
Field pea	6	0
Lentil	3	0
Linseed	4	4
Mustard	1	0
Berseem	1	1

Crop	*Zayad* Varieties Total	JNKVV
Soybean	3	3
Urid bean	2	0
Mung bean	5	0

Fig 15.1 Varieties under Breeder Seed Production during 2020-21

CHAPTER 16

Biodiversity Audit of Field Crops

RS Shukla
Department of Plant Breeding and Genetics
College of Agriculture, Jabalpur

Biodiversity in Cereals:

Table 16.1 Number of accession of cereals

Crop	Wheat		Rice	Oat	Total
Species	***aestivum***	***durum***	02	01	06
Varieties					
JNKVV notified	56		18	07	81
JNKVV denotified	-		-	-	
Other University of MP	01		-	-	1
Other State	08		60	-	68
Primitive cultivars	20		10	-	30
Other Country	·01		15	-	16
Total	89		103	07	199
Accession					
Indigenous	4741		110		4851
Exotic	550		20		570
Total	5291		130		5421
Genetic stock	07		-		7
Advance lines/strain	300		55	54	409
Breeding materials	7653		350	-	8003
Land races	60		25	-	85
Farmers varieties	-		400	-	400
Transgenic (GM)	-		-	-	
Wild relatives (sp)	04		01	-	5
Total	**18781**		**1299**	**69**	**20146**

Biodiversity in pulses:

Table 16.2 Number of accession of pulses

Crop	Chickpea	Pigeonpea	Mungbean	Uridbean	Lentil	Field pea	Total
Species	2		01	01	01	01	06
Varieties						-	
JNKVV notified	20		03	03	02	05	33
JNKVV denotified	3		-	-	-	-	03
Other University of MP	7		-	-	-	-	07
Other State	55		15	16	14	28	128
Primitive cultivars	-		-	-	-	-	
Other Country	3		-	-	-	-	03
Total	90		18	19	16	33	176
Accession							
Indigenous	1257		25	26	-	54	1362
Exotic	215		-	-	-	-	215
Total	1472		25	26	-	54	1577
Genetic stock	196		-	-	-	-	196
Advance lines/ strain	262		10	08	41	15	336
Breeding materials	410		40	40	-	42	532
Land races	6		-	-	-	110	116
Farmers varieties	-		-	-	-	10	10
Total	**2346**		**50**	**48**	**41**	**177**	**2662**

Plate 16.1 Nursery of rice

Biodiversity inoilseeds:

Table 16.3 Number of accession of oilseeds

Crop	Soybean	Linseed	Sesame	Niger	Total
Species	1	3	1	1	06
Varieties		-			
JNKVV notified	11	9	9	5	34
JNKVV denotified	-	-	-	-	
Other University of MP	9	-	-	-	09
Other State	15	15	51	10	91
Primitive cultivars	-	-	-	-	
Other Country	-	-	-	-	
Total	**35**	**24**	**60**	**15**	**134**
Accession					
Indigenous	352	3240	1450	1200	6142
Exotic	63	2	50	-	115
Total	**415**	**3242**	**1500**	**1200**	**6357**
Genetic stock	1	2657	-	-	2658
Advanced lines/strain	22	210	10	-	242
Breeding materials	99	301	20	-	420
Land races	1	-	-	-	
Farmers varieties	-	-	-	-	
Transgenic (GM)	-	-	-	-	
Wild relatives (sp)		2	3	-	05
Weedy relatives(sp)	-	-	-	-	
Wild relatives	1	4	10	-	15
Total	**115**	**3174**	**43**	-	**3332**

Plate 16.2 Breeder seed production of soybean

Biodiversity in Fodder Crops

Table 16.4 Number of accession of fodder crops

Crop	Berseem	Oat	Ricebean	Maize	Sorghum	*Vicia sativa*	Total
Species	01	01	01			01	04
Varieties							
JNKVV notified	03	07	02			01	13
JNKVV denotified	-	-	-			-	
Other University of MP	-	-	-			-	
Other State	-	-	-			-	
Primitive cultivars	-	-	-			-	
Other Country	-	-	-			-	
Total	03	07	02			01	13
Accession							
Indigenous	-	54	-			-	54
Exotic	-	-	-			-	
Total		-	-			-	
Advance lines	22	-	19			06	47
Total	**22**	**54**	**19**			**06**	**101**

Plate 16.3 Forage garden, JNKVV, Jabalpur

Utilization of biodiversity

Table 16.5 Number of accession utilized

Crop	Accession	Evaluated	Characterized	Documented	Utilized	Made available
Cereals						
Wheat	6625	471	10	-	200	50
Rice	130	130	130	-	45	-
Pulses						
Chickpea	1472	500	132	-	400	55
Pigeonpea						
Mungbean	120	120	120	120	120	-
Uridbean	40	40	40	35	20	-
Field pea	54	54	54	52	30	01
Oilseeds						
Soybean	415	415	200	-	10	-
Linseed	3242	3242	2612	2612	80	35
Sesame	1500	1200	1200	500	150	800
Niger	1200	1000	1000	-	100	1200
Mustard						
Fodder						
Berseem	74	54	10	-	06	-
Oat	150	121	16	-	12	-
Rice bean	40	19	02	-	02	-
Vicia sativa	06	06	02	-	-	-

CHAPTER 17

Biodiversity Audit of Medicinal and Aromatic Plants

Gyanendra Tiwari, Sachin Nagre and Satyendra Thakur
Department of Plant Physiology,
College of Agriculture, JNKVV, Jabalpur

Madhya Pradesh has one of the richest diversity of plants particularly of medicinal and aromatic herbs amongst all states in India. The favourable ecosystems harbouring this huge diversity is now degrading gradually due to habitat loss, deforestation, climate change and unscientific use of herbicide in agri-ecosystems and particularly merciless and unscientific exploitation of herbs for medicinal purpose and marketing. Conservation of these medicinal herbs is urgently needed to protect them for future generations. Besides, eradicating any plant in the name of tagged weeds from ecosystem must be curbed strictly otherwise they will be extinct in near future.

The fauna available in the JNKVV Medicinal Garden, Jabalpur are presented in Table 17.1 and familywise representation of plant species depicted in Fig. 17.1 and Plate 17.1.

Plate 17.1 Conservation at Medicinal Garden

Table 17.1 Biodiversity of Medicinal plants

Sr	Botanical Name	Common Name	Family
1.	*Abelmoschus moschatus* (L.) Medic.	Muskdana	Malvaceae
2.	*Abroma augusta (L.) LAF.*	Ulat Kambal	Malvaceae
3.	*Abrus precatorius* L.	Rutti (Lal)	Fabaceae
4.	*Abrus precatorius* L.	Rutti (Safed)	Fabaceae
5.	*Abrus precatorius* L.	Rutti (Kali)	Fabaceae
6.	*Abutilon hirsutum* G.Don.	Kangh (Badi)	Malvaceae
7.	*Abutilon indicum* L. Sweet.	Kanghi	Malvaceae
8.	*Abutilon instia* Willd.	Aila	Malvaceae
9.	*Abutilon tomentosum* W & A.	Raj Bala	Malvaceae
10.	*Acacia auriculiformis* A. Cunn.	Australian Babul	Mimosaceae
11.	*Acacia catechu*	Khair	Mimosaceae
12.	*Acacia concinna* D.C.	Sikakai	Mimosaceae
13.	*Acacia farnaciana* (L.) Willd.	Ghandha Babul	Mimosaceae
14.	*Acacia holoceria*		Mimosaceae
15.	*Acacia leucophaea* Willd.	Safed Babool	Mimosaceae
16.	*Acacia mangium*	Australian teak	Mimosaceae
17.	*Acacia nilotica*	Babool	Mimosaceae
18.	*Acacia pinnata* Willd.	Biswul	Mimosaceae
19.	*Acacia tometosa* Willd.	Lal Sai babool	Mimosaceae
20.	*Acalypha hispida* Burm.	Bahtali	Euphorbiaceae
21.	*Acalypha indica L*	Khokali, Kuppi	Euphorbiaceae
22.	*Acalypha spicata*		Euphorbiaceae
23.	*Acalypha tricolor*	Khokli	Euphorbiaceae
24.	*Achilea millefolium*	Yarrow	Amaranthaceae
25.	*Achyranthus aspera* L.	Adhajhara, Apamarg, Latjira	Amaranthaceae
26.	*Achyranthus bidenta*. Blume	Bada Adhajhara.	Amaranthaceae
27.	*Acorus calamus* L.	Bacha	Acoraceae
28.	*Adansoinia digitata* L.	Gorakh imli, Kalptaru, baobab	Malvaceae
29.	*Adhatoda vasica* Nees.	Basaka, Adusa	Acanthaceae
30.	*Adiantum pedatum* L.	Maidenhair fern	Pteridaceae
31.	*Adiantumcapillus-veneris* L.	Hansraj, Fern	Pteridaceae
32.	*Adina cordifolia*	Haldu	Rubiaceae
33.	*Aegle marmelos* L.Corr.	Bel	Rutaceae
34.	*Aerva lanata* (L.) Juss.	Gorakhbooti/Gorakhganja	Amaranthaceae
35.	*Aeschynomene aspera* L.	Sola	Fabaceae
36.	*Aeschynomene indica* L.	Phulan	Fabaceae
37.	*Agathis spp.*	Kauri	Araucariaceae
38.	*Agave americana* L.	Kantala	Agavaceae
39.	*Agave cantata* Roxb.		Agavaceae
40.	*Agave sisalana* Perr.	Adinath, Sisal	Agavaceae
41.	*Ageratum conyzoides* L.	Mahakaua	Asteraceae
42.	*Ageratum houstonium* Mill.	Floss flower/mist flower	Asteraceae
43.	*Ailanthus excelsa* Roxb.	Arua	Simaroubaceae
44.	*Ailanthus triphysa* (Dennst.) Alston.	Gora neem	simaroubaceae
45.	*AIlium porum*	Ban Lahsun	Alliaceae
46.	*Albizia amara* (Roxb) Boir.	Siris	Mimosaceae
47.	*Albizia lebbek* L.	Kala Siris	Mimosaceae

Sr	Botanical Name	Common Name	Family
48.	*Albizia odoratissima* (L.F.) Bench.	siris	Mimosaceae
49.	*Albizia procera* Roxb. Bench.	Gurar (Safed siris)	Mimosaceae
50.	*Alcasia macrorihiza* (L.) Schott.	Giant Taro	Araceae
51.	*Alhagi camelorum* Fisch.	Javasa	Fabaceae
52.	*Alisma plantago*	Jalkela	Alismataceae
53.	*Alium cepa* L.	Piyaj	Alliaceae
54.	*Allamanda cathartica*	Allamanda	Apocynaceae
55.	*Allamanda cathartica* L.	Allamanda	Apocynaceae
56.	*Allium sativum* L.	Lahsun	Alliaceae
57.	*Alocasia cucullata* (lour) Schott.	Giant Taro	Araceae
58.	*Alocasia indica* (Roxb) Schott.	Mankanda	Araceae
59.	*Aloe vera* (Syn.A.barbadensis)	Kumari, Gwar Patha	Asphodelaceae
60.	*Alpenia officinarum* Hance.	Kulanjan	Zingiberaceae
61.	*Alpinia calcarata*	Kulanjanbhed	Zingiberaceae
62.	*Alpinia galangal* (L.) Sw.	Bara kulanjan/ Greater Galangal	Zingiberaceae
63.	*Alpinia sanderae* Sand.	Kulanjanbhed	Zingiberaceae
64.	*Alpinia speciosa* (Wendl.) K. Schum.	Pannag champa	Zingiberaceae
65.	*Alstonia scholaris* L.R.Br.	Satwin, Saptaparni	Apocynaceae
66.	*Alternanthera sessilis* L.	Gatighusi / Matyakshi	Amaranthaceae
67.	*Althaea officinalis* L.	Gulkhera	Malvaceae
68.	*Althaea rosea* (L.) Cav.	Gulkhera	Malvaceae
69.	*Alysicarpus longifolium* Wight. & Arn.	Shevra	Fabaceae
70.	*Amaranthus blitum* L.var. oleracea Duthie.	Choulai	Amaranthaceae
71.	*Amaranthus cruentus* L.	Ramdana	Amaranthaceae
72.	*Amaranthus spinosus* L.	Kantilichouli	Amaranthaceae
73.	*Amaranthus tricolor* L.	Lal sag	Amaranthaceae
74.	*Ambrosia trifida*	Ragweed	Asteraceae
75.	*Ammannia baccifera* L.	Dadmari	Lythraceae
76.	*Ammi majus* L.	Ammi	Apiaceae
77.	*Ammi visnaga*	Honey plant	Apiaceae
78.	*Amomum aromaticum* Roxb.	Barielaichi	Zingiberaceae
79.	*Amorphophallus campanulatus* (Roxb)	Suran	Araceae
80.	*Amorphophallus sylvesticus* Kunth.	Jangli Jamikhand	Araceae
81.	*Anacardium occidentale* L.	Kaju	Anacdiaceae
82.	*Anagallis arvensis* L.	Krishna nil, Jokmari	Primulaceae
83.	*Anamirta cocculus* (L.) Wight & Arn	Kakmari	Menispermaceae
84.	*Ananas comosus* (L) Merr.	Ananas	Bromeliaceae
85.	*Andrographis paniculata*	Kalmegh, Kadu chiraya	Acanthaceae
86.	*Anethum graveolens* L.	Soya	Apiaceae
87.	*Anethum sowa* Kurz.	Sowa	Apiaceae
88.	*Anisomeles indica* (L.) O.Kuntz.	Kalabhangra	lamiaceae
89.	*Annona cherimolia* Mill.	Hanumanphal	Annonaceae
90.	*Annona muricata* L.	Mamphal	Annonaceae
91.	*Annona reticulata* L.	Ramphal	Annonaceae
92.	*Annona squamosa* L.	Sitaphal	Annonaceae
93.	*Anogeissus accuminata* Roxb.	Kardhai, Dhawra	Combretaceae
94.	*Anogeissus latifolia* (D.C.) Wall ex. Bedd.	Dhawra/gum ghatti	Combretaceae
95.	*Anogeissuspendula* Edgew.	Dhao	

Sr	Botanical Name	Common Name	Family
96.	*Anthocephalus kadambu* Roxb. Miq.	Kadamba	Rubiaceae
97.	*Apium graveolens* L.	Ajmod/salari	Apiaceae
98.	*Arachys hypojea*	Groundnut	Fabaceae
99.	*Araucaria bidwilli*	Spiny Araucaria	Araucariaceae
100.	*Arecacatechu* L.	Supari	Arecaceae
101.	*Argemonealba* Lestib.	Shialkanta	Papaveraceae
102.	*Argemone mexicana* L.	Satyanasi, Shialkanta	Papaveraceae
103.	*Argyreia nurvosa* (Burn E.)	Ghabel, Samandarsokh	Convolvulaceae
104.	*Aristolochia brachiolata* Lamk.	Kiramar	Aristolochiaceae
105.	*Aristolochia indica* L.	Isharmul	Aristolochiaceae
106.	*Artabotrys uncinatus* (Lam.) Mern.	Harichampa	Annonaceae
107.	*Artemisia annua*	Maleria buti	Asteraceae
108.	*Artemisia indica* Willd	Tulsidauna	Asteraceae
109.	*Artemisia maritima* L.	Kirmala	Asteraceae
110.	*Artemisia pallens* Wall.	Davana	Asteraceae
111.	*Artemisia vulgaris* L.	Nagdona	Asteraceae
112.	*A.nilagirica*(Clarke) Pamp.	Nagdona	Asteraceae
113.	*Artocarpus heterophyllus* lamk.	kathal	Moraceae
114.	*Artocarpus lakoocha* Roxb.	Barbal	Moraceae
115.	*Arund donax* L	Baranal/ narhal	Poaceae
116.	*Asclepias curassavica* L.	Kakatundi	Asclepiadaceae
117.	*Asparagus adscendens* Roxb.	Safed musli	Liliaceae
118.	*Asparagus filicinus* Ham.	Alli Palli	Liliaceae
119.	*Asparagus gonocladus* Baker.	Shakakul	Liliaceae
120.	*Asparagus officinalis* L.	Halyun	Liliaceae
121.	*Asparagus plumosus* Baker.	Satawar	Liliaceae
122.	*Asparagus racemosus* Willd.	Satmali, Satwar	Liliaceae
123.	*Asparagus racemosus* willd.	Satavar	Liliaceae
124.	*Asparagus sarmentosus* L.	Satawar (Flat Cladode)	Liliaceae
125.	*Asparagus spengeri* Regel.	Satawar	Liliaceae
126.	Asphodelus *tenuifoli.* L.	Banpiyaji, pola.	Asphodelaceae
127.	*Asteracantha longifolia* L.	Talimkhana	Acanthaceae
128.	*Astragales heratensis*	Indian tragacanth	Fabaceae
129.	*Atropa belladona*	Belladona	Solaneceae
130.	*Avena sativa* L.	Jai	Poaceae
131.	*Averrhoea carambola* L.	Safed Kachnar	Averrhoeaceae
132.	Azanja lampas	Bankapas	Malvaceae
133.	*Azardirachta Indica*	Neem	Meliceae
134.	*Bacopa monneiri* (L) Penn.	Brahmi	Scrophulariaceae
135.	*Balanites aegyptiaca* (L.) Del.	Hingot	Balaniteceae
136.	*Baliospermum montanum* (Willd) Arg.	Danti, Jangli Jamalgota	Euphorbiaceae
137.	*Baliospermum montanum* (Willd).	Dantimul	Euphorbiaceae
138.	*Bambusa arundinacea* Retz.	Bans	Bambusaceae
139.	*Bambusa fulcata* Hort.	Bans	Bambusaceae
140.	*Bambusa polymorpha* Murno.	Betua	Bambusaceae
141.	*Bambusa tulda* Roxb.	Peka Chau	Bambusaceae
142.	*Bambusa vulgaris* Schrader.	Bansni	Bambusaceae
143.	*Barlaria alba* Lodd.	Bajradanti Safed	Acanthaceae

Sr	Botanical Name	Common Name	Family
144.	*Barlaria cristana* L.	Bajradanti	Acanthaceae
145.	*Basella alba* L.	Poi	Basellaceae
146.	*Bauhinia blackii*	Kala kachnar	Caesalpiniaceae
147.	*Bauhinia malabarica* Roxb.	Amli	Caesalpiniaceae
148.	*Bauhinia purpurea* L.	Lal Kachnar	Caesalpiniaceae
149.	*Bauhinia tomentosa* L.	Pilikachnar	Caesalpiniaceae
150.	*Bauhinia vahlii* Wight & Arn.	Mahul, Malihjhan	Caesalpiniaceae
151.	*Bauhinia variegate* L.	Kandla, Asta	Caesalpiniaceae
152.	*Bauhinia variegate* L.	Kachnar	Caesalpiniaceae
153.	*Begonia* L.	*Begonia*	Begoniaceae
154.	*Benincasa hispida* (Thunb.) Cong.	Petha kaddu, Safed Kumda	Cucurbitaceae
155.	*Berberis aristata* D.C.	Daru Hardra	Berberidaceae
156.	*Beta vulgaris*	Chukandar	Chenopodiaceae
157.	*Biden pilosa* L.	Samarakakari	Asteraceae
158.	*Biophytum sensitivum* D.C.	Mukkrootti	Oxalidaceae
159.	*Bixa orellana* L.	Sinduri	Bixaceae
160.	*Blepharis edulis* Pers.	Utangan	Acanthaceae
161.	*Bluemea balsaifera* (L) D.C.	Kakronda	Asteraceae
162.	*Bluemea lacera* D.C.	Kakronda	Asteraceae
163.	*Boerrhavia diffusa*	Punanava	Nyctaginaceae
164.	*Boerrhavia repends*	Punanava safed	Nyctaginaceae
165.	*Bombax ceiba* L.	Semal	Bombacaceae
166.	*Borassus flabellifer L.*	Toddypalm	Arecaceae
167.	*Boswellia serrata*	Salai	Burseraceae
168.	*Boswellia serrata* Roxb.	Salai	Berseraceae
169.	*Brassica juncea* L.	Rai	Brassicaceae
170.	*Bridelia squamosa* Gehrm.	Kassai	Euphorbiaceae
171.	*Brownea coccinea*	Brownea	Fabaceae
172.	*Bryonia laciniosa* L.	Shivalingi	Cucurbitaceae
173.	*Bryophyllum pinnata*	Paranbeej.	Crassulaceae
174.	*Buchanania lanzan* Spreng.	Chironji	Anacardiaceae
175.	*Bursera panicillata* (Sesse & Moc ex D.C.)	Lavender Tree	Burseraceae
176.	*Butea monosperma* Lamk. Tanbert.	Palas	Fabaceae
177.	*Butea superba* Roxb.	Palas lata	Fabaceae
178.	*Cactus grandiflorus*	Thuhar	Cactaceae
179.	*Caesalpinia bonducella* fleming	Sagargoti	Caesalpiniaceae
180.	*Caesalpinia coriaria* willd.	Dibidibi	Caesalpiniaceae
181.	*Caesalpinia crista* Thumb.	Gataran	Caesalpiniaceae
182.	*Caesalpinia digyna* Rottle.	Bakerimul	Caesalpiniaceae
183.	*Caesalpinia pulcherrima* Swartz.	Gultarra	Caesalpiniaceae
184.	*Caesalpinia sappan* L.	Patang	Caesalpiniaceae
185.	*Caesulia axillarisi* Roxb.	Bilja	Asteraceae
186.	*Caladium seguinum*	American arum	Araceae
187.	*Calandrinia ciliata*	Redmaids	Portulacaceae
188.	*Calendula officinalis* L.	Zergul	Asteraceae
189.	*Callicarpa macrophylla*	Priyangu	Verbenaceae
190.	*Callistemon lanceolatus*	Bottlebrush	Myrtaceae
191.	*Calotropis gigantia* (L) R. Br.	Ak, Madar	Asclepiadaceae

Sr	Botanical Name	Common Name	Family
192.	*Calotropis procera* (Ait.) R.Br.	Safed AK	Asclepiadaceae
193.	*Camellia sinensis* (L.)	Chaya	Theaceae
194.	*Cananabis sativa*	Bhang	Cananabinaceae
195.	*Canarium commune* L.	Jangli Badam	Berseraceae
196.	*Canarium strictum* Roxb.	Kaladamar	Berseraceae
197.	*Canavalia ansiformis* D.C.	Sakhar sem	Fabaceae
198.	*Canavlia gludiata* (Jacq.) D.C.	Barasem	Fabaceae
199.	*Cannabis indica*	Bhang	Cannabeceae
200.	*Canscora decussate* Schutt.	Shankhapushpi	Gentieanaceae
201.	*Capparis deciduas* Forsk	Karer, Karira	Capparidaceae
202.	*Capparis grandis*		Capparidaceae
203.	*Capsicum annum* Linn.	Mirch	Solanaceae
204.	*Cardiospermum halicacabum* L.	Kanphuti	Sapindaceae
205.	*Careya arborea* Roxb.	Kumbhi.	Barringtoniaceae
206.	*Carica Papaya* L.	Papita	Caricaceae
207.	*Carissa carandas* L.	Karonda	Apocynaceae
208.	*Carissa grandiflora*	Bara Karonda	Apocynaceae
209.	*Carissa spinarum* L.	Karonda	Apocynaceae
210.	*Carrisa spinosa*	Karonda	Apocynaceae
211.	*Carthamus tinctorius* L.	Kusum	Asteraceae
212.	*Carum carvi* L.	Shiajira	Apiaceae
213.	*Caryota urens* L.	Mari, Toddy	Arecaceae
214.	*Cassia absus* L.	Chaksu	Caesalpiniaceae
215.	*Cassia alata* L.	Dadmardan	Caesalpiniaceae
216.	*Cassia angustifolia* Vahl.	Senai, Senna	Caesalpiniaceae
217.	*Cassia artimisoides* Gaudich.Ex. Dc.		Caesalpiniaceae
218.	*Cassia auriculata* L.	Tarwar	Caesalpiniaceae
219.	*Cassia fistula* L.	Amaltas	Caesalpiniaceae
220.	*Cassia glauca* Lamk.		Caesalpiniaceae
221.	*Cassia grandis* L.F.		Caesalpiniaceae
222.	*Cassia javanica* L.		Caesalpiniaceae
223.	*Cassia mimosifolia*		Caesalpiniaceae
224.	*Cassia seracea*	Chakoda Bhed	Caesalpiniaceae
225.	*Cassia siamea*	Kasod tree	Caesalpiniaceae
226.	*Cassia sophora* L	Kasaundi	Caesalpiniaceae
227.	*Cassia tora* L.	Chakoda Powar	Caesalpiniaceae
228.	*Cassytha filifomis* L	Amarbel	Cassythaceae
229.	*Casuarina equisetifolia* J.R. & G.	Gangli Jhau	Casuarinaceae
230.	*Catharanthus roseus* L.G. Don.	Sada suhagan	Apocynaceae
231.	*Ceiba pentandra*	Apok	Bombacaceae
232.	*Celastruspaniculatus* Willd.	Malkangi	Celastraceae
233.	*Celosiaargentia* L	Safed murga	Amaranthaceae
234.	*Celosia cristata* L.	Morshikha	Amaranthaceae
235.	*Centaurea cynanus* L.	Corn flower/Bachelors button	Asteraceae
236.	*Centella asiatica*	Bramhi	Hydrocotylaceae
237.	*Centipeda minima* (L.) A.Br. Aschers	Nachhikni	Asteraceae
238.	*Centratherum anthelminticum*	Somraj	Asteraceae
239.	*Cephalandra indica*	Ban kundara	Cucurbitaceae

Sr	Botanical Name	Common Name	Family
240.	*Cestrum aurantiacum* Linidl.		Solanaceae
241.	*Cestrum diurnum*	Din-ka-raja	Solanaceae
242.	*Cestrum nocturnum* L.	Rat-ki-rani	Solanaceae
243.	*Cestrum parqui* L.	Rat-ki-rani	Solanaceae
244.	*Chenopodium album* L.	Bathua	Chenopodiaceae
245.	*Chenopodium murate* L.	Khartua	Chenopodiaceae
246.	*Chlorophytum arundinaceum*	Safed Musli Bhed	Liliaceae
247.	*Chlorophytum borivllianum*	Safed musli	Liliaceae
248.	*Chlorophytum laxum*	Safed Musli Bhed	Liliaceae
249.	*Chlorophytum tuberosum*	Safed musli	Liliaceae
250.	*Chlorophytum variegates*		Liliaceae
251.	*Chloroxylon switenia* D.C.	Bhirra	Flindersiaceae
252.	*Chrysanthemum coronarium* L.	Gudaki	Asteraceae
253.	*Chrysanthemum indicum* L.	Guldadi	Asteraceae
254.	*Chukrosia tabularis A. Juss.*	Chikrasi	Malvaceae
255.	*Cicca acida* (L.) Merr.	Haarphrodi	Euphorbiaceae
256.	*Cichorium endive* L.	Kasni	Asteraceae
257.	*Cichorium intybus*	Chikori	Asteraceae
258.	*Cinnamomum camphora* (L.) Nees.	Mushkapur	Lauraceae
259.	*Cinnamomum zeylanica*	Dalchini	Lauraceae
260.	*Cinnamonum tamala*	Tezpatra	Lauraceae
261.	*Cissampelos pareira* L.	Patha, Nirbissi	Menispermaceae
262.	*Cissus quadrangularis* L.	Hadjor	Vitaceae
263.	*Cissus repens* Lamk.	Harjora	Vitaceae
264.	*Citrullus colocynthis* L.	Indrayan	Cucurbitaceae
265.	*Citrullus lanatus*	Tarbuj	Cucurbitaceae
266.	*Citrullus lanatus* var. fistulosus	Tinda	Cucurbitaceae
267.	*Citrus aurantifolia* (Charistum.) Swingle.	Nimbu	Rutaceae
268.	*Citrus limon* (L.) Barm	Bara nimbu	Rutaceae
269.	*Citrus medica* L.	Bijaura	Rutaceae
270.	Citrus reticulata	Blanco. – Santra	Rutaceae
271.	*Citrus sinensis*	Mosambi	Rutaceae
272.	*Clarkia elagans* Dougl.		Onagraceae
273.	*Cleistanthus collinus* Roxb. Benth.	Garari	Euphorbiaceae
274.	*Clematis goourisana* Roxb.	Muxda	Ranunculaceae
275.	*Cleome brachycarpa* Vahl.	Ponkal	Cleomaceae
276.	*Cleome simplifolia* Hook. F. &T.	Shoth ki buti	Cleomaceae
277.	*Cleome spinosa* L.	Hurhur	Cleomaceae
278.	*Cleome viscosa* L	Hurhur	Cleomaceae
279.	*Clerodendrum indicum* (L.) Kuntaz.	Bharangi	Verbenaceae
280.	*Clerodendrum inerme*	Lanjal	Verbenaceae
281.	*Clerodendrum infortunatum* Linn.		Verbenaceae
282.	*Clerodendrum japanicum* (Thunb.) Sweet.	Attardana	Verbenaceae
283.	*Clerodendrum phlomidis* L.F.	Arni, Agnimanth	Verbenaceae
284.	*Clerodendrum serratum* (L.) Spreng.	Bhurarngi	Verbenaceae
285.	*Clerodendrum thomsonmiae*	Bleeding heart vine	Verbenaceae
286.	*Clitoria mariana* L.	Aparajita	Fabaceae
287.	*Clitoria ternatea* L.	Aparajita	Fabaceae

Sr	Botanical Name	Common Name	Family
288.	*Coccinia grandis* Syn. Cephalandra indica	Kundru	Euphorbiaceae
289.	*Cocculus hirsutus*	Jamti ki bel	Menispermaceae
290.	*Cochlospermum religiosperm* (l.) Alston.	Pili kapas, Kumbi	Cochlosperamaceae
291.	*Cocus nucefera* L.	Nariyal	Arecaceae
292.	*Coffea arabica* L.	Kafi	Rubiaceae
293.	*Coix gigantea* Koen ex Roxb.	Gurlu	Poaceae
294.	*Coix lacrima* Jobi.	Garru	Poaceae
295.	*Coleus aromaticus*	Pasan bhed	Lamiaceae
296.	*Coleus barabatus* Benth.		Lamiaceae
297.	*Coleus blumei* Benth.		Lamiaceae
298.	*Colocasia esculanta* (L.)	Arvi, Ghunga	Araceae
299.	*Commelina bengalensis* Wall.	Kankaua	Commelinaceae
300.	*Commelina oblique* Harn	Kanjura	Commelinaceae
301.	*Commiphora bhandari* Com. Nov.	Guggul	Burseraceae
302.	*Commiphora wightii* (Arnott.)	Guggul	Burseraceae
303.	*Convolvlus argentens* Wall	Bhokdi	Convolvlaceae
304.	*Convolvlus arvensis* L.	Hiranchari	Convolvlaceae
305.	*Convolvlus microphyllus* Sieb.	Shankpushpi	Convolvlaceae
306.	*Convolvlus pluricaulis*	Shank Pushpi	Convolvlaceae
307.	*Convolvlus scamonia* L		Convolvlaceae
308.	*Coptis teeta*	Mamira	Ranunculaceae
309.	*Corchorus acutangulus* Lamk.	Chendchdi	Tiliaceae
310.	*Corchorus antichorus* Roeusch.	Chhoti Chench	Tiliaceae
311.	*Corchorus astuan*	Chench	Tiliaceae
312.	*Corchorus capsularis* L.	Jute	Tiliaceae
313.	*Corchorus fascicularis*	Magarmust	Tiliaceae
314.	*Corchorus olitorius* L.	Chench	Tiliaceae
315.	*Corchorus tridens*	Chench	Tiliaceae
316.	*Corchorus trilocularis*	Chench	Tiliaceae
317.	*Cordia dischotoma* var. Wallichi	Lassora	Ehretiaceae
318.	*Cordia macleoidii*	Dahiman	Ehretiaceae
319.	*Cordia sebestena* L.	Lal lasoora	Ehretiaceae
320.	*Coriandrum didymus*		Apiaceae
321.	*Coriandrum sativum* L.	Dhania	Apiaceae
322.	*Costus igneous*	Insulin Plant	Costaceae
323.	*Costus speciosus* Smith.	Pohkar mul/Keokand	Costaceae
324.	*Cotula anthimoides* L.	Babuna	Asteraceae
325.	*Cotulahemisphaerica*		Asteraceae
326.	*Crataevanurvala* Buch Ham.	Barna/Barun	Capparidaceae
327.	*Crinumasiaticum* L.	Sukhdarshan	Amaryllidaceae
328.	*Crinumdefixum* Ker. Gawl.		Amaryllidaceae
329.	*Crinumlatifolium* L		Amaryllidaceae
330.	*Crinum powelli* Hort. Var album Hort.	White crinum	Amaryllidaceae
331.	*Crinumzeylanicum* L.	Sukhdarshan	Amaryllidaceae
332.	*Crocus sativus* L.	Keshar	Iridaceae
333.	*Crotalaria juncea* L.	San	Fabaceae
334.	*Crotalaria medicagenia* Lamk.		Fabaceae
335.	*Crotolaria mucroonata* Desf.	Junjhuna	Fabaceae

Sr	Botanical Name	Common Name	Family
336.	*Crotolaria retusa* L.	Ghunghunia	Fabaceae
337.	*Crotolaria sericea* L.	Jhunjhunia	Fabaceae
338.	*Crotonaromaticus* L		Euphorbiaceae
339.	*Crotonbonplan* dianum		Euphorbiaceae
340.	*Crotonoblongifolius* Roxb.	Chuka	Euphorbiaceae
341.	*Crotontiglium* L.	Jamalgota	Euphorbiaceae
342.	*Cryptolepis buchanani* R & S.	Karanta	Periplocaceae
343.	*Cucubitamaxima* Duch.	Kashifal	Cucurbitaceae
344.	*Cucumisanguria* L.	Gherkin	Cucurbitaceae
345.	*Cucumismelo* L.	Kharbuja	Cucurbitaceae
346.	*Cucumismelo* L. var. Ogrestis	Meki	Cucurbitaceae
347.	*Cucumismelo* var. *mommeordica*Duth. & Full.	Phunt kachra	Cucurbitaceae
348.	*Cucumismelo* var. *utilissimus*Duth. & Full.	Kakri	Cucurbitaceae
349.	*Cucumissativa* L.	Khira	Cucurbitaceae
350.	*Cuminumcyminum* L.	Jira	Apiaceae
351.	*Cupressuslusitanica* Mill.		Cupressaceae
352.	*Cupressustorulosa* D. Don.	Devidiar	Cupressaceae
353.	*Curculigolatifolia*	Kali musli	Hypoxidaceae
354.	*Curculigolatifolia* Dry	Musli	Hypoxidaceae
355.	*Curculigoorchioides* Gaertn.	Kalimusli	Hypoxidaceae
356.	*Curcumaangustifolia* Roxb	Amahaldi	Zingiberaceae
357.	*Curcumaaromatica* Salisb.	Jungli Haldi	Zingiberaceae
358.	*Curcumacaesia* Roxb.	Kali Haldi	Zingiberaceae
359.	*Curcumadomestica* Vahl.	Haldi	Zingiberaceae
360.	*Curcumazedoaria* (Christm.) Rose.	Kachura	Zingiberaceae
361.	*Cuscutachinesis*	Amarbel	Cuscutaceae
362.	*Cuscutaepathymum*	Amarbel	Cuscutaceae
363.	*Cuscutaepilinum*	Amarbel	Cuscutaceae
364.	*Cuscutahyaline* Roth.	Amarbel	Cuscutaceae
365.	*Cuscutareflexaa* Roxb.	Amarbel	Cuscutaceae
366.	*Cyamopsistetragonoloba* (L.) Taub.	Guar	Fabaceae
367.	*Cyanaracolymus*		Asteraceae
368.	*Cyanotisaxillaris* Roem.	Soltraj	Commelinaceae
369.	*Cycascircinalis* L.	Jangali madan-must-ka-phal	Cycadaceae
370.	*Cycasrevolute* Thunb.		Cycadaceae
371.	*Cycaszamia*	Cycas	Cycadaceae
372.	*Cymbopogoncaesius* (Nees) Stapf.	Ginger grass	Poaceae
373.	*Cymbopogoncitratus* D.C. Stapf.	Sugandh rohisha/agiaghas	Poaceae
374.	*Cymbopogonflexuosus*	Lemongrass	Poaceae
375.	*Cymbopogonjwarancusa* Jones Scults.	Lamjak	Poaceae
376.	*Cymbopogonmartini* Roxb. Wats.	Palma rosa	Poaceae
377.	*Cymbopogonnardus* (L) Renole.	Ganjini	Poaceae
378.	*Cymbopogonpendulous* Wats.	Agia Ghas	Poaceae
379.	*Cymbopogonschoenaznthus* L. Spreng.	Rosa-ghas	Poaceae
380.	Cymbopogon winterianus Jowitt.	Citronella	Poaceae
381.	*Cynara scolymus* L.	Hathichuk	Asteraceae
382.	*Cynodon dactylon* Pers.	Dub	Poaceae

Sr	Botanical Name	Common Name	Family
383.	*Cyperus articulates* L.	Gondla	Cyperaceae
384.	*Cyperus bulbosus*	Gondla	Cyperaceae
385.	*Cyperus esculentus* L.	Chichoda	Cyperaceae
386.	*Cyperus rotundus* L.	Gondla	Cyperaceae
387.	*Cyperus scariosus* R.Br.	Nagar motha	Cyperaceae
388.	*Cyphomandra betacea*	Tomato tree	Solanaceae
389.	*Daedalacanthus roseus* T. Anders.	Dasmuli	Acantheceae
390.	*Daemia extensa* Br.	Utaran bel	Asclepidaceae
391.	*Dalbergia latifolia*	Kala shisam	Fabaceae
392.	*Dalbergia sisso* Roxb.	Shisam	Fabaceae
393.	*Datura innoxia* Mill.	Sadha Dhatura	Solanaceae
394.	*Datura metel* L	Kala Dhatura	Solanaceae
395.	*Datura strantonium* L	Dhatura	Solanaceae
396.	*Daucus carota* L	Gajar	Apiaceae
397.	*Delonix regia* (Boj.) Raf.	Gulmohar	Caesalpiniaceae
398.	*Delphinium brunonianum* Rogu.	Nefri	Ranunculaceae
399.	*Dendrocalamus strictus* Nees.	Bans	Poaceae
400.	*Dendrophthoe falcata*	Banda	Loranthaceae
401.	*Desmodium diffusum* D.C.	Shalparni	Fabaceae
402.	*Desmodium gangaticunt* D.C.	Shalparni	Fabaceae
403.	*Desmodium polycarpus* D.C	Shalparni	Fabaceae
404.	*Desmodium tomentosa* Heyne.		Fabaceae
405.	*Desmodium triflorum* (L.) D.C.	Kudalia	Fabaceae
406.	*Desmostachia bipinnata* (L) Stapf.	Kusha	Poaceae
407.	*Dianthus caryophylhts* L	Clovepink	Caryophyllaceae
408.	*Dichrostachys cinerea*	Vartuli	Mimosaceae
409.	*Digera arvensis* Forsk.	Latmuharia	Amaranthaceae
410.	*Digera muricata* (L) Mart.	Lesua	Amaranthaceae
411.	*Dillenia indica* L	Chalta	Dilleniaceae
412.	*Dioscorea alata* L.	Khumalu	Dioscoreaceae
413.	*Dioscorea bulbifera* L.	Ralalu	Dioscoreaceae
414.	*Dioscorea daemon*	Baichandi	Dioscoreaceae
415.	*Dioscorea deltoidea* Wall. ex Kunth.		Dioscoreaceae
416.	*Dioscorea esculenta* (Lour).Burkill.	Susnialu	Dioscoreaceae
417.	*Dioscorea floribunda*		Dioscoreaceae
418.	*Dioscorea hamiltonii* Hook f.		Dioscoreaceae
419.	*Dioscorea pentaphylla* L	Kantalu	Dioscoreaceae
420.	*Diospyros kaki* L	Tendu ka Bhed	Ebenaceae
421.	*Diospyros lotus* L.	Tendu ka bhed	Ebenaceae
422.	*Diospyros melanoxylon* Roxb.	Tendu	Ebenaceae
423.	*Diospyros montena*	Bis tendu	Ebenaceae
424.	*Diospyros tomentosa* Roxb.	Tendu kak	Ebenaceae
425.	*Dodonaea viscosa* (L.) Jacq.	Vilatmenhdi, Aliyar	Sapindaceae
426.	*Dolichos falcatus*		Fabaceae
427.	*Dolichos uniflorus* Lam	Kulthi	Fabaceae
428.	*Drosera indica* L	Mukhjali	Droseraceae
429.	*Dudymocarpus pedicillata* R. Br.	Pathar Phodi	Gesneriaceae
430.	*Echinopes echinatus* D.C.	Utkatara	Asteraceae

Sr	Botanical Name	Common Name	Family
431.	*Eclipta alba* Hassk.	Bhrangraj	Asteraceae
432.	*Elaeies guinensis*	Oilpalm	Arecaceae
433.	*Elaeodendron glaucum* Pers.	Jamrasi	Celastraceae
434.	*Eleocarpus sphaericus* Gaertn.	Rudrakh	Eleocarpaceae
435.	*Elephantopus scaber*	Gobhi, Goujaban	Asteraceae
436.	*Elettaria cardamomum*	Choti elaichi	Zingiberaceae
437.	*Embelia ribes* Burm F.	Baiberang	Myrsinaceae
438.	*Emblica officinalis*	Amla	Euphorbiaceae
439.	*Enicostimma hyssopifolium* (Wild.)	Nami (Chotachirayta)	Gentianaceae
440.	*Equisetum arvens* L. Laden	Field horse tail	Equisetaceae
441.	*Erianthus arundinaceus*	Sarkanda	Poaceae
442.	*Erigeron canadensis* L.	Jaramupriya	Asteraceae
443.	*Eriolaena quinquelocularis* Wt.	Dhamun	Sterculiaceae
444.	*Eruca vesicaria* (L.) Car.	Taramira	Brassicaceae
445.	*Ervthrina variegate* L.	Dadap	Fabaceae
446.	*Erythrina blakei* Hort.ex.Parker.	Pangara	Fabaceae
447.	*Erythrina indica* Lam.	Hadua	Fabaceae
448.	*Eucalyptus citriodora* Hook.	Nilgiri	Myrtaceae
449.	*Eucalyptus globulus* Labill.	Nilgiri	Myrtaceae
450.	*Eucalyptus melanopfiloia* F.Muell.	Nilgiri	Myrtaceae
451.	*Eucalyptus rostrata* Schlecht.	Nilgiri	Myrtaceae
452.	*Eugenia uniflora*	Tropical cherry	Myrtaceae
453.	*Eupatorium triplinerve* Vahl.	Ayapana	Asteraceae
454.	*Euphorbia antiquorum* L		Euphorbiaceae
455.	*Euphorbia dracunculoides*	Dudhi	Euphorbiaceae
456.	*Euphorbia geniculata* Orteg	Dudhi	Euphorbiaceae
457.	*Euphorbia hirta*.L.	Laldudhi	Euphorbiaceae
458.	*Euphorbia hypercifolia*		Euphorbiaceae
459.	*Euphorbia microphylla*		Euphorbiaceae
460.	*Euphorbia nervifolia* Auct.Pl.non L.	Sehund	Euphorbiaceae
461.	*Euphorbia pulcherima* Willd.ex. Koltz.	Lalpata	Euphorbiaceae
462.	*Euphorbia thymifolia*	Chholi Dudhi	Euphorbiaceae
463.	*Euphorbia tirucalli* L.	Konpal	Euphorbiaceae
464.	*Euryale ferox*	Makhana	Eurylaceae
465.	*Evolvulus alsinoides* Wall.	Vishu kranta	Convolvulaceae
466.	*Excoecaria bicolor*	Laila majnu plant	Euphorbiaceae
467.	*Ferula galbeniflua*	Gandhbiroza	Apiaceae
468.	*Ficus bengalensis* L.	Bargad	Moraceae
469.	*Ficus benjamina*	Chilkan	Moraceae
470.	*Ficus carica* L	Anjir	Moraceae
471.	*Ficus elastica* Roxb.	Attabar	Moraceae
472.	*Ficus glomerata* Roxb.	Gular	Moraceae
473.	*Ficus hispidu* L.f.	Kathumbar	Moraceae
474.	*Ficus infectoria* Roxb.	Pilkan	Moraceae
475.	*Ficus krishnae* C. DC.	Krishna Bargad	Moraceae
476.	*Ficus palmata* Forsk.	Anjiri	Moraceae
477.	*Ficus retusa* L	Kamrup	Moraceae
478.	*Ficus rumphii* Blume.	Pakar	Moraceae

Sr	Botanical Name	Common Name	Family
479.	*Flacourtia cataphrcta* Roxh.	Talaspatri	Flacourtiaceae
480.	*Flacourtia indica* (Burm) F.Merr.	Baichi	Flacourtiaceae
481.	*Fragaria vesca* L.	Strawberry	Rosaceae
482.	*Fumaria indica*	Pokharmul	Fumariaceae
483.	*Fumaria of cinalis* L.	Pitpapra	Fumariaceae
484.	*Funiculum vulgare* Mill.	Sonf	Apiaceae
485.	*Gardenia campanulata* Roxb.		Rubiaceae
486.	*Gardenia gummifera* L.F.	Dikamnli	Rubiaceae
487.	*Gardenia jasminoides* Ellis.	Gandhraj	Rubiaceae
488.	*Gardenia latifolia* Ait.	Papra	Rubiaceae
489.	*Gardenia resinifera* Roth.	Dikamali	Rubiaceae
490.	*Geranium nepalense* Sweet.	Bhanda	Geraniaceae
491.	*Gliricidia maculata* (H.B.K.) Steud.	Gliricidia	Fabaceae
492.	*Gloriosa superba* L.	Kalihari	Liliaceae
493.	*Glycosmis pentaphylla* Retz.D.C.	Ban Nimbu	Rutaceae
494.	*Glycyrrhiza glabra* L.	Mulehathi	Fabaceae
495.	*Gmelina arborea* L.	Khamer	Verbenaceae
496.	*Gomphrena celoseodes*		Amaranthaceae
497.	*Gomphrena pungens*		Amaranthaceae
498.	*Gossypium arboreum*	Badikapas	Malvaceae
499.	*Gossypium hirsutum*	Kapas	Malvaceae
500.	*Grevillea robusta* A. Cunn.	Silver Oak	Proteaceae
501.	*Grewia oppositifolia* Buch.	Bhemal/Gursakri	Malvaceae
502.	*Grewia populifolia* Vahl		Malvaceae
503.	*Grewia subinaequalis* D.C.	Phalsa	Malvaceae
504.	*Grewia tenure*	Chabeni	Malvaceae
505.	*Gymnema sylvestre* (Raz.) R.Br.	Gurmar	Asclepiadaceae
506.	*Gynandropsis gynandra*	Hurhur	Cleomaceae
507.	*Hedychium coronarium* Koenig. ex.Retz.	Ginger lily	Zingiberaceae
508.	*Hedychium flavum* Oxb.	Yellow ginger lily	Zingiberaceae
509.	*Hedychium spicattan* Buch.Ham.ex.Smith	Kapur Kachri	Zingiberaceae
510.	*Helianthus annuus* L.	Surajmukhi	Asteraceae
511.	*Helianthus tuberosus* L	Hath chuk	Asteraceae
512.	*Helicteris isora*	Marorphali	Sterculiaceae
513.	*Heliotroipium ellipticum* Ledebour.	Nikatti	Boraginaceae
514.	*Heliotropium indicum*	Hatisura	Boraginaceae
515.	*Heliotropium strigosum* Willd.	Chitiphul	Boraginaceae
516.	*Hernidesmus indicus* R.Br.	Anantmul	Periplocaceae
517.	*Heteropogon contortus* (L.) P.Beauv.	Kasurghas, Lumps	Poaceae
518.	*Hevea brasiliensis*(Willd.ex.Adr.de.Juss) Muel.Arg.	Rubber	Euphorbiaceae
519.	*Hibiscus liliaceus* L.	Chelwa	Malvaceae
520.	*Hibiscus mulabilis* L.	Gulaijaib	Malvaceae
521.	*Hibiscus rosa* sinensis L.	Ambari	Malvaceae
522.	*Hibiscus rosa* subdarifa L.	Rata, Lal Amhari	Malvaceae
523.	*Hibiscus surraliensis* L.	Ranbhindi	Malvaceae
524.	*Hibiscus tiliaceus* L.	Chelwa	Malvaceae
525.	*Hibiscus vitifolius* L	Bankapas	Malvaceae

Sr	Botanical Name	Common Name	Family
526.	*Hiptage berighalensis* (L.) Kurz.	Madhwilata	Malpighiaceae
527.	*Holarrhena antidysentrica*	Kutaj, Indrajav	Apocyneceae
528.	*Holoptelea integrifolia* (Roxb).	Planch, Chironji Papari	Ulmaceae
529.	*Holostemma rheedei* Wall.	Ghirbel	Asclepiadaceae
530.	*Hopea odorata* Roxb.		Dipterocarpaceae
531.	*Hydnocarpus kurzii* Warb.	Chaulmogra	Flacourtiaceae
532.	*Hygrophyla spinosa* T.Abders Buchtiam.	Talmakhana	Acathaceae
533.	*Hyocyamus muticus*	Khurasani ajwain	Solanaceae
534.	*Hyocyamus niger* L.	Khurasani ajwain	Solanaceae
535.	*Hypophorbe lagenicaulis*	Bottle palm	Arecaceae
536.	*Hyptis suaveolens* L.	Vilayti Tulsi	Lamiaceae
537.	*Impatiens balsamica* L.	Gulmehdi	Balasaminaceae
538.	*Imperata arundinacea*	Kansi	Poaceae
539.	*Indigofera linifolia* Retz.	Torki	Fabaceae
540.	*Indigofera trita* L.F.	Baikharia	Fabaceae
541.	*Inula racemosa*	Pokharmul	Asteraceae
542.	*Ipoemoea nil* (L). Roth.	Kaladana	Cunvolvulaceae
543.	*Ipomoea alba* L.	Dudhikalmi	Cunvolvulaceae
544.	*Ipomoea aquatica* Forsk.	Naribhaji	Cunvolvulaceae
545.	*Ipomoea biloba* Forsk.	Vidhara	Cunvolvulaceae
546.	*Ipomoea Dasysperma* Jacq.		Cunvolvulaceae
547.	*Ipomoea eriocarpa* R.Br.	Boota	Cunvolvulaceae
548.	*Ipomoea hederacea* Jacq.	Kaladana	Cunvolvulaceae
549.	*Ipomoea hederifolia* L.	Staripomoea	Cunvolvulaceae
550.	*Ipomoea hispida*	Bhawar	Cunvolvulaceae
551.	*Ipomoea mauritiana* Jacq.	Bilaikand	Cunvolvulaceae
552.	*Ipomoea obsewea* Ker	Bachgandha	Cunvolvulaceae
553.	*Ipomoea palmata* Forsk.	Railway creeper	Cunvolvulaceae
554.	*Ipomoea paniculata*	Vidarikand	Cunvolvulaceae
555.	*Ipomoea pestigridis* L.	Languli Lata	Cunvolvulaceae
556.	*Ipomoea purpurea* (L) Roth.		Cunvolvulaceae
557.	*Ipomoea tricolour* Cav.		Cunvolvulaceae
558.	*Ipomoea turbinata* Lag.	Guiliano Michai	Cunvolvulaceae
559.	*Ipomoea uniflora* Rooem. & Schult.		Cunvolvulaceae
560.	*Ipornoea batatas* Forsk.	Shakarkand	Cunvolvulaceae
561.	*Ixora coccinea* L.	Kangas, Rajana	Rubiaceae
562.	*Ixora parviflora* Vahl.	Kotgandlial	Rubiaceae
563.	*Jacaranda mimosifolia* D.Don.	Nili Gulmohar	Bignoniaceae
564.	*Jasminum angustifolia* Vahl.	Banmalika	Oleaceae
565.	*Jasminum arborescens* Roxb.	Chameli	Oleaceae
566.	*Jasminum auriculatum* Vahl.	Juhi	Oleaceae
567.	*Jasminum calophyllum* Wallet Dc.		Oleaceae
568.	*Jasminum flexile* Vahl	Multi	Oleaceae
569.	*Jasminum frunflorum* L.	Chameli	Oleaceae
570.	*Jasminum humile* L.	Peeli chameli	Oleaceae
571.	*Jasminum multiflorum*	Kunidphul	Oleaceae
572.	*Jasminum officinale* L.	Motia	Oleaceae
573.	*Jasminum sanbac* (L.) Ait	Mogra bela	Oleaceae

Sr	Botanical Name	Common Name	Family
574.	*Jatropha curcas* L.	Safed arandi	Euphorbiaceae
575.	*Jatropha glandulifera* Roxb.	Jungli arandi	Euphorbiaceae
576.	*Jatropha gossypifolia* L.	Bherenda	Euphorbiaceae
577.	*Jatropha multifida* L		Euphorbiaceae
578.	*Jatropha padaraefolia* Andr.		Euphorbiaceae
579.	*Jatropha podagrica* Hook.		Euphorbiaceae
580.	*Juniperus commuris* L. Hauber.		Cuppressaceae
581.	*Jussiaea repens* L.	Jalnonia	Onagraceae
582.	*Jussiaea tenella* Burmf.	Jalnonia	Onagraceae
583.	*Jussiaea uffruticosa* L.	Jalnonia	Onagraceae
584.	*Justicia diffusa* Willd.		Acanthaceae
585.	*Justicia gendarussa* L.F. Neel.	Nirgundi	Acanthaceae
586.	*Justicia procumbens* L.	Udar puchhi	Acanthaceae
587.	*Kalanchoe integra* (Mcedic) Kuntz.	Pannbiji	Crassulaceae
588.	*Kalanchoe laciniata* D.C.		Crassulaceae
589.	*Kalanchoe pinnata*	Hamsagar	Crassulaceae
590.	*Kalanchoe tigrinum*		Crassulaceae
591.	*Kigelia pinnata* D.C.	Jharfha noos	Bignoniaceae
592.	*Kochia indica* Wight.	Buikauraro	Chenopodiaceae
593.	*Kochia scoparia* Schrad.		Chenopodiaceae
594.	*Kydia calycina*	Pola	Malvaceae
595.	*Kyllinga triceps* Roxb.	Nirbis	Cyperaceae
596.	*Lablab purpureus*	Sem	Fabaceae
597.	*Lactuca indica* L.	Lettuce	Asteraceae
598.	*Lactuca remotifolia* D.C.		Asteraceae
599.	*Lactuca runcinata* D.C.		Asteraceae
600.	*Lactuca saliva* L. Salad		Asteraceae
601.	*Lagenaria vulgaris* Ser.	Kadwitumbi	Cucurbitaceae
602.	*Lagerstromia indica*	Lendia	Lythraceae
603.	*Lagerstromia parviflora*	Dhaura	Lythraceae
604.	*Lagerstromia speciosa* (L.) Pers.	Jarul	Lythraceae
605.	*Lathyrus aphuca* L.	Janglimatar	Fabaceae
606.	*Lathyrus odoratus*	Sweet pea	Fabaceae
607.	*Lathyrus sativus* L.	Kheshari	Fabaceae
608.	*Launaea aspleniifolia* Hook f.	Bangobhi	Asteraceae
609.	*Launaea niulicaulus* Hook f.	Ban Gohhi	Asteraceae
610.	*Lavendula angustifolia* Mill.	Lavender	Lamiaceae
611.	*Lavendula bipinnata* Kuntaze.	Lavender	Lamiaceae
612.	*Lawsonia inermis* L.	Henna, Mehdi	Lythraceae
613.	*lberis amara* L	Candytuft	Brassicaceae
614.	*lchnocarpus fruitescens* R.Br.	Anantmul, Dudhilata	Apocynaceae
615.	*Leonotis nepetiolia*	Bara Guma	Lamiaceae
616.	*Lepidagathis cristra* Willd.	Bukhar Jadi	Acanthaceae
617.	*Lepidium sativiun* L.	Chansur	Brassicaceae
618.	*Leucas aspera*	Chhota Halkusa	Lamiaceae
619.	*Leucas aspera* (Willd.) Sperng.	Chhota Halkusa	Lamiaceae
620.	*Leucas cephalotes* (Roth.) Sperng.	Goma	Lamiaceae
621.	*Leucas lavendulifolia* Rees.	Guma	Lamiaceae

Sr	Botanical Name	Common Name	Family
622.	*Licopodium clavatum* L.	Nagbeli	Lycopodiaceae
623.	*Lilium candidum* L.	Modonna Lily	Liliaceae
624.	*Lilium Nanteuni* Wall.		Liliaceae
625.	*Lilium tigrinum*	Tiger lily	Liliaceae
626.	*Limnophylla aromatica* (Lam.) Merrill.	Kuttra	Scrophulaviaceae
627.	*Limnophylla indica* (L.) Druce.	Amragandha	Scrophulaviaceae
628.	*Limnophylla rugosa* (Roth) Merr.		Scrophulaviaceae
629.	*Limonia acidissima* L.	Kaitha	Rutaceae
630.	*Lindenburgia indica* Kuntz.		Scrophulaviaceae
631.	*Lindernia crustacea* (L.) Muell.		Scrophulaviaceae
632.	*Lindernia oppositifolia* Retz.	Mukhurjee	Scrophulaviaceae
633.	*Lippia nodiflfora* Rich.	Jalpippli	Verbanceae
634.	*Litchi chinensis* (Gaertn.) Sonn.	Litchi	Sapindaceae
635.	*Litsea polyantha* Jurs.	Menda lakadi	Lauraceae
636.	*lndigofera glandulosa*	Jurjhuru	Fabaceae
637.	*lndigofera hirsuta* L		Fabaceae
638.	*lndigofera suffruticosa* Mill	Vilayiti Neel	Fabaceae
639.	*lndigofera tinctoria* L	Neel	Fabaceae
640.	*Lolium temulentum* L.	Snelum	Poaceae
641.	*Loranthus globosus* Roxb.	Banda Chhota	Loranthaceae
642.	*Loranthus longiflorus* Rerr.	Banda Bara	Loranthaceae
643.	*Luffa accutangula* Roxb.	Torai	Cucurbitaceae
644.	*Luffa aegyptiaca* Mill.	Ghiya Torai	Cucurbitaceae
645.	*Luffa echinata* Roxb.	Ghagrahela	Cucurbitaceae
646.	*Lycopodium cernuum* Linn.		Lycopodiaceae
647.	*Madhuca india* J.F. Emel.	Mahua	Sapotaceae
648.	*Magnolia grandiflorai* L.	Anda Champ	Magnoliaceae
649.	*Malachra capitata* L.	Ranbhindi	Malvaceae
650.	*Mallotus esculenta* Granth	Tapioca, Cassava	Euphorbiaceae
651.	*Mallotus philippensis* Lamk Muel.	Rohani, Kamela	Euphorbiaceae
652.	*Malus pumilla* Mill.	Apple	Rosaceae
653.	*Malva parvifolia*	Panirak	Malvaceae
654.	*Malva rotundifolia*	Khubasi	Malvaceae
655.	*Malvastrum coromandelianum* (L.)	Garcke-Khareti	Malvaceae
656.	*Manihot esculenta*	Saakarkhand, Sabudhana	Euphorbiaceae
657.	*Manilkara kauki* L.Dub.	Khirni	Sapotaceae
658.	*Mansoa alliaceae*	Garlic vine	Bignoniaceae
659.	*Maranta arundenacea* L.	Arrowroot/Tikhur	Marantaceae
660.	*Marsdenia tenacissima* Roxb.Moon.	Jiti, chit.	Asclepiadaceae
661.	*Marsilea minuta* L.	Chaepati	Marsileaceae
662.	*Martynia annua* L.	Bichua	Martyniaceae
663.	*Marus nigra* L.	Krishna toot	Moraceae
664.	*Matricaria chamomilla* L.	Chamomile	Asteraceae
665.	*Medicago hispidea* Gaetn.	Karmetha	Fabaceae
666.	*Medicago saliva*	Vilayati gawuth	Fabaceae
667.	*Melia azadirach* L.	Bakain	Meliaceae
668.	*Melilotus alba* Medik.	Safed Senji	Fabaceae
669.	*Melitotus parviflora* Desf.	Peeli senji	Fabaceae

Sr	Botanical Name	Common Name	Family
670.	*Mentha arvensis* L.	Mentha	Lamiaceae
671.	*Mentha piperita* L.	Pipermint	Lamiaceae
672.	*Mentha spicata* L.	Spearmint	Lamiaceae
673.	*Mentha viridis*	Pudina	Lamiaceae
674.	*Merrimia dissecta* (Jacq.) Hallrerf.		Convolvulaceae
675.	*Merrimia emarginata* Hallier	Musakani	Convolvulaceae
676.	*Merrimia tubrosa* (L.) Rendle.	Richkand-trivritta	Convolvulaceae
677.	*Mesua ferrea* L.	Nagkeshar	Guttiferae
678.	*Metroxylon sagu* Roxb.	Sago	Arecaceae
679.	*Michelia champaca*	Ganj	Magnoliaceae
680.	*Miliusa tomentosa* Roxb ex.J.Sinclair	Kari	Annonaceae
681.	*Millingtonia hortensis* L.F.	Akash neem	Vignoniaceae
682.	*Millusa ovalifolia* Kurz.		Annonaceae
683.	*Mimosa hamata* Willd.	Aar	Mimosaceae
684.	*Mimosa pudica* L.	Lajwanti	Mimosaceae
685.	*Mimusops elengi* L.	Maulshree	Sapotaceae
686.	*Mirabilis jalapa* L.	Gulabas	Nyctaginaceae
687.	*Mitragyna parvifolia* Roxb. Kurth.	Kern, Kaddam	Rubiaceae
688.	*Mollugo cerviana* (L.) Ser.	Parpat	Molluginaceae
689.	*Mollugo hirta* Thumb.	Okharadi	Molluginaceae
690.	*Mollugo lotoides* L.	Gandhibuti	Molluginaceae
691.	*Mollugo nudicaulis* Lam.		Molluginaceae
692.	*Mollugo oppositifolia* L.	Jim	Molluginaceae
693.	*Mollugo pentaphylla* L.		Molluginaceae
694.	*Mollugo stricta*	Pittapapada	Molluginaceae
695.	*Momordica balsamina* L	Mokha Bhatkarela	Cucurbitaceae
696.	*Momordica charantia* L.	Karela	Cucurbitaceae
697.	*Momordica cochinchinensis*	Kakrol male	Cucurbitaceae
698.	*Momordica cymbelaria* Fenze.	Kadochi	Cucurbitaceae
699.	*Momordica dioca* Roxb.	Kareladhar/Jangli Karela	Cucurbitaceae
700.	*Momordica longifolia*	Khirni	Cucurbitaceae
701.	*Momordica tubrosa* Cogn.	Kadavanchi	Cucurbitaceae
702.	*Morinda citrifolia* L.	Bartund,	Rubiaceae
703.	*Moringa oleifera*	Sainjna	Moringaceae
704.	*Morus alba* L.	Shahtoot	Moraceae
705.	*Morus serrata* Roxb.	Kimu, Himu	Moraceae
706.	*Mucuna bracteata* D.C.	Kewanch	Fabaceae
707.	*Mucuna capitata* Wight. & Arn.	Kewanch	Fabaceae
708.	*Mucuna pruriens* (L) D.C.	Kewanch	Fabaceae
709.	*Murraya koenigii* (L.) Spreng.	Mithi Neem	Rutaceae
710.	*Murraya paniculata* (L.) Jack.	Kamini	Rutaceae
711.	*Musa paradisiaca* L.	Kela	Musaceae
712.	*Mussaenda roxburghii* Hook f.		Rubiaceae
713.	*Nardostachys grandiflora* D.C.	Jata mansi	Veleriaceae
714.	*Nasturtium officinalis* R.Br.	Brahmi sag	Brassicaceae
715.	*Nelumbo nucifera*	Kamal	Nelumbonaceae
716.	*Neptunia prostrata* (Lamk).	Bail Lajalu	Mimosaceae
717.	*Nerium indicum* Mill.	Kaner	Apocynaceae

Sr	Botanical Name	Common Name	Family
718.	*Nerium oleander* L.		Apocynaceae
719.	*Nicotiana plumbagoginifolia*	Jangli tambaku	Solanaceae
720.	*Nicotiana tabacum*	Tambaku	Solanaceae
721.	*Nigella domascena* L.	Love-in-a mist	Ranunculaceae
722.	*Nigella sativa* L.	Kalongi Mungrela	Ranunculaceae
723.	*Nyctanthes arbortristis* L.	Parijat, Harsingar	Nyctanthaceae
724.	*Nymphaea alba*L.	Pandharenkamal white water lily	Nymphaeaceae
725.	*Nymphaea esculenta* Roxb.	Chhota sundi	Nymphaeaceae
726.	*Nymphaea nouchali* Burm F.	Kanval	Nymphaeaceae
727.	*Nymphaea stellalata* Willd.	Indian blue water lily/ Nilkamal	Nymphaeaceae
728.	*Nymphaea tetragona* Georgi.	Pigmi water lily.	Nymphaeaceae
729.	*Ocimum basilicum* Sims.	Basil, tulsi	Lamiaceae
730.	*Ocimum canum* Sims.	Mamri	Lamiaceae
731.	*Ocimum graticimum* L.	Ramtulsi	Lamiaceae
732.	*Ocimum kilmandscharicum* Guerke.	Kapurtulsi	Lamiaceae
733.	*Ocimum sanctum* L.	Tulsi	Lamiaceae
734.	*Oenothera lamarkiana*	Evening Primrose	Onagraceae
735.	*Oldenlandia biflora* L		Rubiaceae
736.	*Oldenlandia corymbosa* L.	Pittpapda	Rubiaceae
737.	*Oldenlandia diffusa* Willd. Roxb.	Snake-Needle Grass	Rubiaceae
738.	*Oldenlandia herbacea* Roxb.	Slender Diamond Flower	Rubiaceae
739.	*Oldenlandia hispida* Benth		Rubiaceae
740.	*Oldenlandia umbellata*	Chay root	Rubiaceae
741.	*Onosma bractatum* Wall.	Goejaban, Shakhahuli	Boraginaceae
742.	*Onosma echioides* L.	Ratanjot	Boraginaceae
743.	*Operculina turpenthum* (L.) S. Manso.	Nisodh	Convolvulaceae
744.	*Opuntia cochinellifera* Mill.	Nagfani (Spineless)	Cactiaceae
745.	*Opuntia dellenii*. Haw.	Nagfani	Cactiaceae
746.	*Opuntia vulgaris* Mill.	Nagfani	Cactiaceae
747.	*Origanum majorana*	Marwa/ Tulsi Marwa	Lamiaceae
748.	*Orobanche aegyptiaca* Pers.		Orobanchaceae
749.	*Oroxylum indicum*	Shivnag	Bignoniaceae
750.	*Ougeinia oojeinensis* (Roxb.) Hochr.	Tinsa	Fabaceae
751.	*Oxalis carniculata* L.	Khattibooti/Chuktripati	Oxalidaceae
752.	*Oxalis latifolia* H.B. & K.	Khatti Buti	Oxalidaceae
753.	*Oxystelma sccamone* (L.)	Dudhilata	Asclepiadaceae
754.	*Paederia foetida* L.	Gandhali	Rubiaceae
755.	*Pandanus andamanensium*	Kewda, Ketki	Pandanaceae
756.	*Pandanus furcatus* Roxb.		Pandanaceae
757.	*Pandanus odoratissimus* F.	Keora	Pandanaceae
758.	*Pandanus utilis* Bory.	Keora	Pandanaceae
759.	*Panicum anitidotale* Retz.	Deenanath Ghas	Poaceae
760.	*Papaver somniferum* L.	Afim	Papaveraceae
761.	*Parkia biglandulosa*	Parkia	Mimosaceae
762.	*Parkinsonia aculeata* L.	Vilayati babul	Caesalpiniaceae
763.	*Parthenium hysterophorus L.*	Gajarghass	Asteraceae
764.	*Paspalum scrohiculatum* L.	Kodo	Poaceae
765.	*Passiflora edulis* Sims.	Jhumakalata	Passifloraceae

Sr	Botanical Name	Common Name	Family
766.	*Passiflora foetida* L.	Kaurava Pandava	Passifloraceae
767.	*Passiflora incarnata* L		Passifloraceae
768.	*Pavonia odorata* Willd.	Sugandhwala	Malvaceae
769.	*Pedalium murex* L	Bara Gokhra	Pedaliaceae
770.	*Peganum hermala*	Hermal	Zygophyllaceae
771.	*Pelargonium graveolens* (L.) Herit.	Rosegeranium	Geraniaceae
772.	*Pellaea calomelanos* Link.	Pella	Sinopteridaceae
773.	*Peltophorum pterocarpum*	Copper Pod	Caesalpiniaceae
774.	*Pennisetum typhoides* (Burm. F.) Stapf.	Bajra	Poaceae
775.	*Pergularia daemia* Forsk.	Utranbel	Asclepiadaceae
776.	*Perilla fruitiscens*	Bhasinda	Lamiaceae
777.	*Peristrophe bicalyculata* Nees.	Atrilal, Kakjangha	Acanthaceae
778.	*Peuraria tubrosa* Roxb. ex. Willd.	Vidarikand	Fabaceae
779.	*Phaseolus trilobatus* (L.)Schreb.	Banmumg	Fabaceae
780.	*Phaseolus vulgaris* L.	Bakla	Fabaceae
781.	*Phoenix acaulis* Roxb.	Khajur	Arecaceae
782.	*Phoenix sylvestris* (L) Roxb.	Khajur	Arecaceae
783.	*Phragmitis communis* Trin.	Reed, Ramsir	Poaceae
784.	*Phragmitis karka* (Retz.)Trin ex. Steud.	Barru Narkul	Poaceae
785.	*Phyllanathus simplex* Retz.	Bhuamla	Euphorbiaceae
786.	*Phyllanthgus madraspaterri*	Ranavali	Euphorbiaceae
787.	*Phyllanthus urinaria* L.	Hazarmani, Bhuamala	Euphorbiaceae
788.	*Physalis alkekenji* L.	Tomato Strawberry	Solanaceae
789.	*Physalis minima* L.	Jangli Rasbhari, Chirpota	Solanaceae
790.	*Physalis peruviana* L.	Rasbhari	Solanaceae
791.	*Pimenta dioca* (L) Merrill.	Allspice tree may	Myrtaceae
792.	*Pimpinella anisurn* L.	Suonf	Apiaceae
793.	*Pinus roxburghii*	Chir	Pinaceae
794.	*Piperbetle* L.	Pan	Piperaceae
795.	*Piperlongum* L.	Pipar, piplamul	Piperaceae
796.	*Piper nigrum*	Kalimirch	Piperaceae
797.	*Pistia stratiotes* L.	Jalkhumbhi	Araceae
798.	*Pitchecolobium dulce* Benth.	Vilayati imli	Mimosaceae
799.	*Plantago ovta* Forsk.	Isabgol	Plantaginaceae
800.	*Pluchia lanceolata* Oliver & Hiern.	Rasna, Vilaikand	Asteraceae
801.	*Plumbago indica* L.	Lal chitrak	Plumbaginaceae
802.	*Plumbago zeylanica* L.	Chitrak	Plumbaginaceae
803.	*Plumeria alba* L.	Safed champa	Apocynaceae
804.	*Plumeria rubra* forma acuminate	Khairchampa	Apocynaceae
805.	*Plumeria rubra* L.	Lalchampa	Apocynaceae
806.	*Pogostemon cabin* (Blanco) Benth.	Patchauli	Lamiaceae
807.	*Pogostemon patchouli* Hook. F.	Patchauli	Lamiaceae
808.	*Polyalthea longifolia* Benth.	Nakali Ashok Devdaru	Annonaceae
809.	*Polyalthea Suberosa* Thw.	Chamkhirni	Annonaceae
810.	*Polygonum amphibium* L.		Polygonaceae
811.	*Polygonum amplex* caule D. Don		Polygonaceae
812.	*Polygonum hydropiper* L.	Waterpepper	Polygonaceae
813.	*Polygonum multiflorum* Thumb.		Polygonaceae

Sr	Botanical Name	Common Name	Family
814.	*Polygonum punctatum* Ham.		Polygonaceae
815.	*Pongamia pinnata* L. Pierre.	Karanj	Fabaceae
816.	*Portulaca grandiflora Hook.*		Portulacaceae
817.	*Portulaca oleracea*	Kulfa	Portulacaceae
818.	*Pothos cathcarti* Schott.		Araceae
819.	*Premna mucronata* Roxb.	Bakar Basota	Verbenaceae
820.	*Premna obtusifolia.*	Agetha, Arni	Verbenaceae
821.	*Prunus domestica* L.	Alubukhara	Rosaceae
822.	*Psidium guajava L.*	Amrud	Myrtaceae
823.	*Psoralea corylifolia* L.	Babachi	Fabaceae
824.	*Pterocarpus marsupium* L.	Beejasar	Fabaceae
825.	*Pterocarpus santalinus* L.	Redsenders	Fabaceae
826.	*Pterospermum acerifolium* Wild.	Kanak champa	Sterculiaceae
827.	*Pterospermum canescens* Roxb.	Muchkand	Sterculiaceae
828.	*Punica granatum* L.	Anar	Punicaceae
829.	*Purtanjiva roxburghii* Wall.	Purtanjiva	Euphorbiaceae
830.	*Pygmaeopremnha herbacea*	Bharangi	Verbenaceae
831.	*Pyrostegia venusa*	Golden shower	Bignoniaceae
832.	*Quisqualis indica L.*	Rangoon creeper	Combetaceae
833.	*Radermachera xylocarpa*	Padri tree	Bignoniaceae
834.	*Randia spinosa* Poir.	Mainphal	Rubiaceae
835.	*Ranunculus sceleratus* L.	Jaldhania	Ranunculaceae
836.	*Raphanus sativus* L.	Muli	Brassicaceae
837.	*Rauvolfia serpentine* L. Benth	Sarpagandha	Apocynaceae
838.	*Rauvolfia teraphylla* L.	Barachandrika	Apocynaceae
839.	*Rhus parviflora*		
840.	*Rhynchosia capitata* D.C.	Paparo	Fabceae
841.	*Rhynchosia minima* (L.) D.C.	Paparo	Fabceae
842.	*Ricinus communis* L.	Arandi (Chhoti)	Euphorbiaceae
843.	*Rivinia humilis*	Blood berry, Pigion berry	Protiveraceae
844.	*Rorirppa indica* (L.) Hern.	Khubkalan	Brassicaceae
845.	*Rosa alba* L	Gulab	Rosaceae
846.	*Rosa banksiae* R.Br.	Lata Gulab	Rosaceae
847.	*Rosa bourboniana* Desportes.	Desi gulab/ Baramasi gulab	Rosaceae
848.	*Rosa braunonii* Lindl	Kunja, Karer climbing rose	Rosaceae
849.	*Rosa centifolia* L	Gulab	Rosaceae
850.	*Rosa chinensis* Jacq.	Dcshi Gulab	Rosaceae
851.	*Rosa damascena* Mill.	Damask rose/Fasti Gulab	Rosaceae
852.	*Rosa gallica* L.	French rose	Rosaceae
853.	*Rosa india* Hook F.		Rosaceae
854.	*Rosa multiflora* Thunb.	Climbing rose	Rosaceae
855.	*Rosmarinus* officinalis L.	Rosmari	Lamiaceae
856.	*Rubia cordifolia* L. Sansu. Hook.R.	Manjit	Rubiaceae
857.	*Ruellia tubrosa* L.	Pathrikat	Acanthaceae
858.	*Rumex acetocella* L.	Chuk	Polygonaceae
859.	*Rumex acetosa* L.	Chuka	Polygonaceae
860.	*Rumex dentatus* L	Ban Palak	Polygonaceae
861.	*Rungia pectinata* (L.) Nees.	Kharmor	Acanthaceae

Sr	Botanical Name	Common Name	Family
862.	*Rungia repens* (L.) Nees.	Kharmor	Acanthaceae
863.	*Ruta graveolens* L.	Sitab, Sadab	Rutaceae
864.	*Saccharum munja* Roxb.	Munj	Poaceae
865.	*Saccharum officinarum* L.	Ganna	Poaceae
866.	*Saccharum spontaneum* L.	Kans	Poaceae
867.	*Sagittaria sagitifolia* L.	Chhotakut	Alismataceae
868.	*Salsola baryosma* (R. & S.) Dandy.	Loonuk	Chenopodiaceae
869.	*Salvadora oleoides* Decne.	Barapilu	Salvadoraceae
870.	*Salvadora persica* L.	Chhota pilu	Salvadoraceae
871.	*Salvadora plebeia* Br.	Kamarkas	Salvadoraceae
872.	*Salvia aegyptiaca* L.	Salbia	Lamiaceae
873.	*Salvia hispanica*	Chiaseeds	Lamiaceae
874.	*Salvia officinalis* L	Salbia, sefakuss	Lamiaceae
875.	*Salvia sclarez* L	Clarysage	Lamiaceae
876.	*Salvia splendens* Ker Gawl.	Sailbia	Lamiaceae
877.	*Sansevieria roxburghiana* Schutt.	Marul, Murwa	Agavaceae
878.	*Sansevieria zeylanica* Willd.	Murwa	Agavaceae
879.	*Santalum album* L.	Chandan	Santalaceae
880.	*Sapindus emerginatus* Vahl.	Ritha	Sapindaceae
881.	*Sapindus mukorossi* Gaertn.	Ritha	Sapindaceae
882.	*Saraca asoca* (Roxb.) de.Wild.	Sita Ashok	Caesalpiniaceae
883.	*Sarcostemma acidum*	Somlata	Apocynaceae
884.	*Schelichera oleosa* (Lour) Oken.	Kosum	Sapindaceae
885.	*Scindapsus officinalis*Schott. (Syn Pothos officinalis)	Gajpipal	Araceae
886.	*Scirpus grossus* L.F.	Kaseru	Cyperaceae
887.	*Scirpus lacustris* L.	Clubrush	Cyperaceae
888.	*Sclera pergracillis* Nees. Kunth.	Lemon scented cyprus	Cyperaceae
889.	*Scoparia dulcis* L.	Meethi Buti Seeds/ planting	Flacourtiaceae
890.	*Semecarpus anacardium* L.F.	Bhilwan	Anacardiaceae
891.	*Senecio vulgaris* L.		Asteraceae
892.	*Sesamum indicum* L.	Til/Kala Til	Pedaliaceae
893.	*Sesbania aculeata* Pers.	Dhuincha	Fabaceae
894.	*Sesbania bispinosa* Jacq.W.F.	Jayanti	Fabaceae
895.	*Sesbania grandiflora* (L.) Pair.	Agastoya	Fabaceae
896.	*Sesbania sesban* (L.) Merr.	Jayanti	Fabaceae
897.	*Setaria italica* (L.) Beauv.	Kangni	Poaceae
898.	*Shorea robusta* Gaertn.	Sal	Dipterocarpaceae
899.	*Sida alba* L. SynS. Spinosa L.	Nagbala	Malvaceae
900.	*Sida carpinifolia* Mast	Bariagn	Malvaceae
901.	*Sida cordata* (Burms. F.) Borssum	Bhumibala	Malvaceae
902.	*Sida cordifolia* L.		Malvaceae
903.	*Sida rhombifolia* L.Var.*rhombodea*Roxh. Mast.	Mahabala, safed beriala	Malvaceae
904.	*Silybum marianum*	Muilk thistle	Asteraceae
905.	*Sisymbrium irio* L.	Khukalan	Brassicaceae
906.	*Smilax china* Linn.	Chobchini	Smilacaceae
907.	*Smilax zeylanica* L.	Ramdataun	Smilacaceae

Sr	Botanical Name	Common Name	Family
908.	*Smithia sensitiva* Ait	Odabrini	Fabaceae
909.	*Solanum grandiflorum*		Solanaceae
910.	*Solanum indicum*	Makoi	Solanaceae
911.	*Solanum lacianiatum*	Jangli bhata	Solanaceae
912.	*Solanum pseudo* capsicum	Golmirch	Solanaceae
913.	*Solanum sissymbrifolium*	Golmirch	Solanaceae
914.	*Solanum stramonifolium* Jacq.	Rambegun	Solanaceae
915.	*Solanum surattense* Burmf.(Syn. *S. xanthocarpam*)	Kateli/ Bhatkatai	Solanaceae
916.	*Solanum torvum* Swartz.	Titbaigun	Solanaceae
917.	*Solanum viarum* Dunal.	Jangli Bhatkatai	Solanaceae
918.	*Solidago canadensis* L.	Golden rod	Asteraceae
919.	*Sonchus arvensis*	Sahadevi Badi	Asteraceae
920.	*Sonchus oleraceus* L	Daukad	Asteraceae
921.	*Soymida febrifuga*& Juss.	Rohan	Meliaceae
922.	*Spathodea campanulata* Beauv.	Fountain tree	Bignoniaceae
923.	*Spatholobus roxburghii*Benth.Syn. *Butea parviflora*	Palas bel	Fabaceae
924.	*Spermadictylon sauveolens* Roxb.	Mahabel	Rubiaceae
925.	*Sphaeranthus. indicus*	Gorakhmundi	Asteraceae
926.	*Spilanthus oleracea* Murr.	Akarkara	Asteraceae
927.	*Spilanthus paniculata* Wall ex. Dc.	Akarkara	Asteraceae
928.	*Spinacia oleracea*	palak	Chenopodiaceae
929.	Sterculia villosa Roxb.	Chandan	Sterculiaceae
930.	*Sterculia wrens* Roxb.	Gulu, Kullu	Sterculiaceae
931.	*Stereospermum coleus*		Bignoniaceae
932.	*Stereosperum suaveolens* D.C.	Paral, Padaria	Bignoniaceae
933.	*Strychnos nux-vomica* L.	Kuchla	Strychnaceae
934.	*Strychnos potatorum* L.F.	Nirmal	Strychnaceae
935.	*Swertia chirata* (Roxb. ex.flem.)	Chirayata	Gentianaceae
936.	*Swietenia mahagoni* (L) Jack.	Mahogoni	Meliaceae
937.	*Symplocas racemosa* Roxb	Lodh	Symplocaceae
938.	*Syzygium aromaticum*	laung	Myrtaceae
939.	*Syzygium aromaticus*	Loung	Myrtaceae
940.	*Syzygium cumini* (L.) Skeels.	Jamun	Myrtaceae
941.	*Syzygium fruiticosum* Rook D.C.	Jangli Jamun/Kat Jamun	Myrtaceae
942.	*Syzygium samaragense*	Water apple	Myrtaceae
943.	*Tabebuia pentaphylla* (L) Hemsl.	Basant ki rani	Bignoniaceae
944.	*Tabebuia rosea* Bertol D.C.	Basant ki Rani	Bignoniaceae
945.	*Tabebuia spectahilis* N.Chils.	Basand ki rani	Bignoniaceae
946.	*Tabernaemontanadivaricata* (L.) R.Br.	Chandni	Apocynaceae
947.	*Tacoma stuns* (L) H.B. & K.	Yellow bell	Bignoniaceae
948.	*Tagetes erecta* L.	Genda	Asteraceae
949.	*Tagetes lucidu* Cav	Genda	Asteraceae
950.	*Tagetes minuta* L.	Gendi	Asteraceae
951.	*Tagetes patula* L.	Genda	Asteraceae
952.	*Tagetes tenuifolia* Cav	Striped marigold	Asteraceae
953.	*Tamarindus indica* L.	lmli	Mimosaceae

Sr	Botanical Name	Common Name	Family
954.	*Tamarix troupii* Hole.	Jhau, Makin	Tamaricaceae
955.	*Taxodium distichum* L.Rich.	Taxodium	Taxodiaceae
956.	*Tecomella undulata* G.Don.	Rugtrora, Rohida	Bignoniaceae
957.	*Tectona grandis* L.F.	Sagon	Verbenaceae
958.	*Tephorosia candida* Roxb D.C.	Lashtia	Fabaceae
959.	*Tephrosia purpurea* (L) Pers.	Sarphonka	Fabaceae
960.	*Tephrosia tinctoria* Pers.		Fabaceae
961.	*Tephrosia vilosa* Pers.	Swetsurfunka	Fabaceae
962.	*Terminalia arjuna* (Roxb) Wight.	Ajun	Combretaceae
963.	*Terminalia bellirica* (Gaertn) Roxb.	Bahera	Combretaceae
964.	*Terminalia catappa* L.	Jangli badam	Combretaceae
965.	*Terminalia chebula* Retz.	Harra	Combretaceae
966.	*Terminalia tomentosa*	Saja, Asan	Combretaceae
967.	*Thalictrum foliolosum* D.C.	Pseuedo Mameera	Ranunculaceae
968.	*Theobroma cacao L.*	Cocoa	Sterculiaceae
969.	*Thespesia popuMea* (L.)	Paraspipal	Malvaceae
970.	*Thevetia peruviana* (Pers.) K. Schum	Pilikaner	Apocynaceae
971.	*Thuja occidentalis* L.	Mayurpankh	Cupressaceae
972.	*Thuja orientalis* L.	Mayur pankh	Cupressaceae
973.	*Thunbergia alata* Boj. Exsim.	*Black eye susan*	Thunbergiaceae
974.	*Thunbergia faragrans* Roxb.		Thunbergiaceae
975.	*Thunbergia grandiflora* Roxb.	Kukua-lata	Thunbergiaceae
976.	*Thymus serpyllum* L.	Jangli ajwain	Lamiaceae
977.	*Tinospora cordifolia* (Wind) Meers.	Amrita, Gloe	Menispermaceae
978.	*Tinospora crispa* (L) Miers. ex Hook. F. Thoms.	Gurch	Menispermaceae
979.	*Tinospora sinensis* (Lour.) Merr.	Gurch	Menispermaceae
980.	*Toona ciliata*	Tun	Meliaceae
981.	*Trachyspermum ammi* (L) Spraoue.	Ajwain	Apiaceae
982.	*Trachyspermum roxburghianum* (Oic) Gnaib.	Ajmud	Apiaceae
983.	*Tradescantia discolour* L Herit.	Bugnell	Commelinaceae
984.	*Trapa natans* L.	Singhara	Trapaceae
985.	*Triadica sebifera* (L.)Small.	Chinese tallow tree	Euphorbiaceae
986.	*Trianthema portulacastrum*	Viskhapara	Aizoaceae
987.	*Tribulus terestris* L.	Gokharu	Zygophyllaceae
988.	*Trichodesma amplexicanle* Roth.	Aundh Fali/ Gojaban	Boraginaceae
989.	*Trichodesma indicum* Br.	Aynahuli	Boraginaceae
990.	*Trichodesma zeylanicum* Br.	Jalsiris	Boraginaceae
991.	*Tricholepis glabrrima* D.C.	Brahmudandi	Asteraceae
992.	*Trichosanthes anguina* L	Chachinga, Chachinda	Cucurbitaceae
993.	*Trichosanthes cucumerina* L.	Jangli chachinda	Cucurbitaceae
994.	*Trichosanthus dioica Roxb.*	Parwal	Cucurbitaceae
995.	*Trichosanthus laciniosa* Wall.	Rambel	Cucurbitaceae
996.	*Tridax procumbens*	Ekdandi, Gudaria	Asteraceae
997.	*Trigonella foenum* graecum L.	Methi	Fabaceae
998.	*Triumfetta pentandra* A. Rich.	Chikatyo	Tiliaceae
999.	*Triumphetta rhomboidea* Jacq.	Rakat kadawan	Tiliaceae

Sr	Botanical Name	Common Name	Family
1000.	*Tylophora fasciculata* Buch. Ham.		Asclepiadaceae
1001.	*Tylophora indica* (Burm.f.)	Antamul	Asclepiadaceae
1002.	*Typha angusta* Bory. & Chaub.	Pater, Ramban	Typhaceae
1003.	*Uraria lobata* L	Pitharan	Fabaceae
1004.	*Uraria picta* Desv.	Prishinparni	Fabaceae
1005.	*Urena lobata* L var. Cinuata King.	lotloti	Malvaceae
1006.	*Urena repanda* Roxb ex.sm.	Lotloti	Malvaceae
1007.	*Urenia lobata* L.	Lapetua	Malvaceae
1008.	*Urginea indica* (Roxb) Kunth	Jaglipyaj, Jangli Khanda	Liliaceae
1009.	*Urtica dioca* L.	Bichu booti	Urticaceae
1010.	*Valeriana jatamanshi* Janes.	Muskhbala	Valeriaceae
1011.	*Vallisnaria spiralis* L.	Kai	Hydrocharitaceae
1012.	*Vanda parviflora*	Vanda	Orchidaceae
1013.	*Vanda tessellata*	Vanda, rasna	Orchidaceae
1014.	*Ventilago denticulata* Willd.	Keoti, Raidhani	Rhamnaceae
1015.	*Ventilago madraspatana* Gaertn.	Bikakali bel	Rhamnaceae
1016.	*Verbascum thaspus* L	Jangli tambaku	Scrophulaiaceae
1017.	*Verbena officinalis* L	Pamukh	Verbenaceae
1018.	*Vernonia anthelmintica* Willd.	Kalijiri	Asteraceae
1019.	*Vernonia cinerea* (L) Less.	Sahadevi	Asteraceae
1020.	*Vetiveria zizanioides* (L.) Mash.	Khas	Poaceae
1021.	*Vicia hirsuta* (L.) S.F.	Akari	Fabaceae
1022.	*Vicia sativa* L.	Akari	Fabaceae
1023.	*Victoria regia* lindl.	Kaamal	Euryalaceae
1024.	*Vigna aconitifolia* (Jacq.) Marechal.	Moth	Fabaceae
1025.	*Vigna trilobata* (L) Verdcourt.	Mugam	Fabaceae
1026.	*Vigna umbellata* (Thumb) Ohwi & Ohashi.	Sutri	Fabaceae
1027.	*Vinca major* L.	Periwinkle	Apocynaceae
1028.	*Vinca minor* L.	Periwinkle	Apocynaceae
1029.	*Viola odorata*	Banafasha	Violaceae
1030.	*Viola serpens* Wall ex.Ging.	Banafasha	Violaceae
1031.	*Vitex glabra* R.Br.	Shersa	Verbenaceae
1032.	*Vitex negundo* L.	Nirgundi	Verbenaceae
1033.	*Vitex peduncularis* Wall.	Shilgiri	Verbenaceae
1034.	*Vitis cetosa*	Dokarbela	Vitaceae
1035.	*Vitis indica* L	Jangli angur	Vitaceae
1036.	*Vitis latifolia* Roxb	Gobil. Panibel	Vitaceae
1037.	*Vitis pedata* Vahi.	Godhapadi	Vitaceae
1038.	*Vitis trifolia* L.	Ramchana, Panikisanbhalu	Vitaceae
1039.	*Volutarella divericata* Benth.	Brahmdani	Asteraceae
1040.	*Withania somnifera* Dunal.	Ashwagandha	Solanaceae
1041.	*Woodfordia fruitcosa* L. Kurtz.	Dhawai	Lythraceae
1042.	*Wrightia tinctoria* R.Br.	Mitha indrajou, Dudhi	Apocynaceae
1043.	*Wrightia tomentosa* (Roxb.) Roem. & Schul.	Dudhi	Apocynaceae
1044.	*Xanthium strumarium* L.	Adhashishi	Asteraceae
1045.	*Yucca aloifolia* L.	Adannath	Agavaceae
1046.	*Yucca filanentosa* L.	Adannath	Agavaceae

Sr	Botanical Name	Common Name	Family
1047.	*Yucca gloriosa*	Adannath Bhed	Agavaceae
1048.	*Zingiber cassumunar* Roxb.	Jangli Adrak	Zingiberaceae
1049.	*Zingiber officinalis* Rose.	Adrak	Zingiberaceae
1050.	*Ziziphus mauritiana*	Ber	Rhamnaceae
1051.	*Ziziyphus nummularia*	Jharber	Rhamnaceae
1052.	*Ziziyphus xylopyra* Wills.	Kat ber/Ghontber	Rhamnaceae
1053.	*Zizyphus oenopila* Mill.	Makor	Rhamnaceae

Family wise representation of plant species

Total 1053 plant species representing 164 families have been conserved in the medicinal garden. Ten families have presentation of more than 15 species with maximum representation of 81 species by family Fabaceae.

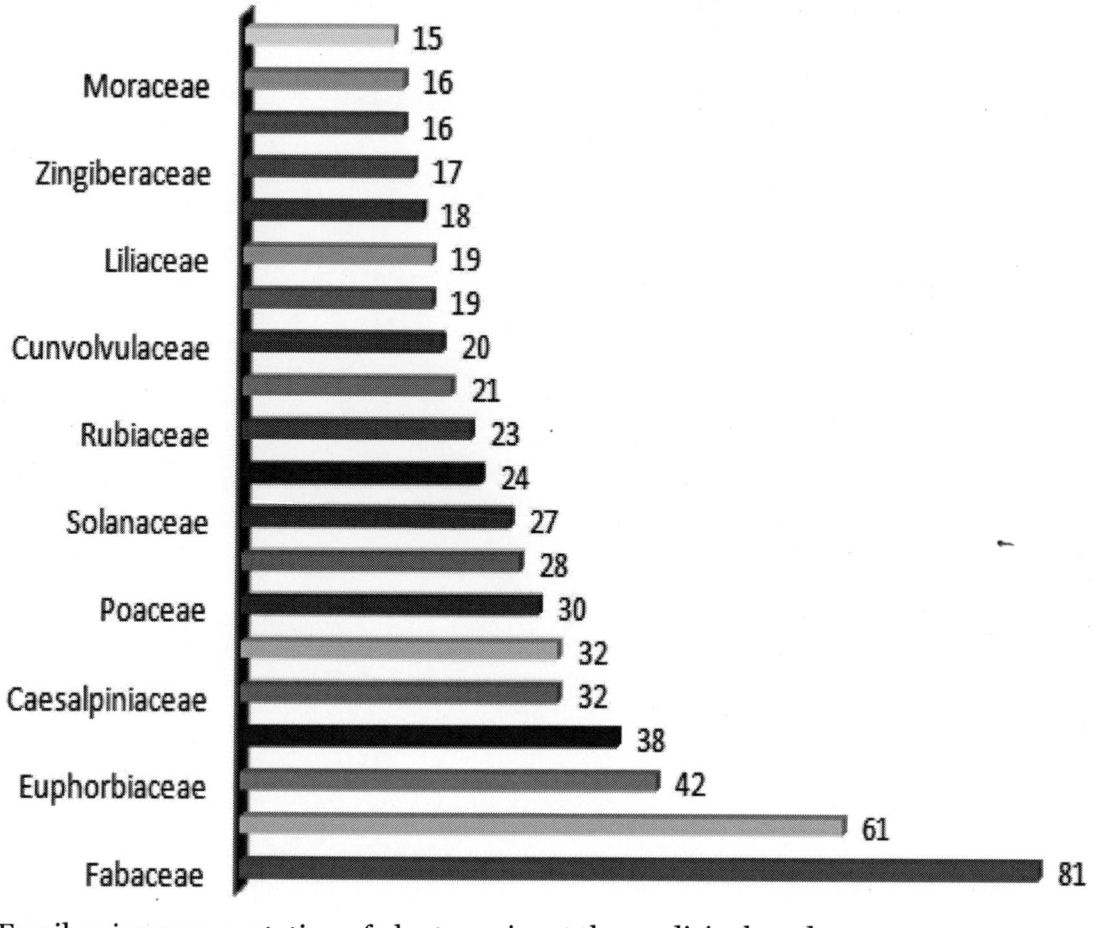

Fig 17.1 Family wise representation of plant species at the medicinal garden

CHAPTER 18

Biodiversity Audit of Mushroom

Vijay Kumar Yadav,
Jayant Bhatt and Gitanshu Dinkwar
Department of Plant Pathology
College of Agriculture, Jabalpur

Mushrooms are cosmopolitan, heterotrophic, beneficial higher fungi that are quite specific in their nutritional and ecological requirements. Mushrooms help in maintaining the ecosystem by catalyzing the mineralization of organic matter. They have been used as food ever since the hunting-and-gathering stage of our prehistoric forebears. Archaeological evidences reveal edible species associated with people living 13,000 years ago in Chile (Rojas *et al.*, 1995). They were included in the diet by Greeks and Romans since ancient times. Romans regarded them as food of God while Chinese termed them as elixir of life.

Diversification in any farming system imparts sustainability. Mushrooms are one such component that not only impart diversification but also help in addressing the problems of quality food, health and environmental related issues. Commercial production of edible mushrooms represents unique exploitation of the microbial technology for the bioconversion of the agricultural, industrial, forestry and household wastes into nutritious food (mushrooms). Indoor cultivation of mushrooms, utilizing the vertical space, is regarded as the highest protein producer per unit area and time. Hi-tech horticulture venture has a promising scope to meet the food shortages, without undue pressure on land.

Edible mushrooms are a delicious source of food all over the world. The consumption of edible mushrooms is increasing, even in the developed world, due to awareness for their appreciable content of proteins and vitamins B, C, and D as well as a higher content of trace minerals. They are low in calorie with negligible starch, sugars and fats and have a high nutritional value, almost twice that of any vegetable or fruit. Apart from their use as food, sufficient evidences suggest that many species contain substances that may prevent or reduce the chances of contracting cancer, heart diseases, diabetes, and viral infections (Oei, 1991). Mushrooms are known to produce many kind of bioactive compounds, generally linked with the mycelial cell wall, that help in enhancing immunity against carcinogens (Ramesh and Pattar, 2010). In addition, many edible and non-edible mushrooms have been used for medicinal purposes (Wasser and Weis, 1999; Hobbs, 1995).

Mushroom cultivation in India is of recent origin. In the 1961, ICAR funded a project on cultivation technology of button mushroom at Solan which led to establishment of a UNDP project with FAO experts. Research on mushrooms at JNKVV was started in the year 1973 with successful cultivation of Paddy Straw mushroom (*Volvariella volvacea*) at Jabalpur. Overall, 19 mushroom species are recorded from the university campus in which 12 are edible, 4 are medicinal and 3 species are non-edible (Fig 18.1).

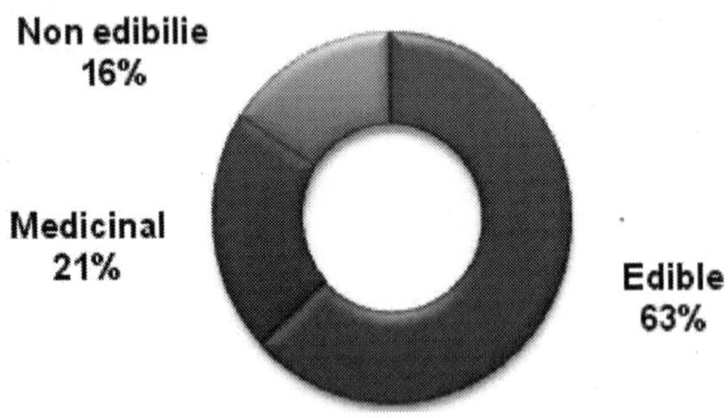

Fig 18.1 Types of mushroom reported from JNKVV campus Jabalpur

At JNKVV Jabalpur four commercially cultivated edible mushrooms viz., Button, White Milky, Oyster (Plate 18.1) and Paddy straw are cultivated experimentally with the highest species diversity in oyster mushrooms.

Plate 18.1 Oyster with basidiospores

Among all the edible mushrooms listed in Table 18.1, *Termitomyces* is the only mushroom which cannot be cultivated. It only grows in association with termites in their nests dependent on the organic matter brought by the termites. Basically, *Termitomyces* aresaprobic, but symbiotic with termites. The fungus helps the termites to degrade the plant-derived material (e.g., wood, dry grass, and leaf litter) on which they live. During rainy season, basidiocarps of this mushroom are gathered by the local inhabitants for consumption, as well as for sells. It carries great ecological and socio-economic significance with high nutritional value viz., rich in various nutritional and nutraceutical components such as proteins, fibers, carbohydrates, low amount of fats, minerals, phenolics, flavonoids, ascorbic acid, β carotene, lycopene, alkaloids (Fig. 18.1 and 18.2).

Plate 18.2 Termitomyces

Table 18.1 List of mushroom reported form JNKVV campus Jabalpur

	Scientific name	Common Name	Use
1.	*Agaricus bisporus*	Button	Edible
2.	*Calocybe indica*	Milky Mushroom	Edible
3.	*Hypsizygus ulmarius*	Elm Oyster	Edible
4.	*Pleurotus djamor*	Pink Oyster	Edible
5.	*Pleurotus eous*	Oyster	Edible
6.	*Pleurotus flabellatus*	Oyster	Edible
7.	*Pleurotus membranaceus*	Oyster	Edible
8.	*Pleurotus ostreatus*	Oyster	Edible
9.	*Pleurotus sajor-caju*	Grey Oyster	Edible
10.	*Pleurotus florida*	White Oyster	Edible
11.	*Termitomyces clypeatus*	Termitomyces	Edible
12.	*Volvariella volvacea*	Paddy straw	Edible
13.	*Auricularia polytricha*	Cloud ear fungus	Medicinal
14.	*Ganoderma applanatum*	Reishi mushroom	Medicinal
15.	*Ganoderma lucidum*	Reishi mushroom	Medicinal
16.	*Hericium erinaceus*	Monkey head mushroom	Medicinal
17.	*Coprinus* sp.	Inky Cap	Non edible
18.	*Lepiota cristata*	Stinking dapperling	Non edible
19.	*Lycoperdon* sp.	Puffballs	Non edible

References:

Hobbs, CH. 1995. Medicinal mushrooms: an exploitation of tradition, healing and culture. Botanica press. *Santacruz*. pp 219-251.

Oei P. 1991. Manual on mushroom cultivation. Techniques and opportunity for commercial applications in developing countries. CTA. pp 21-26.

Ramesh C and Pattar MG. 2010. Antimicrobial properties, antioxidant activity and bioactive compounds from six wild edible mushrooms of Western Ghats of Karnataka, India. Pharmacognosy Research 2: 107-112.

Rojas C and Mansur E. 1995. Ecuador: General information on non-timber products in Ecuador. In: Memory, expert consultation on non-timber forest products for Latin America and the Caribbean. Forest Series. Santiago Chile. pp. 208-223.

Wasser, SP and Weis, AL. 1999. Medicinal properties of substances occurring in higher basidiomycetes mushrooms: current perspectives (review). International Journal of Medicinal Mushrooms 1: 31-62.

CHAPTER 19

Biodiversity Audit of Weeds

PB Sharma
Department of Agronomy
College of Agriculture, Jabalpur

Total 149 weeds documented at Jabalpur campus of the University represent 43 families with the maximum 26 representation of family Poaceae followed by 12 from Fabaceae and 11 from Asteraceae. Out of total documented weeds, 70 % weeds are dicot and remaining 30% monocot. 46% weeds are of Kharif and 32% of Rabi season whereas remaining 22 percent perennial/ from water bodies/ parasite (Table 19.1).

Table 19.1 Weeds reported from Jabalpur campus

Sr	Scientific name	Family	Common name	Cotyledon	Season	Density
1.	*Abelmoschus fioulneus*	Malvaceae	White wild musk mallow	Dicot	Kharif	Medium
2.	*Abelmoschus manihot*	Malvaceae	Sunset hibiscus	Dicot	Kharif	Medium
3.	*Abelmoschus moschatus*	Malvaceae	Musk mallow	Dicot	Kharif	Medium
4.	*Achyranthes aspera* L.	Amaranthaceae	Prickly chaff flower	Dicot	Kharif	Medium
5.	*Acrachne racemosa*	Poaceae	Goose grass	Monocot	Kharif	Low
6.	*Ageratum conyzoides*	Asteraceae	Goat weed	Dicot	Perennial	Medium
7.	*Agropyron repens*	Asteraceae		Dicot		Medium
8.	*Alternanthera philoxeroides*	Amaranthaceae	Alligator weed	Dicot	Kharif	Medium
9.	*Alternanthera sessilis*	Amaranthaceae	Creeping chaff weed	Dicot	Kharif	Medium
10.	*Alternanthera* spp	Amaranthaceae	Sessile joy weed			High
11.	*Amaranthus spinosus*	Amaranthaceae	Spiny pigweed	Dicot	Kharif	Medium
12.	*Amaranthus viridis*	Amaranthaceae	Smooth pigweed	Dicot	Kharif	Medium
13.	*Ammannia baccifera*	Lythraceae	Blistering ammannia	Dicot	Kharif	Medium
14.	*Anagallis arvensis*	Primulaceae	Blue pimpernel	Dicot	Rabi	Medium
15.	*Antennaria neglecta*	Asteraceae	Field pussy toes			Low
16.	*Argemone mexicana*	Papaveraceae	Mexican poppy	Dicot	Rabi	Medium
17.	*Asphodelus tenuifolius*	Asphodelaceae	Wild onion	Dicot	Rabi	Medium
18.	*Avena fatua*	Poaceae	Wild oat	Monocot	Rabi	Medium
19.	*Avena ludoviciana*	Poaceae	Winter wild oat	Monocot	Rabi	Medium
20.	*Azolla filiculoides*	Salviniaceae	Water fern	Dicot	Aquatic	Medium
21.	*Brachiaria mutica*	Poaceae	Para grass	Monocot		High
22.	*Brachiaria reptans*	Poaceae	Running grass	Monocot	Kharif	Medium
23.	*Caesulia axillaris*	Asteraceae	Pink node flower	Dicot	Kharif	Medium
24.	*Calotropis procera*	Apocynaceae	Maddar root	Dicot	Perennial	Medium
25.	*Carthamus oxyacantha*	Asteraceae	Wild safflower	Dicot	Rabi	Medium
26.	*Cassia toria*	Fabaceae	Coffee weed/ Sickle pod	Dicot	Kharif	Medium
27.	*Ceratophyllum demersum*	Ceratophyllaceae	Hornwort	Dicot	Aquatic	Medium
28.	*Chenopodium album*	Amaranthaceae	Common lambsquarter	Dicot	Rabi	Medium
29.	*Chenopodium murale*	Amaranthaceae	Nettle leaf	Dicot	Rabi	Medium
30.	*Cichorium intybus*	Asteraceae	Blue daisy	Monocot	Rabi	Medium
31.	*Cirsium arvensis*	Asteraceae	Canada thistle	Dicot	Perennial	Medium
32.	*Citrullus colocynthis*	Cucurbitaceae		Dicot	Kharif	
33.	*Cleome viscosa*	Cleomaceae	Yellow Spider Flower	Dicot	Kharif	Medium
34.	*Commelina benghalensis*	Commelinaceae	Day flower	Dicot	Kharif	Medium
35.	*Commelina diffusa*	Commelinaceae	Spreading day flower	Dicot	Kharif	Medium
36.	*Commelina erecta*	Commelinaceae	Slender day flower	Dicot	Kharif	Medium
37.	*Convolvulus arvensis*	Convolvulaceae	Field bind weed	Dicot	Perennial	Medium
38.	*Conyza stricta*	Poaceae	Erect horse weed	Monocot	Kharif	Medium
39.	*Corchorus tridens*	Tiliaceae	Wild jute	Dicot	Kharif	Medium

Sr	Scientific name	Family	Common name	Cotyledon	Season	Density
40.	*Coronopus didymus*	Brassicaceae	Garden cress	Monocot	Rabi	Medium
41.	*Crotalaria medicaginea*	Fabaceae	Rattle pod	Dicot	Kharif	Medium
42.	*Cucumis hardwickii*	Cucurbitaceae		Dicot	Kharif	Low
43.	*Cuscuta reflexa*	Convolvulaceae	Dodder	Monocot	Parasitic	Medium
44.	*Cynodon dactylon*	Poaceae	Bermuda grass	Monocot	Perennial	Medium
45.	*Cyperus rotandus*	Poaceae	Motha	Monocot	Perennial	Medium
46.	*Cyperus compressus*	Poaceae	Hedgehog sedge	Monocot	Kharif	Medium
47.	*Cyperus iria*	Poaceae	Rice flat sedge	Monocot	Kharif	Medium
48.	*Dactyloctenium aegyptium*	Poaceae	Crow foot grass	Monocot	Kharif	Medium
49.	*Datura stramonium*	Solanaceae	Jimson weed	Dicot	Kharif	Medium
50.	*Digera arvensis*	Amaranthaceae	False amaranth	Dicot	Kharif	Medium
51.	*Digera muricata*	Amaranthaceae	False amaranth	Dicot	Kharif	Medium
52.	*Digitaria sanguinalis*	Poaceae	Crab grass	Monocot	Kharif	Medium
53.	*Echinochloa colona*	Poaceae	Water grass	Monocot	Kharif	Medium
54.	*Echinochloa crusgalli*	Poaceae	Barnyard grass	Monocot	Kharif	Medium
55.	*Eclipta alba*	Asteraceae	False daisy	Dicot	Kharif	Medium
56.	*Eichhornia crassipes*	Pontederiaceae	Water hyacinth	Dicot	Aquatic	Medium
57.	*Eleusine indica*	Poaceae	Indian goose grass	Monocot	Kharif	Medium
58.	*Eragrostis pilosa*	Poaceae	Love grass	Monocot	Kharif	Medium
59.	*Eragrostis tenella*	Poaceae	Feather love grass	Monocot	Kharif	Medium
60.	*Euphorbia heterophylla*	Euphorbiaceae	Painted spurge	Dicot	Kharif	Medium
61.	*Euphorbia hirta*	Euphorbiaceae	Pill pod spurge	Dicot	Kharif	Medium
62.	*Euphorbia microphylla*	Euphorbiaceae	Euphorbia microphylla	Dicot	Kharif	Medium
63.	*Euphorbia prostrata*	Euphorbiaceae	Petty spurge	Dicot	Kharif	Medium
64.	*Fumaria parviflora*	Papaveraceae	Fumitory	Monocot	Rabi	Medium
65.	*Gnaphalium purpureum*	Asteraceae	purple cud weed	Monocot	Rabi	Medium
66.	*Heliotropium indicum*	Boraginaceae	/Devil weed	Dicot	Kharif	Medium
67.	*Hibiscus ovalifolius*	Malvaceae	Rope fiber	Dicot	Kharif	Medium
68.	*Hydrilla verticillata*	Hydrocharitaceae	Hydrilla	Monocot	Aquatic	Medium
69.	*Indigofera linnaei*	Fabaceae	Creeping indigo	Dicot	Kharif	Medium
70.	*Ipomoea carnea*	Convolvulaceae	Blush morning glory	Dicot	Perennial	Medium
71.	*Ipomoea nil*	Convolvulaceae	Blue morning glory	Dicot	Kharif	Medium
72.	*Ipomoea pestigridis*	Convolvulaceae	Tiger foot morning glory	Dicot	Kharif	Medium
73.	*Ischaemum rugosum*	Poaceae	Wrinkle grass	Monocot	Kharif	Medium
74.	*Lactuca serriola*	Asteraceae		Dicot		
75.	*Lantana camara*	Verbenaceae	Lantana	Dicot	Perennial	Medium
76.	*Lathyrus sativus*	Fabaceae	Grass pea	Dicot	Rabi	Medium
77.	*Lathyrus aphacaora*	Fabaceae	Meadow pea	Dicot	Rabi	Medium
78.	*Leptochloa chinensis*	Poaceae	Red sprangle top	Monocot	Kharif	Medium
79.	*Lolium temulentum*	Poaceae	Poison rye grass	Monocot	Rabi	Medium
80.	*Loranthus micranthus*	Orobanchaceae	Loranthus	Dicot	Parasitic	Medium
81.	*Lagascea mollis*	Asteraceae				
82.	*Malva parviflora*	Malvaceae	Little mellow	Dicot	Rabi	Medium
83.	*Marsilea hirsuta*	Marsileaceae	Four leaf water clover	Dicot	Aquatic	Medium
84.	*Medicago denticulata*	Fabaceae	Toothed bur clover	Dicot	Rabi	Medium
85.	*Medicago polymorpha*	Fabaceae	California bur clover	Dicot	Rabi	Medium
86.	*Medicago truncatula*	Fabaceae	Barrel clover	Dicot	Rabi	Medium
87.	*Melilotus alba*	Fabaceae	White sweet clover	Dicot	Rabi	Medium
88.	*Melilotus indicus*	Fabaceae	Yellow sweet clover	Dicot	Rabi	Medium

Sr	Scientific name	Family	Common name	Cotyledon	Season	Density
89.	*Mollugo pentaphylla*	*Molluginaceae*	Carpet weed	Dicot	Kharif	Medium
90.	*Mollugo verticillata*	*Molluginaceae*	Green carpet weed	Dicot	Kharif	Medium
91.	*Momordica blasamina*	Cucurbitaceae		Dicot	Kharif	
92.	*Momordica cochinchinesis*	Cucurbitaceae		Dicot	Kharif	
93.	*Momordica diocia*	Cucurbitaceae		Dicot	Kharif	
94.	*Myriophyllum aquaticum*	Haloragaceae	Parrot's feather	Monocot	Aquatic	Medium
95.	*Nymphaea alba*	Nymphaeaceae	White water lily	Dicot	Aquatic	Medium
96.	*Nymphoides cristata*	Menyanthaceae	Crested Floating Heart	Dicot	Aquatic	Medium
97.	*Oenothera laciniata*	Onagraceae	Cutleaf evening primrose	Dicot	Rabi	Medium
98.	*Opuntia dillenii*	Cactaceae	Prickly pear	Dicot	Perennial	Medium
99.	*Orobanche aegyptiaca*	Orobanchaceae	Egyptian broomrape	Dicot	Parasitic	Medium
100.	*Oryza sativa var. fatua*	Poaceae	Wild rice	Monocot	Kharif	Medium
101.	*Oxalis corniculata*	Oxalidaceae	Indian sorrel	Dicot	Kharif	Medium
102.	*Parthenium hysterophorus*	Asteraceae	Wild carrot weed	Dicot	Perennial	Medium
103.	*Paspalum distichum*	Poaceae	Knot grass	Monocot	Kharif	Medium
104.	*Paspalum vaginatum*	Poaceae	Seashore paspalum	Monocot	Kharif	Medium
105.	*Phalaris minor*	Poaceae	Canary grass	Monocot	Rabi	Medium
106.	*Phragmites australis*	Poaceae	Common reed	Monocot	Aquatic	Medium
107.	*Phyllanthus niruri*	Phyllanthaceae	Gulf leaf flower	Dicot	Kharif	Medium
108.	*Physalis minima*	Solanaceae	Ground cherry	Dicot	Kharif	Medium
109.	*Pistia stratiotes*	Araceae	Water lettuce	Dicot	Aquatic	Medium
110.	*Poa annua*	Poaceae	Sweet grass	Monocot	Rabi	Medium
111.	*Polypogon monspeliensis*	Poaceae	Beard grass	Dicot	Rabi	Medium
112.	*Portulaca oleracea*	Portulacaceae	Purselane	Dicot	Kharif	Medium
113.	*Potamogeton perfoliatus*	Potamogetonaceae	Pond weed	Dicot	Aquatic	Medium
114.	*Ranunculus muricatus*	Ranunculaceae	Rough seed buttercup	Dicot	Rabi	Medium
115.	*Ranunculus sceleratus*	Synonyms	Cursed buttercup	Dicot	Rabi	Medium
116.	*Rumex dentatus*	Polygonaceae	Sour dock	Dicot	Rabi	Medium
117.	*Rumex hymenosepalus*	Polygonaceae	Canaigre dock	Dicot	Rabi	Medium
118.	*Rumex spinosus*	Polygonaceae	Dock/Sorrel	Dicot	Rabi	Medium
119.	*Saccharum spontaneum*	Poaceae	Tiger grass	Monocot	Perennial	Medium
120.	*Sagittaria subulata*	Alismataceae	Arrowhead	Monocot	Aquatic	Medium
121.	*Salvinia minima*	Salviniaceae	Water spangles	Dicot	Aquatic	Medium
122.	*Saponaria vaccaria*	Caryophyllaceae	Cow cockle	Dicot	Rabi	Medium
123.	*Setaria verticillata*	Poaceae	Rough bristle foxtail	Monocot	Kharif	Medium
124.	*Setaria viridis*	Poaceae	Green foxtail	Monocot	Kharif	Medium
125.	*Sida acuta*	Malvaceae		Dicot	Perennial	
126.	*Silene conoidea*	Caryophyllaceae	Forked catchfly	Dicot	Rabi	Medium
127.	*Sisymbrium irio*	Brassicaceae	Wild mustard	Dicot	Rabi	Medium
128.	*Solanum dulcamara*	Solanaceae	Night shade	Dicot	Rabi	Medium
129.	*Solanum nigrum*	Solanaceae	Black nightshade	Dicot	Kharif	Medium
130.	*Sonchus arvensis*	Asteraceae	Perennial saw thistle	Dicot	Rabi	Medium
131.	*Sorghum halepense*	Poaceae	Johnson grass	Monocot	Perennial	Medium
132.	*Spergula arvensis*	Caryophyllaceae	Corn spurry	Monocot	Rabi	Medium
133.	*Striga lutea*	Loranthaceae	Witch weed	Dicot	Parasitic	Medium

Sr	Scientific name	Family	Common name	Cotyledon	Season	Density
134.	*Trianthema monogyna*	Aizoaceae				
135.	*Trianthema portulacastrum*	Aizoaceae	Desert horse purslane	Dicot	Kharif	Medium
136.	*Tribulus terrestris*	Zygophyllaceae	Puncture vine	Dicot	Kharif	Medium
137.	*Trichosanthes cucumerina*	Cucurbitaceae		Dicot	Kharif	Medium
138.	*Trichosanthes lobata*	Cucurbitaceae		Dicot	Kharif	Medium
139.	*Trichosanthes palmata*	Cucurbitaceae		Dicot	Kharif	Medium
140.	*Tridax procumbens*	Asteraceae	Night lamp	Dicot	Kharif	Medium
141.	*Trigonella polycerata*	Fabaceae	Wild fenugreek	Dicot	Rabi	Medium
142.	*Typha angustifolia*	Typhaceae	Narrow leaf cattail	Monocot	Aquatic	Medium
143.	*Typha latifolia*	Typhaceae	Cattail	Monocot	Aquatic	Medium
144.	*Veronica agrestis*	Plantaginaceae.	Green field-Speedwell	Dicot	Rabi	Medium
145.	*Vicia hirsuta*	Fabaceae	Hairy vetch	Dicot	Rabi	Medium
146.	*Vicia sativa*	Fabaceae	Akari	Dicot	Rabi	Medium
147.	*Wild Helianthus*	Asteraceae	Wild sunflower	Dicot	Rabi	Medium
148.	*Wolffia arrhiza*	Araceae	Duckweed	Dicot	Aquatic	Medium
149.	*Xanthium strumarium*	Asteraceae	Cockle bur	Dicot	Kharif	Medium

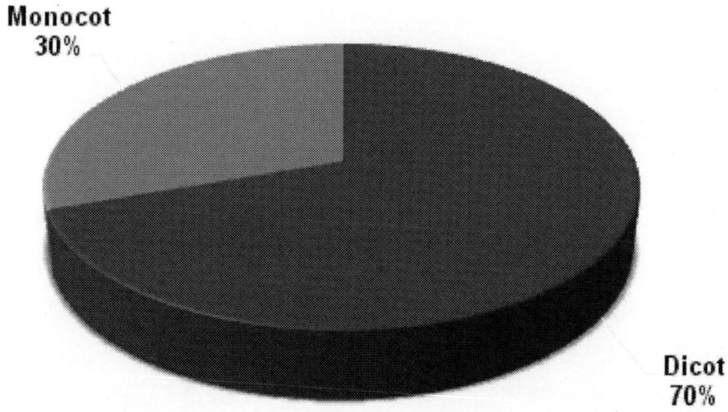

Fig 19.1 Diversity in Weeds for Cotyledon

Out of 149 reported weeds from campus Jabalpur, 70% are dicot and 30% monocot (Fig 19.1).

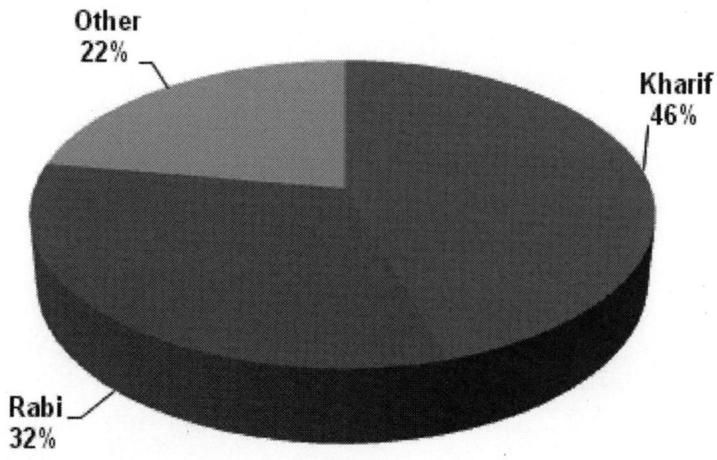

Fig 19.2 Diversity in Weeds for season

Out of 149 reported weeds at Jabalpur campus, 46% are of Kharif and 32 % are of Rabi season (Fig 19.2) whereas 22% are others.

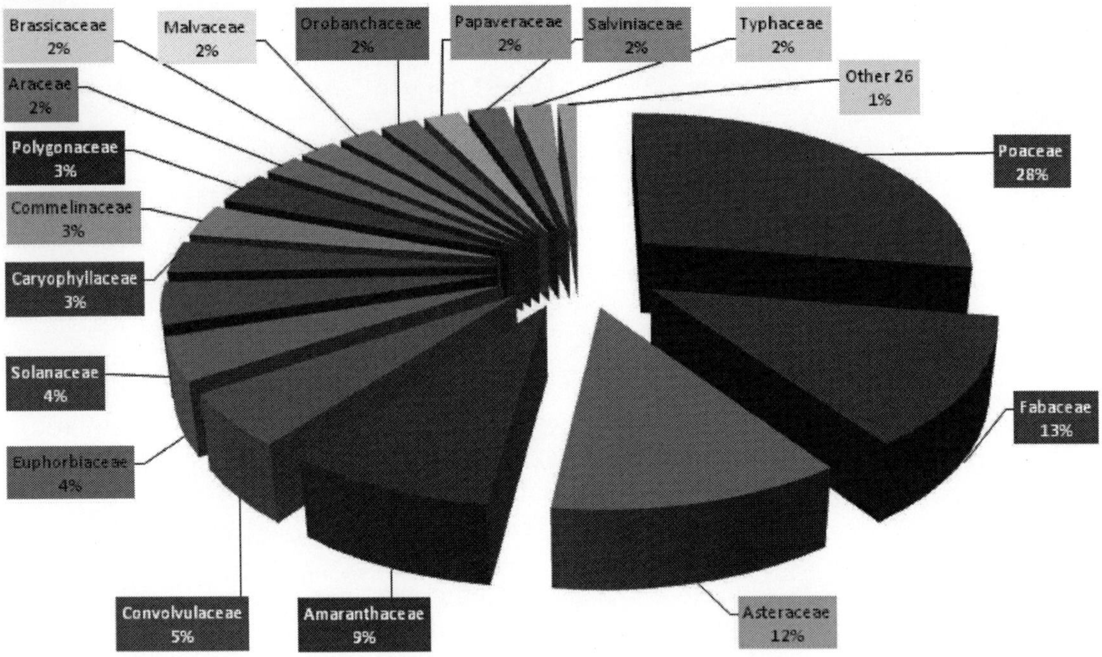

Fig 19.3 Percent representation of weeds in terms of families at Jabalpur

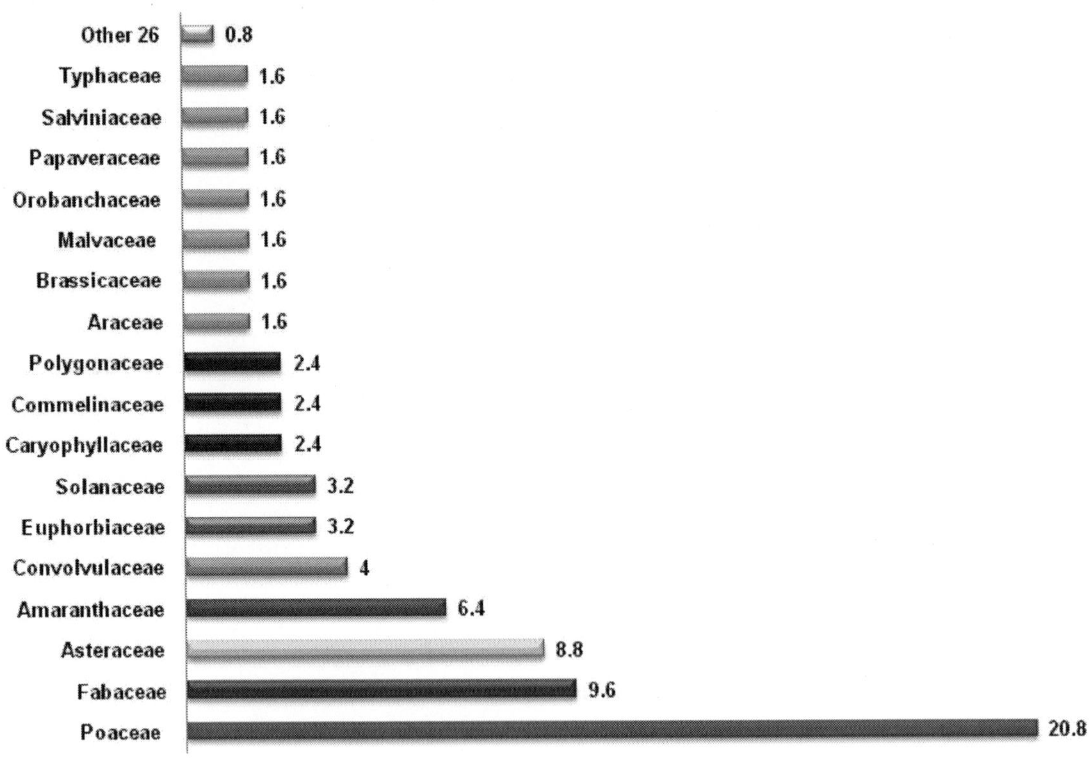

Fig 19.4 Diversity for Families in Weeds at Jabalpur

Representation of 43 families for weeds has been reported form the Jabalpur campus of the University. The maximum number of weeds are form the family Poaceae (28%), followed by Fabaceae (13%), Asteraceae 912%), Amaranthaceae (9%) and others (Fig 19.3 and 19.4).

CHAPTER 20

Biodiversity Audit of Insect, Mites, Spiders, Birds and Rodents

SB Das, Jagrati Upadhyay, Kumari Pragya and Subrata Goswami
Department of Entomology
College of Agriculture, JNKVV, Jabalpur

Biodiversity, which is a contraction of "biological diversity", refers to the variety of life forms within an ecosystem, biome or the entire globe. It includes the complex assemblage of plants, animals and even the microorganisms that works together in an ecosystem like an intricate web to maintain the balance of life. Biodiversity is one of the measures of the health of biological system. It is a multidimensional concept encompassing all aspects of biological variation *viz.*, genetic variability within individuals and among populations of different species, the composition and abundance of species within a particular environment (community level) and the variety of ecological roles played by species in the environment (ecosystem level).

The JNKVV, Jabalpur campus situated between 22^{0} 49' and 24^{0} 8' North latitude and 78^{0} 21' and 80^{0} 58' East longitude and at an altitude of 411.78 m above mean sea level, is spread across an area of 543.54. Jabalpur has a humid subtropical climate and receives an average annual rainfall of around 1386 mm. The campus has an anthropogenically modified ecosystem (mostly an agroecosystem) where a wide range of crops are cultivated round the year. Apart from the crops, numerous herbs, shrubs and trees are present within the campus representing a rich diversity of flora.

Over the years several researchers have pointed out many criteria for a good biodiversity indicator. But practically all these requirements can't be combined in reality, because there is always compromise between the inherent complexity of biodiversity and the simplicity of what in fact is affordably measurable. For monitoring the local species richness, rapid biodiversity assessment based on identification of morphospecies is an affordable tool. Though local species richness *i.e.,* alpha diversity is not always considered a relevant aspect of measuring biodiversity on its entirety, but ecological resilience and ecosystem services like pollination, natural suppression of pests *etc.* are often very much intertwined with local species richness. So, an overall assessment about the diversity of organisms and their functional positions can provide an idea about the health, functioning and sustainability of the concerned ecosystem while at the same time, may be instrumental in better understanding the biodiversity, quantifying its loss and formulating strategies for its conservation and effective utilization.

In all 571 fauna have been reported from JNKVV campus, Jabalpur that consisits of 502 insects, 42 spiders, 6 rodents and 5 mites (Table 20.1).

Table 20.1 Fauna recorded in JNKVV campus, Jabalpur

S.No.	Fauna	Number
I.	Insects	502
II.	Spiders	42
III.	Rodents	06
IV.	Mites	05
		555
V.	Insectivorous Birds	01
VI.	Weeds as insect host	10
VII.	Diseases of insects	03
VIII.	Entomopathogenic nematodes	02
		16
		571

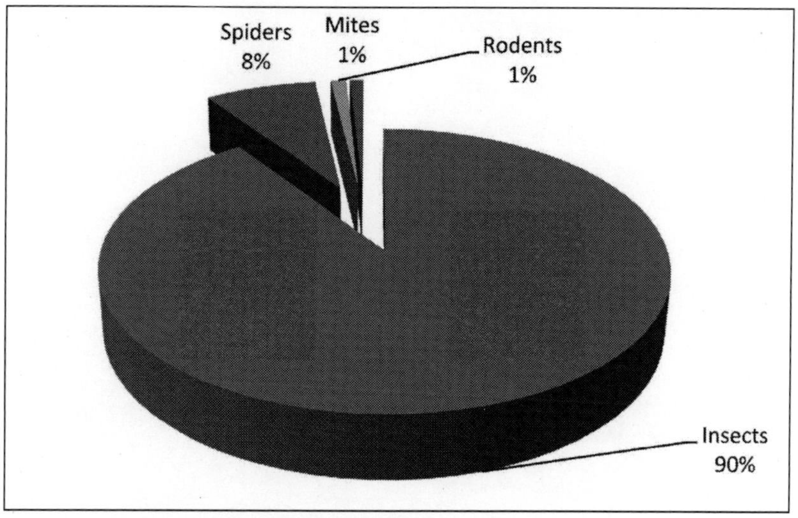

Fig 20.1 Status of arthropods and rodentsin JNKVV campus, Jabalpur

In an attempt to explore the average local species richness of arthropods along with the rodents, insectivorous birds, entomopathogenic microbes, nematodes and alternative weed hosts at JNKVV, Jabalpur campus; a total of 571 different morphospecies were recorded. Arthropods, the most diverse groups of animals in terrestrial ecosystem, were the most numerous here counting upto 559 in numbers. Out of this 551, 502 belonged to the class Insecta and 47 belonged to Arachnida (42 spiders and 5 mites). The dominance of insects (90%) can be correlated to their greater reproduction potential, survival potential, better adaptability and colonization of high diversity of niche environment (Fig.20.1 and 20.2).

Ten weeds were also identified as insect hosts thereby permitting the off-season survival and continuous perpetuation of the insects. Three biotic resistance factors against the insects, each from different taxa, were recorded too *viz.,* 1 insectivorous bird, 3 entomopathogenic microbes and 2 entomopathogenic nematodes (Table 20. 2).

Table 20.2 Distribution of fauna recorded in JNKVV campus, Jabalpur

S.No.	Group	Fauna	Nos.
I.	Insects	Insect Pests	312
		Insect Predators	98
		Insect Parasites	74
		Insect Pollinators	07
		Insects of Industrial Importance	05
		Hyperparasites	04
		Mushroom insect pests	02
			502
II.	Mites	Mites	05
			507
III.	Arachnids	Spider Predators	42
IV.	Rodentia	Rodents	06
			555
V.	Avians	Insectivorous Birds	01
VI.	Plants	Weeds as insect host	10
VII.	Diseases	Diseases of insects	03
VIII.	Nematodes	Entomopathogenic nematodes	02
			16
			571

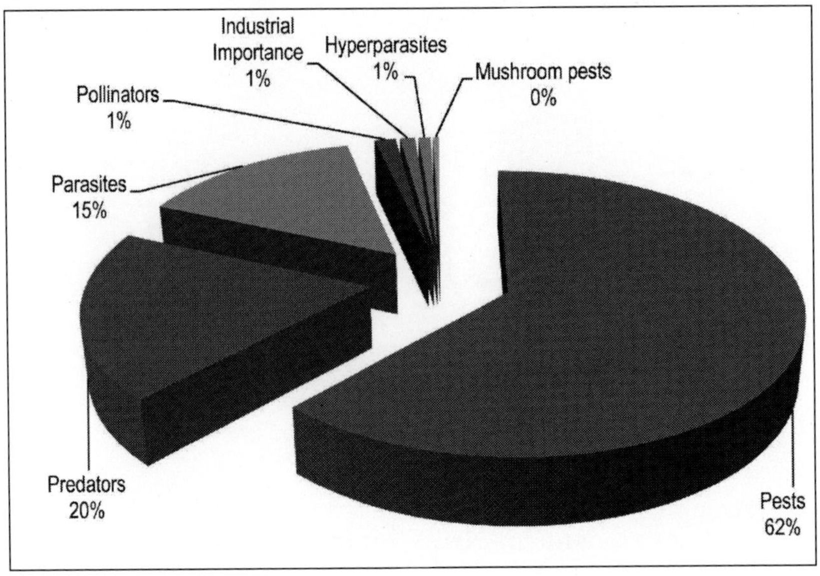

Fig 20. 2 Status of insects in JNKVV campus, Jabalpur

Arthropods constitute the largest proportion of species richness at any spatial scale. So, they are more representative of extensive organismal biodiversity than any other group of organisms. Arthropods, especially insects, play a major role in ecosystem services. The insect fauna recorded from the JNKVV campus belonged to several different guilds with different ecological requirements. Thus, the degree of interspecific competition was less and such a sound diversity of insects is coexisting. Majority (62%) of the insects recorded was pests of different agricultural crops and 2 insect pests of mushroom were also documented. But the simultaneous presence of large proportion of intrinsic regulating forces, within the same taxon, in the form of arthropod predators (98 insects and 42 spiders) and parasite (74 insects) are indicative of a functional natural balance to some extent. The phytophagous insects damage the crops and in turn serve as the food source of the carnivorous ones. Thus, there is scope for establishment of dynamic equilibrium of eating (the crops) and being eaten (by the natural enemies). However, presence of 4 hyperparasitic insects which devour other parasitic species were noted too. Besides, one of the most important mutualistic interactions between insects and flowering plants, *i.e.*, entomophily was also observed in the campus. 7 different insect pollinators were recorded round the season. Thus, two of the most prominent and best understood arthropod-provided regulating ecosystem services, *i.e.*, pollination and pest suppression were rendered by 1% and 35% of the total insects respectively.

While arthropods contribute to production indirectly via regulating services, they also generate products like honey, silk and lac *etc.* or can be directly used as livestock feed or human food. These come under the provisioning ecosystem services that contribute directly and materially to products used by humans. Five such insects of industrial importance providing provisioning services were recorded.

Out of 502 insects, 312 are pests, 98 are predators, 74 are parasites, 07 are pollinators, 05 are of industrial importance, 04 hyperparasites and 02 are pests on mushroom have been reported from JNKVV campus, Jabalpur.

Table 20.3 Arthropods, rodents *etc.* reported from College of Agriculture, Jabalpur

S.No.	Scientific name	Common name	Order	Family	Host
	Insect pests				
1.	*Acanthiophilus helianthi* Rossi	Capsule fly	Diptera	Tephritidae	Safflower
2.	*Achaea janata* (L.)	Castor semi looper	Lepidoptera	Noctuidae	Castor, Rose
3.	*Acherontia styx* (Westwood)	Til hawk moth	Lepidoptera	Sphingidae	Sesame, Potato
4.	*Acheta domestica* (Linnaeus)	Cricket	Orthoptera	Gryllidae	Fodder Grasses
5.	*Acyrthosiphon pisum* Harris	Pea aphid	Hemiptera	Aphididae	Pea
6.	*Agonoscelis nubilis* (Fab.)	Stink bug	Hemiptera	Pentatomidae	Okra, Uridbean
7.	*Agrius convolvuli* (Linn.)	Hawk moth	Lepidoptera	Sphingidae	Sweet Potato, Sunflower, Soybean
8.	*Agrotis ipsilon* Hufnagel	Black cutworm	Lepidoptera	Noctuidae	Chickpea, Linseed, Wheat, Onion, Potato
9.	*Aiolopus thalassinus* (Fab.)	Grasshopper	Orthoptera	Acrididae	Grasses
10.	*Aleurocanthus woglumi* Ashby	Citrus black fly	Hemiptera	Aleyrodidae	Citrus
11.	*Altica oleracea* (Linnaeus)	Steel blue beetle	Coleoptera	Chrysomelidae	Rice
12.	*Altica sp.*	Flea beetle	Coleoptera	Chrysomelidae	Niger
13.	*Amata cyssea* (Cramer)	Handmaiden moth	Lepidoptera	Erebidae	Polyphagous
14.	*Amata* sp.		Lepidoptera	Arctiidae	Fodder Crops
15.	*Amerila astrea* Drury		Lepidoptera	Erebidae	Polyphagous
16.	*Amerila eugenia* Fab.		Lepidoptera	Erebidae	
17.	*Amrasca bigutulla bigutulla* (Ishida)	Cotton leaf hopper	Hemiptera	Cicadellidae	Brinjal, Chilli, Cotton, Niger, Okra, Tomato
18.	*Amrasca devastans* Dist.	Jassid	Hemiptera	Jassidae	French Bean, Okra, Tomato
19.	*Amritodus atkinsoni* (Leth.)	Mango hopper	Hemiptera	Cicadellidae	Mango
20.	*Amsacta lactinea* (Cramer)	Groundnut red hairy caterpillar	Lepidoptera	Arctidae	Castor, Groundnut, Sunflower
21.	*Amsacta moorei* Butler	Red hairy caterpillar	Lepidoptera	Arctidae	Sun Hemp, Maize, Sorghum
22.	*Anarsia ephbippias* May	Bud worm	Lepidoptera	Gelechiidae	Soybean
23.	*Anomis flava* (Fabricius)	Green semilooper	Lepidoptera	Erebidae	Cotton, Okra
24.	*Anoplonemis phasiana* (Fabricius)	Pod sucking bug	Hemiptera	Pentatomidae	Mungbean
25.	*Antestiopsis cruciata* (Fabricius)	Coffee plant bug/ Jasmine bug	Hemiptera	Pentatomidae	Coffee, Jasmine
26.	*Antigastra catalaunalis* (Duponchel)	Leaf webber/ capsule borer	Lepidoptera	Pyralidae	Sesame
27.	*Apheliona maculosa* (Dist.)	Jassid	Hemiptera	Delphacidae	Soybean
28.	*Aphis craccivora* Koch	Lab Lab aphid	Hemiptera	Aphididae	Pulses, Uridbean
29.	*Aphis gossypii* Glover	Cotton aphid	Hemiptera	Aphididae	Brinjal, Chilli, French Bean, Okra, Tomato, Soybean, Cotton, Ashwagandha (*Withania somnifera)*

S.No.	Scientific name	Common name	Order	Family	Host
30.	*Aphis punicae* Passerini	Pomegranate aphid	Hemiptera	Aphididae	Pomegranate
31.	*Arachnis picta* (Packard)	Painted tiger moth	Lepidoptera	Erebidae	
32.	*Argina astrea* (Durry)	Hairy caterpillars	Lepidoptera	Erebidae	Fodder crops
33.	*Argina cribraria* (Clerck)	Hairy caterpillars	Lepidoptera	Arctiidae	Sunhemp
34.	*Asota caricaei* (Fabr.)	Tropical tiger moth	Lepidoptera	Erebidae	Fodder crops
35.	*Asota ficus* (Fabricius)		Lepidoptera	Noctuidae	Fodder crops
36.	*Aspondylia capsici* Barnes	Chilli gall midge	Diptera	Cecidomyiidae	Chilli
37.	*Aspondylia sesami* (Felt.)	Sesame gall fly	Diptera	Cecidomyiidae	Sesame
38.	*Athalia lugens proxima* (Klug)	Mustard sawfly	Hymenoptera	Tenthridinidae	Mustard
39.	*Athalia* sp.		Hymenoptera	Tenthridinidae	Cruciferous Crops
40.	*Atherigona bituberculata* Malloch	Wheat stem fly	Diptera	Muscidae	Wheat
41.	*Atherigona soccata* Rond.	Sorghum shoot fly	Diptera	Muscidae	Sorghum, Soybean
42.	*Atherigona* sp.		Diptera	Muscidae	Pearl Millet, Kodo, Maize, Sudan Grass, Sudax
43.	*Atractomorpha crenulata* (Saussure)	Grasshopper	Orthoptera	Pyrgomorphidae	Uridbean
44.	*Atractomorpha* sp.	Grasshopper	Orthoptera	Pyrgomorphidae	Polyphagous, Okra
45.	*Aulacophora foveicollis* (Lucas)	Red pumpkin beetle	Coleoptera	Chrysomelidae	Cucurbits, Indian Bean, Pigeonpea, Niger, Uridbean
46.	*Bacterocera cucurbitae* (Coquillet)	Cucurbit fruit fly	Diptera	Tephritidae	Cucurbits
47.	*Bacterocera dorsalis* Hendel	Mango fruit fly	Diptera	Tephritidae	Mango
48.	*Bacterocera rufomaculata* DeGeer	Mango stem borer	Coleoptera	Cerambycidae	Mango
49.	*Bagrada cruciferarum* Kirkaldy	Painted bug	Hemiptera	Pentatomidae	Cabbage
50.	*Bagrada hilaris* (Burmeister)	Painted bug	Hemiptera	Pentatomidae	Okra, Uridbean
51.	*Banisia fenestrifera* (Walker)		Lepidoptera	Thyrididae	
52.	*Bastilla crameri* Moore		Lepidoptera	Noctuidae	
53.	*Bathrips melanicornis* (Shumsher)		Thysanoptera	Thripidae	Brinjal
54.	*Bemisia tabaci* Gennadius	Whitefly	Hemiptera	Aleyrodidae	Brinjal, Chili, Tomato, Coriander, Okra, French Bean, Indian Bean, Mungbean, Uridbean, Pea, Niger, Sesame, Linseed, Soybean, Cotton
55.	*Brevicoryne brassicae* (L.)	Cabbage aphid	Hemiptera	Aphididae	Cabbage
56.	*Cadra cautella* Walker	Almond moth	Lepidoptera	Pyralidae	Storage Insect Pest
57.	*Caliothrips indicus* (Bagnall)	Thrips	Thysanoptera	Thripidae	Linseed, Mung, Pea, Urid, Pigeonpea,
58.	*Calliteara grotei* (Moore)		Lepidoptera	Erebidae	
59.	*Callosobruchus analis*	Pulse beetle	Coleoptera	Bruchidae	Pulses

S.No.	Scientific name	Common name	Order	Family	Host
60.	*Callosobruchus chinensis* (L)	Pulse beetle	Coleoptera	Bruchidae	Pulses
61.	*Callosobruchus maculatus* (Fabricius)	Pulse beetle	Coleoptera	Bruchidae	Pulses
62.	*Callosobruchus phaseoli* (Gylh)	Been weevil	Coleoptera	Bruchidae	Common Bean
63.	*Calocoris angustatus* Leth.	Earhead bug	Hemiptera	Miridae	Sorghum
64.	*Charidotella sexpunctata* (Fab.)	Golden tortoise beetle	Coleoptera	Chrysomelidae	Okra
65.	*Chilo infuscatellus* Snell.	Early shoot borer	Lepidoptera	Pyralidae	Sugarcane
66.	*Chilo partellus* (Swinhoe)	Maize stem borer	Lepidoptera	Pyralidae	Maize, Sorghum, Pearl Millet, Rice, Sugarcane
67.	*Chionaema peregrina* (Walker)		Lepidoptera	Erebidae	Mungbean
68.	*Chrotogonus* sp.	Surface grasshopper	Orthoptera	Pyrgomorphidae	Niger
69.	*Chrotogonus trachypterus* (Blan.)	Grasshopper	Orthoptera	Acrididae	Coriander
70.	*Chrysocraspeda* sp.		Lepidoptera	Geometridae	
71.	*Chrysodeixis acuta* (Walker)	Green semilooper	Lepidoptera	Noctuidae	Soybean
72.	*Chrysodeixis chalcites* (Esper)	Green semilooper	Lepidoptera	Noctuidae	Bean, Tomato, Potato, Groundnut, Soybean
73.	*Chrysomya* sp.	Screw-worm	Diptera	Calliphoridae	Farm Animals
74.	*Cirrhochrista brizoalis* (Walker)		Lepidoptera	Crambidae	
75.	*Clavigralla gibbosa* Spinola	Pod bug	Hemiptera	Coreidae	Pigeonpea
76.	*Cletus bipunctatus* (Herrich – Schaffer)	Spined legume bug	Hemiptera	Coreidae	Okra, Uridbean
77.	*Cnapholocris medinalis* (Guen)	Rice Leaf folder	Lepidoptera	Pyraustidae	Rice
78.	*Cneorane* sp.	Blue beetle	Coleoptera	Chrysomelidae	Soybean
79.	*Cofana spectra distant* (Distant)	White Leaf hopper	Hemiptera	Cicadellidae	Rice
80.	*Condica conducata* (Walker)	Niger caterpillar	Lepidoptera	Noctuidae	Niger
81.	*Contarinia sorghicola* Coq.	Sorghum midge	Diptera	Cecidomyiidae	Sorghum
82.	*Corcyra cephalonica Stainton*	Rice moth	Lepidoptera	Pyralidae	Cereals, Sesame
83.	*Cretonotos gangis* (Linnaeus)	Arctinee moth / Rice tiger moth	Lepidoptera	Arctiidae	Polyphagous, Rice
84.	*Cretonotus jangia* Linn.		Lepidoptera	Arctiidae	
85.	*Crocidolomia binotalis* Zell	Cabbage web worm	Lepidoptera	Pyralidae	Cabbage, Mustard
86.	*Crytopeltis tenuis* (Reuter)	Mirid bug	Hemiptera	Miridae	Sesame
87.	*Cyana peregrina* (Walker)		Lepidoptera	Erebidae	
88.	*Cyrtacanthacris sp.*	Grasshopper	Orthoptera	Acrididae	Pigeonpea
89.	*Cyrtacanthacris tatarica* (Linn)	Grasshopper	Orthoptera	Acrididae	Cotton, Grasses
90.	*Dactynotus carthami* HRL	Aphid	Hemiptera	Aphididae	Niger

S.No.	Scientific name	Common name	Order	Family	Host
91.	*Dacus cucurbitae* (Coquillett)	Melon fruit fly	Diptera	Tephritidae	Chilli, Citrus, Cucurbits, Guava, Musk Melon, Tomato
92.	*Daphnis nerii* (Linnaeus)	Oleander hawk-moth	Lepidoptera	Sphingidae	Feeds On The Nectar Of Flowers Of Jasmine, Oleander, Petunia
93.	*Dasyneura lini* Barnes	Linseed bud fly	Diptera	Cecidomyiidae	Linseed, Sesame
94.	*Dialeurodes citri* (Ashmead)	Citrus blackfly	Hemiptera	Aleyrodidae	Citrus
95.	*Diaphorina citri* Kuwayama	Citrus psylla	Hemiptera	Psyllidae	Citrus
96.	*Dichocrosis punctiferalis* Gn.	Castor capsule borer	Lepidoptera	Crambidae	Castor
97.	*Dicladispa armigera* (Oliv.)	Rice hispa	Coleoptera	Chrysomelidae	Rice
98.	*Digama hearseyana similis* Moore		Lepidoptera	Erebidae	Polyphagous
99.	*Dioxyna sororcula* (Wiedemann)	Capsule fly	Diptera	Tephritidae	Niger
100.	*Drosicha mangiferae* (Green)	Mango mealy bug	Hemiptera	Margarodidae	Mango
101.	*Dudua aprobola* (Meyrick)	Shoot borer	Lepidoptera	Tortricidae	Mango, Rose
102.	*Dysdercus cingulatus* (Fab.)	Red cotton bug	Hemiptera	Pyrrhocoridae	Cotton, Niger, Okra, Pigeonpea, Portia Tree [*Thespesia populnea* (L.) Soland ex.]
103.	*Dysdercus koengii* (Fab.)	Red cotton bug	Hemiptera	Pyrrhocoridae	Coriander, Okra, Niger, Pigeonpea, Shatavari (*Asparagusracemosus*)
104.	*Earias fabia* Cram.	Bollworm	Lepidoptera	Noctuidae	Okra
105.	*Earias insulana* (Boisd.)	Spotted bollworm	Lepidoptera	Noctuidae	Cotton, Okra
106.	*Earias jabia* Stoll		Lepidoptera	Noctuidae	
107.	*Earias vitella* (Fab.)	Fruit borer	Lepidoptera	Noctuidae	Okra
108.	*Ectatops sp.*	Hemiptera	Pyrrhocoridae	Polyphagous	
109.	*Edwardsiana rosae* L.	Jassids	Hemiptera	Cicadellidae	Rose
110.	*Eilema antica* (Walker)		Lepidoptera	Erebidae	
111.	*Emmalocera depresella* Swinh.	Root borer	Lepidoptera	Pyralidae	Sugarcane
112.	*Empoasca fabae* (Harris)	Jassid	Hemiptera	Cicadellidae	Indian Bean, Okra, Pigeonpea, Cotton
113.	*Empoasca devastans* Dist.	Jassid	Hemiptera	Jassidae	Okra
114.	*Empoasca kerri* Pruthi	Jassid	Hemiptera	Jassidae	Mungbean, Soybean, Uridbean
115.	*Epilachna vigintioctopunctata* (Fab.)	Hadda Beetle	Coleoptera	Coccinellidae	Brinjal
116.	*Etiella zinckenella* Treitschke	Pea pod borer	Lepidoptera	Pyralidae	Field Pea, Lentil
117.	*Euchrrysops cnejus* (Fab.)	Blue butterfly	Lepidoptera	Lycaenidae	Uridbean
118.	*Euconocephalus* sp.	Bush cricket	Orthoptera	Tettigoniidae	Grasses
119.	*Eudocima aurantia* (Moore)	Fruit-sucking moth	Lepidoptera	Erebidae	
120.	*Euploea core core* Cramer	Nerium butterfly	Lepidoptera	Nymphalidae	Rose

S.No.	Scientific name	Common name	Order	Family	Host
121.	*Euproctis lunata* (Walker)	Tussock caterpillar	Lepidoptera	Lymantriidae	Castor
122.	*Euproctis similis* (Moore)	Tussock moth	Lepidoptera	Lymantriidae	Mulberry, Rice, Finger Millet
123.	*Euproctis subnotata* (Walker)	Tussock caterpillar	Lepidoptera	Lymantridae	Pigeonpea
124.	*Eupterote* sp.	Hairy caterpillar	Lepidoptera	Eupterotidae	Grasses
125.	*Eureama hecabe var. contubernalis* (Moore)	Yellow and white butterfly	Lepidoptera	Pieridae	Rose
126.	*Euscyrtus concinnus* (de Haan)	Field cricket	Orthoptera	Gryllidae	Rice And Fodder Grasses
127.	*Exelastis atomosa* Walsh.	Plume moth	Lepidoptera	Pterophoridae	Pigeonpea
128.	*Eysarcoris* sp.	New forest shield bug	Hemiptera	Pentatomidae	Sunhemp
129.	*Ferrisia virgata* Ckll.	Striped mealy bug	Hemiptera	Pseudococcidae	Ornamental & Vegetable Crops, Brinjal, Acalypha, Croton, Coccinea Indica
130.	*Flata* spp.	Hopper	Hemiptera	Flatidae	
131.	*Forficula auricularia* (Linnaeus)	Earwig	Dermaptera	Forficulidae	
132.	*Frankliniella sulphurea* Schmutz	Blossom thrips	Thysanoptera	Thripidae	Pigeonpea
133.	*Gastropacha pardale* (Walker)	Brown lapped moth	Lepidoptera	Lasiocampidae	
134.	*Gesonia gemma* (Swinhoe)	Grey semilooper	Lepidoptera	Noctuidae	Soybean
135.	*Glyphodes canthusalis* (Walker)		Lepidoptera	Crambidae	Cotton
136.	*Glyphodes* sp.	Leaf eating caterpillar	Lepidoptera	Crambidae	Jackfruit
137.	*Grammodes geometrica* (Fab.)	Leaf webber	Lepidoptera	Erebidae	Feed On Phyllanthus, Cistus, Diospyros, Ricinus, Rice, Numerous Grasses
138.	*Grapholita critica* (Meyr)	Leaf webber	Lepidoptera	Tortricidae	Pigeonpea
139.	*Gryllotalpa africana* Palisot de Beauvois	Small mole cricket	Orthoptera	Gryllotalpidae	
140.	*Gryllotalpa orientalis* (Burmeister)	Mole cricket	Orthoptera	Gryllotalpidae	Rice
141.	*Gryllus bimaculatus* (De Geer)	Field cricket	Orthoptera	Gryllidae	Fodder Grasses
142.	*Gryllus* sp.	Field cricket	Orthoptera	Gryllidae	Mungbean, Soybean
143.	*Halys serrigera* (Westwood)		Hemiptera	Pentatomidae	Polyphagous
144.	*Haplothrips gowdeyi* (Franklin)	Gold tipped tubular thrips	Thysanoptera	Thripidae	Chilli
145.	*Haptoncus reflexicollis* Motschulskyi		Coleoptera	Nitidulidae	Sunhemp
146.	*Helicoverpa armigera* (Hub.)	American bollworm	Lepidoptera	Noctuidae	Chickpea, Pea, Pigeonpea, Safflower, Chilli, Indian Bean, Okra, Onion, Potato, Tomato Cotton, Rose,
147.	*Hellula undalis* Fab.	Cabbage borer	Lepidoptera	Pyralidae	Cabbage, Cauliflower
148.	*Hexacentrus* sp.	Bush cricket	Orthoptera	Tettigoniidae	Grasses
149.	*Hieroglyphus banian* Fab.	Rice grasshopper	Orthoptera	Acrididae	Rice

S.No.	Scientific name	Common name	Order	Family	Host
150.	*Hieroglyphus oryzivorous* Carl	Grasshopper	Orthoptera	Acrididae	Grasses
151.	*Hippotion celerio* (Linnaeus)	Horned caterpillar	Lepidoptera	Sphingidae	Elephant Yam, Grapevine
152.	*Holotrichia consanguinea* Blanchard	White grub	Coleoptera	Scarabidae	Ground Nut, Soybean, Sugarcane And Chilli
153.	*Hyblaea puera* Cramer	Teak defoliator	Lepidoptera	Hyblaeidae	Indian Cork Tree, Jasmine, Neem, Teak
154.	*Hydaphis coriandri* (Das)	Coriander aphid	Hemiptera	Aphididae	Coriander
155.	*Hylemya cilicrura* (Rond.)	Seed corn maggot	Diptera	Anthomyiidae	Soybean
156.	*Hymenia fasciolis* (Balsam)	Leaf caterpillar	Lepidoptera	Pyralidae	
157.	*Hypomecis umbrosaria* (Hubner)	Umber moth	Lepidoptera	Geometridae	
158.	*Hyposidra talaca* (Walker)	Defoliator	Lepidoptera	Geometridae	Polyphagous Pest- Teak
159.	*Idioscopus niverosparsus* (Leth)	Mango hopper	Hemiptera	Cicadellidae	Mango
160.	*Idioscopus* sp.	Mango hopper	Hemiptera	Cicadellidae	Mango
161.	*Indarbela tetraonis* (Moore)	Citrus shoot &bark caterpillar	Lepidoptera	Cossidae	Citrus, Pomegranate
162.	*Lampides boeticus* (Lin.)	Blue butterfly	Lepidoptera	Lycaenidae	Mungbean, Uridbean
163.	*Lamprosema dieminalis* Guen.	Leaf folder	Lepidoptera	Pyralidae	Soybean
164.	*Lamprosema indicata* (Fab.)	Leaf folder	Lepidoptera	Pyralidae	Mungbean, Soybean
165.	*Lasioptera falcata* Felt	Gall midge	Diptera	Cecidomyidae	Ribbed Gourd, Smooth Gourd
166.	*Laspeyresia pseudonectis* Meyr	Top shoot borer	Lepidoptera	Euscosmidae	Sunhemp
167.	*Leptispa pygmaea* (Baly)	Rice leptispa	Coleoptera	Chrysomelidae	Rice
168.	*Leptocorisa acuta* (Thumb)	Paddy earhead bug or Gundhi bug	Hemiptera	Alydidae	Rice, Gudmar (*Gymnemasylvestre*)
169.	*Leptocorisa oratorius* (Guen.)	Rice ear bug	Hemiptera	Coreidae	Indian bean
170.	*Leucinodes orbonalis* (Guen.)	Shoot & fruit borer	Lepidoptera	Crambidae	Brinjal
171.	*Lipaphis erisymi* (Kalt.)	Mustard aphid	Hemiptera	Aphididae	Cabbage, mustard
172.	*Liriomyza trifolii* Burgess	Serpentine leaf miner	Diptera	Agromyzidae	Cabbage, French bean, Pea, Tomato
173.	*Longitarsus belgaumensis* Jacoby	Flea beetle	Coleoptera	Chrysomelidae	Sunhemp
174.	*Lucilia sericata* Meigen	Blow fly	Diptera	Calliphoridae	Rose
175.	*Maconellicoccus hirsutus* (Green)	Mealy bug	Hemiptera	Pseudococcidae	Okra
176.	*Macroglossum stellatarum* (Linn)	Horned caterpillar	Lepidoptera	Sphingidae	Indian Mulberry (Noni)
177.	*Macrosiphum rosaeiformis* Das	Rose aphid	Hemiptera	Aphididae	Rose
178.	*Madurasia obscurella* Jacoby	Flea beetle	Coleoptera	Olyrocidae	Mungbean, Uridbean
179.	*Marsamia* sp.		Lepidoptera	Crambidae	
180.	*Maruca testulalis* Geyer	Spotted pod borer	Lepidoptera	Pyraustidae	Pigeonpea

S.No.	Scientific name	Common name	Order	Family	Host
181.	*Mecopoda* sp.	Bush cricket	Orthoptera	Tettigoniidae	Grasses
182.	*Megachile anthracina* Smith	Rose leaf cutter bee	Hymenoptera	Megachilidae	Rose
183.	*Megalurothrips usitatus* (Bagnall)	Thrips	Thysanoptera	Thripidae	Pigeonpea
184.	*Melanagromyza hibisci* Spencer	Petiole maggot	Diptera	Agromyzidae	Hibiscus esculentus, Okra
185.	*Melanagromyza obtusa* Malloch	Pod fly	Diptera	Agromyzidae	Pigeonpea
186.	*Melanagromyza sojae* Zehntner	Stem fly	Diptera	Agromyzidae	Soybean
187.	*Melanapsis glomerata* (Green)	Scale insect	Hemiptera	Diaspididae	Sugarcane
188.	*Melanitis leda ismene* (Cramer)	Rice butterfly	Lepidoptera	Nymphalidae	Rice
189.	*Microtermes obesi* Holmgr.	Termite	Isoptera	Termitidae	Maize, wheat
190.	*Mocis undata* Fab.	Brown striped semilooper	Lepidoptera	Noctuidae	Soybean
191.	*Monolepta signata* (Olivier)	White spotted flea beetle	Coleoptera	Chrysomelidae	Okra, uridbean
192.	*Monomorium destructor* Jard	Ants	Hymenoptera	Formicidae	Sesamum, Wheat
193.	*Mylabris pustulata Thunberg*	Blister beetle	Coleoptera	Meloidae	Pigeonpea
194.	*Mylabris vitrata*Fab.	Blister beetle	Coleoptera	Meloidae	Pigeonpea
195.	*Myllocerus maculosus* Desbr.	Grey weevil	Coleoptera	Curculionidae	Soybean
196.	*Myllocerus sp.*	Grey weevil	Coleoptera	Curculionidae	Mungbean, Pigeonpea, Uridbean
197.	*Mythimna separata* Wlk.	Rice armyworm or ear cutting caterpillar	Lepidoptera	Noctuidae	Rice, Maize, Sorghum, Wheat
198.	*Mythimna unipunctata* (Haworth)	Cutworm	Lepidoptera	Noctuidae	Barley, Rice, Sorghum
199.	*Myzus persicae* (Sulzer)	Potato aphid	Hemiptera	Aphididae	Potato
200.	*Necroscia sparaxes* Westwood	Phasmid	Phasmatodea	Lonchodidae (Phasmida)	Forest plant Jamrasi (*Elaeodendron glaucum*)
201.	*Nephotettix nigropictus* (Stal.)	Green leaf hopper	Hemiptera	Cicadellidae	Rice
202.	*Nephotettix virescens* Dist.	Green leaf hopper	Hemiptera	Cicadellidae	Rice
203.	*Nesidiocoris tenuis* (Rent.)	Mirid bug	Hemiptera	Miridae	Sesame, okra
204.	*Nezara viridula* L.	Green sting bug	Hemiptera	Pentatomidae	Brinjal, Okra, Mung, Indian Bean, Pigeonpea, Niger, Safflower, Sesame, Soybean, Uridbean
205.	*Nilaparvata lugens* Stal.	Brown plant hopper	Hemiptera	Delphacidae	Rice
206.	*Nymphula depunctalis* Guen.	Caseworm	Lepidoptera	Pyraustidae	Rice
207.	*Nysius raphanus* (Howard)	False chinch bug	Hemiptera	Lygadiae	Okra
208.	*Obereopsis (= Oberea) brevis* Swed.	Girdle beetle	Coleoptera	Cerambycidae	Urid Bean, Cowpea, Mung Bean, Pigeonpea, Soybean

S.No.	Scientific name	Common name	Order	Family	Host
209.	*Odontotermes obesus* Rambur	Termite	Isoptera	Termitidae	Sugarcane, Wheat
210.	*Oecanthus* sp.	Tree crickets	Orthoptera	Tettigoniidae	Grasses
211.	*Oecophylla smaragdina* (F)	Red ant	Hymenoptera	Formicidae	Mango
212.	*Oocassida pudibunda* Boh.	Tortoise beetle	Coleoptera	Chrysomelidae	Jujube
213.	*Ophiomyia phaseoli* (Tryon)	Stem/ Bean fly	Diptera	Agromyzidae	Soybean, Uridbean
214.	*Orgyia leucostigma* (Smith)	Tussock moth	Lepidoptera	Erebidae	Rose
215.	*Orocious albicinctus* (Distant)	Jassids	Hemiptera	Jassidae	Sesame
216.	*Orseolia oryzae* (WM) Mani	Gall midge	Diptera	Cecidomyiidae	Rice
217.	*Othreis fulonica* (Clerck)	Citrus fruit sucking moth	Lepidoptera	Noctuidae	Citrus
218.	*Otinotus oneratus* W.	Cow bug	Hemiptera	Membracidae	Indian Bean, Pigeonpea
219.	*Oyxa* sp.	Grasshopper	Orthoptera	Acrididae	Grasses
220.	*Oxyrachis taradus* (Fab.)	Cow bug	Hemiptera	Membracidae	Okra
221.	*Pagria kamaraesis* Baly		Coleoptera	Alticidae	Soybean
222.	*Palpita quadristigmalis* (Guenée)	Four spotted palpita moth	Lepidoptera	Crambidae	Ornamental Plant (Jasmine)
223.	*Palpita vitrealis* (Rossi)	Jasmine moth or white pearl moth	Lepidoptera	Crambidae	Jasmine
224.	*Papilio demoleus* (Linn.)	Lemon butterfly	Lepidoptera	Papilionidae	Citrus
225.	*Parapoynx stagnalis* (Zeller)	Rice case worm or rice case bearer	Lepidoptera	Crambidae	Rice
226.	*Parnara mathias* Fab.	Rice skipper	Lepidoptera	Hesperidae	Rice, Indian sago- palm (*Caryota urens* L.)
227.	*Parotis marginata* (Hampson)		Lepidoptera	Crambidae	
228.	*Pectinophora gossypiella* (Saund)	Pink bollworm	Lepidoptera	Gelechiidae	Cotton
229.	*Pempherelus affinis* (Fst.)	Stem weevil	Coleoptera	Curculionidae	Cotton
230.	*Peregrinus maidis* D	Sorghum shoot bug	Hemiptera	Delphacidae	Sorghum
231.	*Pericallia ricini* (F.)	Woolly bear	Lepidoptera	Arctiidae	Castor, Sesame
232.	*Perina nuda* (Fab.)	Hairy caterpillar	Lepidoptera	Erebidae	Jack fruit
233.	*Phenacoccus solenopsis* Tinsley	Mealy bug	Hemiptera	Pseudococcidae	Coriander, Cotton, Ashwagandha (*Withania somnifera*), Gudmar (*Gymnemasylvestre*)
234.	*Phengaris arion* L.	Large blue butterfly	Lepidoptera	Lycaenidae	Rose
235.	*Phthorimaea operculella* Zeller	Potato tuber moth	Lepidoptera	Gelechiidae	Potato
236.	*Phyllocnisti citrella* Stainton	Citrus leaf miner	Lepidoptera	Gracillaridae	Citrus
237.	*Phyllotreta chotanica* Duv.	Flea beetle	Coleoptera	Chrysomelidae	Wheat
238.	*Phyllotreta cruciferae* G.	Flea beetle	Coleoptera	Chrysomelidae	Cabbage, Mustard
239.	*Phytomyza atricornis* Meigen	Pea leaf miner	Diptera	Agromyzidae	Pea
240.	*Pieris brassicae* (L)	Cabbage butterfly	Lepidoptera	Pieridae	Cabbage

S.No.	Scientific name	Common name	Order	Family	Host
241.	*Plecia amplipennis* (Skuse)		Diptera	Bibionidae	Fodder Pest
242.	*Plusia acuta* Wlk.	Cabbage semilooper	Lepidoptera	Noctuidae	
243.	*Plusia chalcites* (Esper)	Tomato looper	Lepidoptera	Noctuidae	Polyphagous Pest Of Family Solanaceae &Brassicaceae.
244.	*Plusia orichalcea* (Fab.)	Cabbage semilooper	Lepidoptera	Noctuidae	Pea, Soybean
245.	*Plutella maculipennis* (Curt.)	Cabbage moth	Lepidoptera	Yponomeutidae	Crucifers
246.	*Plutella xylostella* (L.)	Diamond back moth	Lepidoptera	Yponomeutidae	Cabbage
247.	*Psalis pennatula* (Fab.)	Yellow hairy caterpillar	Lepidoptera	Erebidae	Rice, Sorghum, Sugarcane, Tea
248.	*Psalis securis* Hb.		Lepidoptera	Lymantriidae	
249.	*Pulvinaria psidii* (Maskell)	Scale insect	Hemiptera	Coccidae	Guava
250.	*Pyhhria umbra* Huf.	Rose budworm	Lepidoptera	Noctuidae	Rose
251.	*Pyrausta machaeralis* Walker	Teak skeletoniser	Lepidoptera	Pyralidae	Teak
252.	*Pyrilla perpusilla* Wlk.	Sugarcane leaf hopper	Hemiptera	Lophopidae	Maize, Rice, Sorghum, Sugarcane
253.	*Ragmus importunitas* Dist.	Sunhemp mired	Hemiptera	Miridae	Sunhemp
254.	*Raphidopalpa foveicollis* Lucas	Red pumpkin beetle	Coleoptera	Chrysomelidae	Cucurbits
255.	*Rastrococcus (=Phenacoccusi) iceryoides* Green	Mango coccid	Hemiptera	Coccidae	Mango, China Rose
256.	*Rhipiphorothrips cruentatus* Hood	Grape thrips	Thysanoptera	Thripidae	Rose
257.	*Rhizopertha dominica* Fab.	Lesser grain borer	Coleoptera	Bostrichidae	Storage Pest
258.	*Rhopalosiphum maidis* (Fitch)	Aphid	Hemiptera	Aphididae	Maize
259.	*Riptortus pedestris* Fab.	Bean / Pod bug	Hemiptera	Coreidae	Indian Bean, Uridbean
260.	*Riptortus* sp.	Riptortus	Hemiptera	Coreidae	Pigeonpea
261.	*Riptortus strenuus* (Fabricius)	Black bug	Hemiptera	Coreidae	Soybean
262.	*Saccaricoccus sacchari* (Cockerell)	Mealy bug	Hemiptera	Pseudococcidae	Sugarcane
263.	*Sameodes cancellalis* Zeller		Lepidoptera	Crambidae	Grasses
264.	*Scirpophaga incertulas* Wlk.	Yellow stem borer	Lepidoptera	Pyralidae	Rice
265.	*Scirpophaga nincle* Fab.		Lepidoptera	Pyraustidae	
266.	*Scirpophaga novella* Fab.	Top shoot borer	Lepidoptera	Pyraustidae	Sugarcane
267.	*Scirtothrips dorsalis* Hood	Chilli thrips	Thysanoptera	Thripidae	Brinjal, Chilli, Rose
268.	*Scutiphora pedicellate* (Leach)	Jewel bug	Hemiptera	Scutelleridae	Uridbean
269.	*Sesamia inferens* Walker	Pink stem borer	Lepidoptera	Noctuidae	Maize, Wheat
270.	*Sitophilus oryzae* (Linn.)	Rice weevil	Coleoptera	Curculionidae	Storage Pest, Wheat
271.	*Sitotroga cerealella* (Olivier)	Angoumois grain moth	Lepidoptera	Gelechiidae	Storage Pest

S.No.	Scientific name	Common name	Order	Family	Host
272.	*Sogatella furcifera* Horvath	White backed plant hopper	Hemiptera	Delphacidae	Rice
273.	*Spilarctia* sp.		Lepidoptera	Erebidae	Polyphagous
274.	*Spilarctica (= Spilosoma) obliqua* Walker	Bihar hairy caterpillar	Lepidoptera	Arctiidae	Polyphagous- Castor, Linseed, Niger, Sesame, Soybean, Uridbean
275.	*Spodoptera (=Laphygma) exigua* (Hub.)	Lucerne / Linseed caterpillar	Lepidoptera	Noctuidae	Coriander, Linseed, Soybean
276.	*Spodoptera frugiperda* J.E. Smith	Fall army worm	Lepidoptera	Noctuidae	Maize, Oat
277.	*Spodoptera litura* (Fab.)	Tobacco caterpillar	Lepidoptera	Noctuidae	Cabbage, Cotton, Cucurbits, Chilli, Mungbean, Pea, Potato, Rose, Soybean, Uridbean
278.	*Spoladea recurvalis* (Fab.)	Leaf caterpillar	Lepidoptera	Crambidae	Polyphagous pest - Tuber crop- Koorkan (*Coleus parviflora*)
279.	*Sternochetus mangiferae* (F)	Mango stone or nut weevil	Coleoptera	Curculionidae	Mango
280.	*Stomopteryx subsecivella* Zeller	Leaf miner	Lepidoptera	Gelechidae	Soybean
281.	*Stromatium barbatum* (Fabricius)	Teak trunk borer	Coleoptera	Cerambycidae	Timber pest
282.	*Scutiphora pedicellate* (Kirby)	Jewel bug	Hemiptera	Scutelleridae	Okra
283.	*Sylepta derogata* Fabr.	Cotton leaf roller	Lepidoptera	Pyralidae	Cotton, Okra
284.	*Taeniothrips distalis* Karny	Blossom thrip	Thysanoptera	Thripidae	Pigeonpea
285.	*Taeniothrips nigricornis* Schmutz	Blossom thrips	Thysanoptera	Thripidae	Pigeonpea
286.	*Taeniothrips* sp.	Thrips	Thysanoptera	Thripidae	Mungbean
287.	*Taylorilygus pallidulus* Blanchard	Mirid bug	Hemiptera	Miridae	Niger
288.	*Tetrix subulata* Linn.	Short horn grass hopper	Orthoptera	Tetrigidae	Rice, Grasses
289.	*Tettigonia viridissima* L.	Long horned grasshopper	Odonata	Tettigoniidae	Rose
290.	*Thrips florum* (Schmutz)	Flower thrips	Thysanoptera	Thripidae	Chilli, Sunhemp
291.	*Thrips palmi* (Karny)	Melon thrips	Thysanoptera	Thripidae	Brinjal
292.	*Thrips tabaci* Lindeman	Thrips	Thysanoptera	Thripidae	Coriander, Okra, Onion, Cotton, Soybean
293.	*Thysanoplusia orichalcea* (Fab.)	Green semilooper	Lepidoptera	Noctuidae	Niger, Soybean, Sunflower, Pea, Potato
294.	*Trabala vishnou* (Lefebve)	Rose-myrtle lappet moth	Lepidoptera	Lasiocampidae	Rose, Castor, Jamun, Pomegranate
295.	*Trachys* sp.	Okra leaf miner	Coleoptera	Buprestidae	Okra
296.	*Tribolium castaneum* (Herbst)	Rust red flour beetle	Coleoptera	Tenebrionidae	Storage Pest, Cereal Flours, Sesame
297.	*Tribolium confusum* (Herbst)	Rust red flour beetle	Coleoptera	Tenebrionidae	Storage Pest
298.	*Trichoplusia ni* (Hub.)	Semilooper	Lepidoptera	Noctuidae	Indian Bean
299.	*Trigonidium cicindeloides* (Saussure)	Winged bush cricket	Orthoptera	Trigonidiidae	Uridbean
300.	*Trigonidium* sp.	Swordtail crickets	Orthoptera	Orchidaceae Trigonidiidae	Grasses

S.No.	Scientific name	Common name	Order	Family	Host
301.	*Trigonodes hyppasia* (Cramer)	Semilooper	Lepidoptera	Noctuidae	Polyphagous
302.	*Trogoderma granarium* Everts	Khapra beetle	Lepidoptera	Dermestidae	Storage Pest
303.	*Tryporyza incertulas* Wlk	Rice borer	Lepidoptera	Crambidae	Rice
304.	*Tuta absoluta* (Meyrick)	American leaf miner	Lepidoptera	Gelechiidae	Tomato
305.	*Urentius sentis* Dist	Tingid bug	Hemiptera	Tingidae	Brinjal And Bean
306.	*Uroleucon carthami* (Hille Ris Lambers)	Safflower aphid	Hemiptera	Aphididae	Safflower
307.	*Uroleucon compositae (= Dactynotus carthami)* (Theobald)	Safflower aphid	Hemiptera	Aphididae	Safflower
308.	*Utetheisa ornatrix* (Linnaeus)	Hairy caterpillar	Lepidoptera	Erebidae	Legumes
309.	*Utetheisa pulchella* (Linn.)	Sunhemp hairy caterpillar	Lepidoptera	Arctiidae	Sunhemp
310.	*Virachola isocrates* F.	Pomegranate butterfly	Lepidoptera	Lycaenidae	Pomegranate
311.	*Zamarada excisa* (Hampson)		Lepidoptera	Geometridae	
312.	*Zanna* sp.	Tropical plant hopper	Hemiptera	Fulgoridae	Polyphagous
II	**Mites**				
313.	*Polyphagotarsonemus* latus Banks	Chilli mites	Trombidiformes	Tarsonemidae	Chilli
314.	*Tetranychus* sp.	Red spider mite	Trombidiformes	Tetranychidae	Bean, Brinjal, Onion, Soybean
315.	*Tetranychus neocaledonicus* Andre	Mite	Trombidiformes	Tetranychidae	Rose
316.	*Tetranychus telarius* K.	Red spider mite	Trombidiformes	Tetranychidae	Okra, Onion, Rose
317.	*Tetranychus urticae* Koch	Red spider mite	Trombidiformes	Tetranychidae	Rose
II	**Insect Pollinators**				
318.	*Apis cerana indica* Fab.	Indian bee	Hymenoptera	Apidae	Cauliflower, Okra, Onion
319.	*Apis dorsata* Fab.	Rock bee or giant bee	Hymenoptera	Apidae	Cotton, Mango
320.	*Apis florea* Fab.	Little honey bee	Hymenoptera	Apidae	
321.	*Apis melipona* Illiger	Stingless Bee	Hymenoptera	Apidae	Wide range of tropical plants
322.	*Apis mellifera* Linn.	European honey bee	Hymenoptera	Apidae	Rice, Cashew nut, Tomato
323.	*Daphnis nerii* (Linnaeus)	Oleander hawk-moth	Lepidoptera	Sphingidae	Feed on nectar of variety of flowers, Jasmine, Petunia
324.	*Eupeodes confrater* (Wiedemann)	Hover fly	Diptera	Syrphidae	Maggots predate on aphids. Adults are pollinators
III	**Insect Predators**				
325.	*Adalia bipunctata* Linn.	Two-spot Ladybird	Coleoptera	Coccinellidae	Aphid, Psyllids, Whiteflies
326.	*Amantis* sp.	Mantis	Dictyoptera	Mantidae	Predator
327.	*Anaciaeschna jaspidea* (Burmeister)	Dragonfly	Odonata	Aeshnidae	Predator
328.	*Anax* sp.	Dragon fly	Odonata	Aeshnidae	Rose
329.	*Antilochus cocqueberti* (Fab.)	Predacious bug	Hemiptera	Pyrrhocoridae	Cotton - Nymph of Red Cotton Bug

S.No.	Scientific name	Common name	Order	Family	Host
330.	*Archimantis latistyla*, (Serville)	Large brown mantis	Dictyoptera	Mantidae	Predator of Crickets, Cockroach, Flies
331.	*Ascalaphus sinister* Walker	Owlfly	Neuroptera	Ascalaphidae	
332.	*Ascalaphus* sp.	Owl fly	Neuroptera	Ascalaphidae	Adult on Caterpillars, Grubs
333.	*Brachinus longipalpis* (Wiedemann)	Ground beetle	Coleoptera	Carabidae	General predator
334.	*Brachinus sexmaculatus* Dejan	Bombardier beetles	Coleoptera	Carabidae	General predator
335.	*Brachythemis contaminate* Fab.	Asian Groundling (Dragonfly)	Odonata	Libellulidae	Predator
336.	*Brumus suturalis* F.	Coccinellid beetle	Coleoptera	Coccinellidae	Citrus, Sugarcane
337.	*Brumoides suturalis* F.	Three striped lady bird beetle/ Coccinellid beetle	Coleoptera	Coccinellidae	Okra, Uridbean
338.	*Calosoma* sp.	Large ground beetle	Coleoptera	Carabidae	General predator
339.	*Cantheconidia furcellata* Wolff	Predatory stink bug	Hemiptera	Pentatomidae	*Spodoptera litura* on Soybean, *Oocassida pudibunda*
340.	*Chalenites circumdatus* (Bru.)		Coleoptera	Carabidae	
341.	*Cheilomenes sexmaculatus* (F.)	Six spotted zigzag lady bird beetle	Coleoptera	Coccinellidae	Okra, Rose, Safflower, Soybean, Uridbean
342.	*Chlaenius* sp.	Ground beetle	Coleoptera	Carabidae	Predator of Locust's egg
343.	*Chlaenius circumdatus* (Dejean)	Ground beetle	Coleoptera	Carabidae	Predator
344.	*Chlaenius medioguttatis* (Chaudoir)	Ground beetle	Coleoptera	Carabidae	Lepidopterous larvae
345.	*Chlaenius nigricans* (Riley and Clausen)	Ground beetle	Coleoptera	Carabidae	*Laphgma pyrausta nubilalis* and *Pinusinsiguos* sp.
346.	*Chlaenius panagaeoides* Chaudoir	Ground beetle	Coleoptera	Carabidae	General predator
347.	*Chlaenius pictus* (Choudoir)	Ground beetle	Coleoptera	Carabidae	General predator of Lepidopterous larvae
348.	*Chrysopa madestes* Banks	Chrysopids	Neuroptera	Chrysopidae	Lac Insect, Aphid, Coccods, Mites
349.	*Chrysoperla* sp.	Green lace wing	Neuroptera	Chrysopidae	Lac insect, *Ferrisia virgata* Ckll.
350.	*Chrysoperla carnea* (Stephens)	Green lace wing	Neuroptera	Chrysopidae	Brinjal, Mungbean, Pigeonpea, Rose, Sesame
351.	*Chrysoperla rufilabris*	Green lace wing	Neuroptera	Chrysopidae	Indian bean
352.	*Chrysoperla sillemi* (Esben-petersen)	Lace wing	Neuroptera	Chrysopidae	Aphids, Leaf Hoppers
353.	*Cicindela cardoni* Flr.	Tiger beetle	Coleoptera	Cicindellidae	
354.	*Cicindela flexuosa* (Distant)	Tiger beetle	Coleoptera	Carabidae	General predator of small insects
355.	*Cicindela melancholica* Fab.	Tiger beetle	Coleoptera	Cicindellidae	
356.	*Cicindela venosa* Kollar	Tiger beetle	Coleoptera	Cicindellidae	
357.	*Coccinella reparada* Thumb	Lady bird beetle	Coleoptera	Coccinellidae	

S.No.	Scientific name	Common name	Order	Family	Host
358.	*Coccinella septempunctata* (L)	Seven spotted lady bird beetle	Coleoptera	Coccinellidae	Coriander, Lentil, Okra, Onion, Pigeonpea, Sesame, Soybean, Tomato, Uridbean
359.	*Coccinella transversalis* (Fab.)	Lady bird beetle	Coleoptera	Coccinellidae	Brinjal, Chili, Indian Bean, Okra, Uridbean, Gudmar (*Gymnemasylvestre*), Shatavari (*Asparagusracemosus*)
360.	*Coenagrion* sp.	Damsel fly	Odonata	Coenagrionidae	Monarch Butterfly, Stem Borer, Gall Midge, Leaf-Eating Caterpillar
361.	*Conocephalus* sp.	Bush-cricket	Orthoptera	Tettigoniidae	Lepidopteran eggs
362.	*Creobroter apicalis* Saussure	Praying mantids	Dictyoptera	Hymenopodidae	Predator
363.	*Crocothemis servilia* Drury	Red pepper dragonfly	Odonata	Libellulidae	Predator
364.	*Crosopedophorus elegans* Dejean		Coleoptera	Carabidae	Lepidoptera larvae, Soft body insects
365.	*Cyphon padi* Fleming	Lady bird beetle	Coleoptera	Scirtidae	General predator of small insects
366.	*Deserida lineola* (Macleay)	Carabids	Coleoptera	Carabidae	
367.	*Didymocorypha lanceolata*Fab	Mantid	Dictyoptera	Tarachodidae	
368.	*Diplacodes trivialis* Rambur	Ground skimmer (Dragonfly)	Odonata	Libellulidae	
369.	*Diplocheila polita* (Fab.)	Ground beetle	Coleoptera	Carabidae	
370.	*Diplonychus rusticus* (Fabricius)	Water bug	Hemiptera	Belostomatidae	Feed on aquatic insects
371.	*Dytiscus marginalis* (Linnaeus)	Water beetle	Coleoptera	Dytisadae	Predator
372.	*Ectomocoris cordiger* (Stal.)	Assassin bug	Hemiptera	Reduvidae	
373.	*Ectomocoris ululans* (Rossi)	Assassin bug	Hemiptera	Reduviidae	Predator of caterpillars and small insects
374.	*Elaunon bipartitus* (Kirby)	Earwigs	Dermaptera	Forficulidae	General predator on Lepidopteran larvae
375.	*Eocanthecona furcellata* (Wolff)	Predaceous stink bug	Hemiptera	Pentatomidae	Okra - caterpillars and small insects
376.	*Eublemma amabilis* Salmuller	Predator	Lepidoptera	Noctuidae	Lac Insect
377.	*Eupeodes confrater* (Wiedemann)	Syrphid fly	Diptera	Syrphidae	Maggots predate on aphids. Adults are pollinators
378.	*Geocoris tricolor* Fab.	Big-eyed bug	Hemiptera	Lygaeidae	*Ferrisia virgata* Ckll., Brinjal and Bean pests - Tingid bug, *Urentius sentis* Dist.; Mite, *Tetranychus* sp.; Sorghum shoot bug, *Peregrinus maidis* D.
379.	*Heteronychus* sp.	Black beetle	Coleoptera	Scarabaeidae	General predator of aphids, coccids, whitefly and bugs
380.	*Hydrochara caraboides* (Latreille)	Lesser silver water beetlwe	Coleoptera	Hydrophilidae	General predator of aquatic insects

S.No.	Scientific name	Common name	Order	Family	Host
381.	*Hydropyche* sp.	Caddisfly	Trichoptera	Hydropsychidae	Aquatic predators
382.	*Hyperaspis maindroni Sci. sp.brumoides* Sci.		Coleoptera	Coccinellidae	*Centrococcus insolitus* (Green), *Ferrisia virgata* Ckll. Eggs of *Epilachna 28-punctata* F.
383.	*Illeis cincta* (Fabr.)	Lady bird beetle	Coleoptera	Coccinellidae	Kundru, Okra
384.	*Lasigus niger* (Linn.)	Black garden ant	Hymenoptera	Formicidae	Okra, Uridbean
385.	*Libellua* sp.	Skimmers (Dragonfly)	Odonata	Libellulidae	
386.	*Macrochilustri pustulatus* (Dejean)		Coleoptera	Carabidae	Predator
387.	*Mantis* sp.	Preying mantid	Neuroptera	Mantispidae	Predator
388.	*Mantis religiosa* (Linnaeus)	Preying mantid	Mantodea	Mantidae	Rose, Predator of crickets and beetles
389.	*Menochilus sexmaculatus* Fabr.	Lady bird beetle	Coleoptera	Coccinellidae	Brinjal, Coriander, Okra Mustard- *Aphis gossypii* Glov., *A. craccivora* Koch, *A. maidis*, *Ferrisia varigata* Ckll., *Centrococcus insolitus* (Green), *Tetranychus* sp. (ludeni group) Ashwagandha (*Withania somnifera*), Gudmar (*Gymnemasylvestre*)
390.	*Myrmicaria brunnea* Saunders	Ant	Hymenoptera	Formicidae	
391.	*Myrmicaria sp.*	Ant	Hymenoptera	Formicidae	
392.	*Nehalennia* spp.	Damsel fly	Odonata	Ceriogrionidae	Rose
393.	*Nephus regularis* Sic.	Lady beetle	Coleoptera	Coccinellidae	*Ferrisia virgata* Ckll., *Centrococcus insolitus* (Green),
394.	*Oecophylla* sp.	Red ant	Hymenoptera	Formicidae	Mango- *Rastrococcus* (*Phenacoccusi*)*iceryoides* Green
395.	*Oligota* sp.		Coleoptera	Staphylinidae	Beans, Brinjal - Red spider mite, *Tetranychus* sp.,
396.	*Oncocephalus* sp.	Assassin bug	Hemiptera	Reduviidae	
397.	*Onitis falcutus* (Wulfen)		Coleoptera	Scarabaeidae	Aphids, Bugs, Coccids, Whitefly
398.	*Ophionea indica* (Thunberg)	Ground beetle	Coleoptera	Carabidae	Aphids, Bugs Coccids, Whitefly
399.	*Orthectrum cancellatum*	Dragon fly	Odonata	Aeshnidae	Mungbean
400.	*Orthetrum sabina* (Drury)	Slender skimmer	Odonata	Libellulidae	Predator
401.	*Paederus fuscipes* (Curtis)	Rove beetle	Coleoptera	Staphylinidae	Uridbean
402.	*Pantala flavescens* Fabricus	Dragonfly	Odonata	Rhyothemisidae	Pigeonpea, Lepidopterous, Dipterous and Hymenopterous insects
403.	*Podisus maculiventris* (Say)	Predatory bug	Hemiptera	Pentatomidae	Soybean
404.	*Poecilus cupreus* (Linnaeus)	Ground beetle	Coleoptera	Carabidae	General predator of Lepidopterous larvae

S.No.	Scientific name	Common name	Order	Family	Host
405.	*Prothyma proxima* (Chd.)	Beetle	Coleoptera	Cicindellidae	
406.	*Prothyma* sp.	Tiger beetle	Coleoptera	Carabidae	Colorado potato beetle and small insects
407.	*Pseudagrion microcephalum* Rambur	Damselfly	Odonata	Ceriogrionidae	Pigeonpea
408.	*Pseudohypatopa pulverea Meyrick*		Lepidoptera	Blastobasidae	Lac insect
409.	*Pullus coccidivora* Ayyar	Lady beetle	Coleoptera	Coccinellidae	*Ferrisia virgata* Ckll., *Centrococcus insolitus* (Green)
410.	*Rodolia fumida* Muls.	Coccinelid	Coleoptera	Coccinellidae	*Perissopneumon tamarindus* (Green)
411.	*Scadra annulipes* Reuter	Assassin bug	Hemiptera	Reduviidae	Predator
412.	*Scymnus nubillus* Mils.	Lady bird beetle	Coleoptera	Coccinellidae	
413.	*Sirthenea carinata* (Fabricius)	Assassin bug	Hemiptera	Reduviidae	*Gryllus* sp., Mole cricket
414.	*Sirthenes* sp.	Assassin bug	Hemiptera	Reduvidae	*Orycetes* sp., *Scapterisus* sp., Small insects
415.	*Statilia maculata* (Thunberg)	Asian Jumping mantis	Dictyoptera	Mantidae	
416.	*Stethorus gilvifrons* Muls.	Ladybird beetle	Coleoptera	Coccinellidae	Brinjal, Beans- Red spider mite, *Tetranychus* sp. (ludeni group)
417.	*Stethorus pauperculus* Weise	Lady bird beetle	Coleoptera	Coccinellidae	Okra- aphid, Red spider mite
418.	*Stratiolaelaps scimitus*	Predatory mite	Mesostigmata	Laelapidae	Mushroom- Sciarid / Phorid fly
419.	*Tholymis tillarga* Fab.	Dragonfly	Odonata	Libellulidae	
420.	*Xanthogramma javanna* Wd.	Hoverfly	Diptera	Syrphidae	*Ferrisia virgata* Ckll.
421.	*Zosteria* sp.	Robber fly	Diptera	Asilidae	
422.	*Zuphium olens* (Rossi)	Carabid beetle	Coleoptera	Carabidae	
IV	**Insects of Industrial Importance**				
423.	*Antheraea paphia* Linnaeus	Silk worm	Lepidoptera	Saturniidae	Production of silk
424.	*Apis cerana indica* Fab.	Indian bee	Hymenoptera	Apidae	Honey production and its by-products
425.	*Apis mellifera* Linn.	European honey bee	Hymenoptera	Apidae	Honey production and its by-products
426.	*Bombyx moorei* F. *Meridionalis* F.	Silkworm	Lepidoptera	Bombycidae	Silk worm
427.	*Kerria (=Laccifer) lacca* Kerr.	Lac insect	Hemiptera	Lacciferidae	Lac
V	**Insect Pests of Mushroom**				
428.	*Lycoriella ingénue* Dufour	Sciarid fly	Diptera	Sciaridae	Mushroom
429.	*Megaselia scalaris* (Loew)	Phorid fly	Diptera	Phoridae	Mushroom
VI	**Spiders Predators**				
430.	*Ariamnes simulans* (Cambridge)	Comb-footedspider	Araneae	Therididae	Rice
431.	*Clubiona drassodes* Cambridge	Clubionids	Araneae	Clubionidae	Pigeonpea

S.No.	Scientific name	Common name	Order	Family	Host
432.	*Cyclosa* sp.	Trashline orbweavers	Araneae	Araneidae	Pigeonpea
433.	*Cyrtophora* sp.	Tent-web Spider	Araneae	Araneidae	Pigeonpea
434.	*Dictyna* sp.	Cribellate araneomorph spider	Araneae	Dictynidae	Cowpea, Rice, Pigeonpea
435.	*Drassodes* sp.	Ground Spider	Araneae	Gnaphosidae	Pigeonpea
436.	*Gnaphosa poonaensis* Tikader	Ground Spider	Araneae	Gnaphosidae	Pigeonpea
437.	*Gnaphosa* sp.	Ground Spider	Araneae	Gnaphosidae	Pigeonpea
438.	*Guizygiella indica* (Tikader & Bal)	Asian long-jawed orbweavers	Araneae	Tetragnathidae	Mustard, Pigeonpea
439.	*Guizygiella* sp.	Long-jawed orbweavers	Araneae	Tetragnathidae	Mustard, Onion, Rice, Pigeonpea, Soybean, Wheat
440.	*Hogna lenta*Hentz	Wolf spider	Araneae	Lycosidae	Indian bean
441.	*Hyllus semicupreus* (Simon)	Heavy bodied-jumper	Araneae	Salticidae	Cowpea, Mung, Linseed, Pigeonpea
442.	*Larinia* sp.	Orb-weaver	Araneae	Araneidae	Mustard
443.	*Leucauge decorata* (Blackwall)	Decorative silver orb spider	Araneae	Tetragnathidae	Wheat
444.	*Leucauge* sp.	Long-jawed orb-weaver	Araneae	Tetragnathidae	Rice, Pigeonpea, Soybean
445.	*Marpissa* sp.	Jumping spider	Araneae	Salticidae	Cowpea, Linseed, rice, Uridbean
446.	*Neoscona arabesca* (Walckenaer)	Orb weaver spide	Araneae	Araneidae	Uridbean
447.	*Neoscona ellipticus* (Tikader & Bal)	Spotted orb-weavers	Araneae	Araneidae	Coriander, Pigeonpea
448.	*Neoscona* sp.	Spotted orb-weavers	Araneae	Araneidae	Linseed, Mustard, Rice, Pigeonpea, Wheat, Okra
449.	*Neoscona vigilans* (Blackwall)	Spotted orb-weavers	Araneae	Araneidae	Rice, Pigeonpea
450.	*Olios* sp.	Huntsmanspider	Araneae	Sparassidae	Pigeonpea
451.	*Oxyopes birmanicus*Thorell	Lynx spider	Araneae	Oxyopidae	Garlic, Rice, Pigeonpea, Uridbean
452.	*Oxyopes javanus* Thorell	Lynx spider	Araneae	Oxyopidae	Rice, Okra
453.	*Oxyopes salticus* Hentz	Striped lynx spider	Araneae	Oxyopidae	Soybean, Uridbean
454.	*Oxyopes* sp.	Lynx spider	Araneae	Oxyopidae	Cowpea, Onion, Rice, Pigeonpea
455.	*Pardosa birmanica* Simon	Wolf Spider	Araneae	Lycosidae	Pigeonpea, Soybean
456.	*Peucetia* sp.	Lynx spider	Araneae	Oxyopidae	Mung, Maize, Rice, Soybean
457.	*Peucetia viridans* (Hentz)	Green lynx spider	Araneae	Oxyopidae	Rose
458.	*Phidippus* sp.	Jumpingspider	Araneae	Salticidae	Cowpea, Soybean
459.	*Phintella* sp.	Jumpingspider	Araneae	Salticidae	Okra
460.	*Plexippus* sp.	Jumpingspider	Araneae	Salticidae	Rice, Pigeonpea, Wheat
461.	*Runcinia* sp.	Crabspider	Araneae	Thomisidae	Garlic, Onion
462.	*Synema decoratum* Tikader	Crab spider	Araneae	Thomisidae	Cowpea, Pigeonpea, Soybean
463.	*Synema* sp.	Crab spider	Araneae	Thomisidae	Cauliflower, Mung bean, Soybean
464.	*Telamonia dimediata* (Simon)	Two striped jumping spider	Araneae	Salticidae	Rice, Pigeonpea, Okra

S.No.	Scientific name	Common name	Order	Family	Host
465.	*Tetragnatha* sp.	Long-jawed orb-weaver	Araneae	Tetragnathidae	Rice
466.	*Theridion* sp.	Tangle-web spider	Araneae	Theridiidae	Cowpea, Soybean,
467.	*Thomisus* sp.	Crab spider	Araneae	Thomisidae	Marigold, Sorghum, Soybean
468.	*Thomisus onustus* (Walckenaer)	Crab spider	Araneae	Thomisidae	Okra
469.	*Uloborus jabalpurensis* Bhandari & Gajbe	Cribellate orb weaver	Araneae	Uloboridae	Maize, Pigeonpea
470.	*Uloborus* sp.	Cribellate orb weaver	Araneae	Uloboridae	Coriander, Pigeonpea
471.	*Xysticus* sp.	Spider	Araneae	Thomisidae	Garlic, Soybean
VII	**Rodents**				
472.	*Bandicota bengalensis* (Gray and Hardwicke)	Lesser bandicoot	Rodenta	Muridae	Polyphagous
473.	*Meltada urbanus* Hodgson	Common Indian mouse	Rodenta	Muridae	Polyphagous
474.	*Millardia meltada* (Gray)	Soft fur rat	Rodenta	Muridae	Polyphagous
475.	*Mus booduga* (Gray)	Little Indian field mouse	Rodenta	Muridae	Polyphagous
476.	*Rattus rattus rufescens* Gray	House rat	Rodenta	Muridae	Polyphagous
477.	*Suncus murinus* (Linn.)	Mole	Soricomorpha	Soricidae	Polyphagous

Table 20.4 Insect parasites reported form College of Agriculture, Jabalpur

S.No.	Scientific name	Order	Family	Host
VII	**Insect Parasites**			
478.	*Achrysocharella* sp.	Hymenoptera	Eulophidae	Larva of *Trachys sp.*
479.	*Aenasius advena* Compere	Hymenoptera	Encyrtidae	Adult of *Ferrisia virgata*
480.	*Apanteles* sp.	Hymenoptera	Braconidae	Larva of *S.litura*
481.	*Apanteles* sp.(ultor group)	Hymenoptera	Braconidae	Larva of cabbage web worm, *Crocidolomia binotalis*
482.	*Apanteles obliquae* Wilkinson	Hymenoptera	Braconidae	Larva of cabbage web worm, *Crocidolomia binotalis*
483.	*Apanteles ruficrus*	Hymenoptera	Braconidae	Larva of *Perigea capensis, S. litura*
484.	*Apanteles taragamae* Viereck	Hymenoptera	Braconidae	*Laspeyresia pseudonectis Meyr*
485.	*Asympiesella* sp.	Hymenoptera	Eulophidae	Larva of *Lamprosema indicate*
486.	*Avga* sp.	Hymenoptera	Hormiini	Larva of *Trachys sp.*
487.	*Brachymeria lasus* (Walker)	Hymenoptera	Chalcididae	Larva of *Mocis undata*
488.	*Brachymeria* sp.	Hymenoptera	Chalcididae	Larval- Pupalparasite of *Oocassida pudibunda* Boh.
489.	*Bracon gelechiae* Ashm.	Hymenoptera	Braconidae	Larva of *Lamprosema indicata*
490.	*Bracon hebetor* Say	Hymenoptera	Braconidae	Larval parasite of *Dichocrosis punctiferalis, Eublemma olivacea* and *Hellula undalis.* Brinjal, Castor, Chickpea, Okra, Tomato
491.	*Bracon* s.l.sp.	Hymenoptera	Braconidae	Larva of *Trachys* sp., *Laspeyresia pseudonectis*
492.	*Bracon* spp	Hymenoptera	Braconidae	Larval / pupal paraite of Ber, Brinjal, Cabbage, Chickpea, Guava, Maize, Mango, Mung bean, Okra, Paddy, Sapota, Tomato, *Aspondylia capsici* Barnes

S.No.	Scientific name	Order	Family	Host
493.	*Brevicoryne brassicae* Linn	Hemiptera	Aphididae	Cabbage
494.	*Bulborogas* sp.	Hymenoptera	Braconidae	Brinjal
495.	*Campoletis chloridae* (Uchida)	Hymenoptera	Ichneumonidae	Chickpea, Pigeonpea
496.	*Cantheconidia furcellata* Wolf.	Hemiptera	Pentatomidae	Cabbage web worm, *Crocidolomia binotalis*
497.	*Carcelia* sp.	Diptera	Tachinidae	Larva and pupa of *H. armigera, A. janata on* Chickpea, Pigeonpea
498.	*Chelonus* sp.	Hymenoptera	Braconidae	Maize - *S.exigua*
499.	*Chrysopa madestes Banks*	Neuroptera	Chrysopidae	Pest of Lac insect
500.	*Chrysoperla sp.*	Neuroptera	Chrysopidae	Pest of Lac insect
501.	*Cirrospilus* sp.	Hymenoptera	Eulophidae	Larval parasite of *Trachys sp.*
502.	*Cotesia (=Apanteles)* sp.	Hymenoptera	Braconidae	Bean, Ber, Brinjal, Cabbage, Chickpea, Guava, Mango, Mung bean, Pigeonpea, Rice, Sapota, Sesame
503.	*Dairetiella rapae* (M'Intosh)	Hymenoptera	Braconidae	Parasitoid of Mustard aphid
504.	*Dinarmus* sp.	Hymenoptera	Pteromalidae	*Aspondylia capsici* Barnes Parasitoid on *Callosobruchus* sp.
505.	*Disophrys* sp.	Hymenoptera	Braconidae	Larva of *S.exigua*
506.	*Doryctobracon* sp.	Hymenoptera	Braconidae	Tomato
507.	*Dorylus* sp.	Hymenoptera	Formicidae	Lepidopterous and Dipterous insects
508.	*Echthromorpha agrestoria* Swed.	Hymenoptera	Ichneumonidae	Larva of *Mocis undata*
509.	*Ecphoropsis (=Campoletis) perdistinctus* Vier	Hymenoptera	Ichneumonidae	Larva of *H.armigera*
510.	*Elasmus brevicornis* Gab.	Hymenoptera	Elasmidae	Larva of *Lamprosema indicata*
511.	*Elasmus* sp.	Hymenoptera	Elasmidae	Larva of *Laspeyresia pseudonectis*
512.	*Elasmus* sp. nr. *fumipennis*Cameron	Hymenoptera	Elasmidae	Larva of *Laspeyresia pseudonectis*
513.	*Enicospilus capensis* (Thunberg)	Hymenoptera	Ichneumonidae	Light Trap catches
514.	*Enicospilus purgatus* (Say)	Hymenoptera	Ichneumonidae	Lleaf folder, Stem borer and Lepidopterous insects
515.	*Enicospilus xanthocephalus* Cameron	Hymenoptera	Ichneumonidae	Larva of Cabbage web worm, *Crocidolomia binotalis*
516.	*Eriborus argentecupilosus* (Cameron)	Hymenoptera	Ichneumonidae	Larva of *S.exigua*
517.	*Eucelatoria bryani* Sabrosky	Diptera	Tachinidae	Larva of *H.armigera*
518.	*Eucollidea* sp.	Hymenoptera	Eucoilidae	Pupa of *Melanagromyza hibisci*
519.	*Euderus agromyzae* Gangrade	Hymenoptera	Eulophidae	Pre pupa of *Melanagromyza obtusa*
520.	*Euplectrus* sp.	Hymenoptera	Eulophidae	Larva of *Lamprosema indicata*
521.	*Euplemus tachardiae Howard*	*Diptera*	Eupelmidae	Pest of Lac insect
522.	*Eurytoma* sp.	Hymenoptera	Eurytomidae	Maggot of *Melanagromyza hibisc,* Pupa of*Aspondylia capsici* Barnes
523.	*Exorista sorbillans* Wield.	Diptera	Tachinidae	Larva of *Mocis undata*

S.No.	Scientific name	Order	Family	Host
524.	*Glabridorsum stokesii* (Cameron)	Hymenoptera	Ichneumonidae	Larva of Leaf folder, Lepidopterous insects, Stem borer
525.	*Glyptapanteles* spp. (Wasp)	Hymenoptera	Braconidae	Cabbage, Okra
526.	*Habrobracon brevicornis* (Wesmael)	Hymenoptera	Braconidae	Cabbage
527.	*Homolobus* sp.	Hymenoptera	Braconidae	Light trap
528.	*Leurinion* sp.	Hymenoptera	Hormiini	Larval parasite of *Trachys sp.*
529.	*Macrocentrus* sp.	Hymenoptera	Braconidae	Light trap
530.	*Microplitis* sp.	Hymenoptera	Braconidae	Larva of *Lamprosema indicata*
531.	*Microplitis demolitor* Wilkinson	Hymenoptera	Braconidae	Mung bean, Guava, Sapota
532.	*Microplitis maculipennis* (Szep.)	Hymenoptera	Braconidae	Larva of Castor- *A. janata, S.exigua*
533.	*Misocyclops* sp.	Hymenoptera	Selionidae	Larva of *Lasioptera falcata*
534.	*Myrmicaria brunnea* Saunders	Hymenoptera	Formicidae	Egg of Lepidopteran insects
535.	*Netelia* sp.	Hymenoptera	Ichneumonidae	Light trap catches
536.	*Palexorista solenis* Walk.	Diptera	Tachinidae	Larva of *Cabbage web worm, Crocidolomia binotalis*
537.	*Phanerotoma* sp.	Hymenoptera	Braconidae	Larva of *Lamprosema indicata*
538.	*Pseudohypatopa pulverea* Meyrick	Lepidoptera	Blastobasidae	Lac insect
539.	*Rhogas percunets* Lyle	Hymenoptera	Braconidae	Larva of *A. janata*
540.	*Rhogas* sp.	Hymenoptera	Braconidae	Larva of *Perigea capensis*
541.	*Tachardiaephagus tachardiae Howard*	*Diptera*	Encyrtidae	Pest of Lac insect
542.	*Temelucha* sp.	Hymenoptera	Braconidae	*Lamprosema indicata*
543.	*Tetrastichus* sp.	Hymenoptera	Eulophidae	Larva of *Lamprosema indicata,* Female nymph and adult of *Rastrococcus (Phenacoccus) iceryoides Green,* *Lamprosema indicata* (Fab.), Egg parasite *Oocassida pudibunda* Boh.
544.	*Thelaira macropus* Wield.	Diptera	Tachinidae	Larva of *Mocis undata*
545.	*Utetes anastrephae*	Hymenoptera	Braconidae	Chickpea
546.	*Vespa orientalis* Linnaeus (Wasp)	Hymenoptera	Vespidae	Coleopterous, Dipterous and Lepidopterous, insects
547.	*Xanthopimpla predator* (Fab.)	Hymenoptera	Ichneumonidae	Larva of *Mocis undata*
548.	*Xanthopimpla* sp.	Hymenoptera	Ichneumonidae	Lepidopterous insects
549.	*Yellicones* sp.	Hymenoptera	Braconidae	Light trap catches
550.	*Zele chlorophthalama*	Hymenoptera	Braconidae	Adults/Nymphs of *Myzus persicae,* Larvae *Spodoptera littoralis and S. litura*
551.	*Zelomorpha* sp.	Hymenoptera	Braconidae	Light trap koinobiont endoparasitoids of late instar larvae of noctuid and arctiid moths
VIII	**Hyperparasites**			
552.	*Brachymeria sp.*	Hymenoptera	Chalcididae	Hyper parasite of *Bracon hebetor Say* Larval parasite of *Dichocrosis punctiferalis*
553.	*Eurytoma sp.*	Hymenoptera	Eurytomidae	Hyper parasite of *Bracon hebetor Say, Larval parasite of Dichocrosis punctiferalis*
554.	*Metaphycus* sp.	Hymenoptera	Encyrtidae	Hyper parasite of *Brachymeria* sp.
555.	*Pediobus parvulus* Ferr.	Hymenoptera	Eulophidae	Hyper parasite of *Brachymeria* sp. pupa

Dr. Govind Amarchand Gangarade
Served
8.6.1946 to 8.5.1981
First record of new parasite species reported by Dr. Gangrade
Euderus agromyzae Gangrade (Hymenoptera: Eulophidae)
Pre pupal parsite of *Melanagromyza obtusa*

Table 20.5 Other information related to insect pests reported form College of Agriculture, Jabalpur

	Scientific name	Common name	Order	Family	Host
	Diseases of insects				
556.	*Beauveria bassiana* (Bals.)	Entomopathogenic fungus	Hypocreales	Cordiycipitaceae	Soybean, beneficial microbial
557.	*Lecanicillium lecanii* (Zimm.)	White hallow fungus	Hypocereales	Cordiycipitaceae	Beneficial microbial
558.	*Metarhizium anisopilae* (Metsch.) Sorok	Green muscardine fungus	Hypocereales	Clavicipitaceae	Beneficial microbial
	Entomo-pathogenic nematode				
559.	*Mermis* sp.	Grasshopper nematode	Mermithida	Mermithidae	*Antigastra catalaunalis*
560.	*Strobiliomyia orbata* W.				*S.exigua*
	Weed as insect host				
561.	*Abutilon graveolens* Wight & Arnott	Mallow plant	Malvales	Malvaceae	*Oberea brevis* Swed.
562.	*Aeschynomene indica* (Linn.)	Curly indigo	Fabales	Fabaceae	*O. brevis* Swed., *H.armigera*
563.	*Alysicarpus* spp.	Moneyworts	Fabales	Fabaceae	*O. brevis* Swed.
564.	*Alysicarpus rugousus* (Wild.) DC	Red Moneywort	Fabales	Fabaceae	*O. brevis* Swed.
565.	*Alysicarpus tetragonolobus* Edgew	Red alyce clover	Fabales	Fabaceae	*O. brevis* Swed.
566.	*Desmodium gangetium* DC Pennel	Salparni	Fabales	Fabaceae	*O. brevis* Swed.
567.	*Corchorus acutaangulus* L.	Jute mallow	Malvales	Malvaceae	*O. brevis* Swed.
568.	*Indigofera glandulosa* J.C. Wendl	Asian indigo	Fabales	Fabaceae	*O. brevis* Swed.
569.	*Triumfetta rhombidea* Jacq.	Chinese bur	Malvales	Malvaceae	*O. brevis* Swed.
570.	*Xanthium strumarium* L.	Common cocklebur	Asterales	Asteraceae	*O. brevis* Swed.
	Insectivorous birds				

	Scientific name	Common name	Order	Family	Host
571.	*Dicrurus macrocercus* (Vieillot)	Black drongo	Passeriformes	Dicruridae	Ants, termites, locusts, crickets, beetles, bees, moths and butterflies infesting different crops especially found in Pigeonpea and chickpea

Compilation of the information on insect diversity revealed that 502 species of insects were recorded which represented 16 orders *viz.,* Coleoptera, Dermaptera, Dictyoptera, Diptera, Hemiptera, Hymenoptera, Isoptera, Lepidoptera, Mantodea, Mesostigmata, Neuroptera, Odonata, Orthoptera, Phasmatodea, Thysanoptera and Trichoptera and included 82, 2, 5, 37, 89, 82, 2,138,1,1,10, 14, 24, 1, 13 and 1 species, respectively (Fig. 20.3 and 20.4).

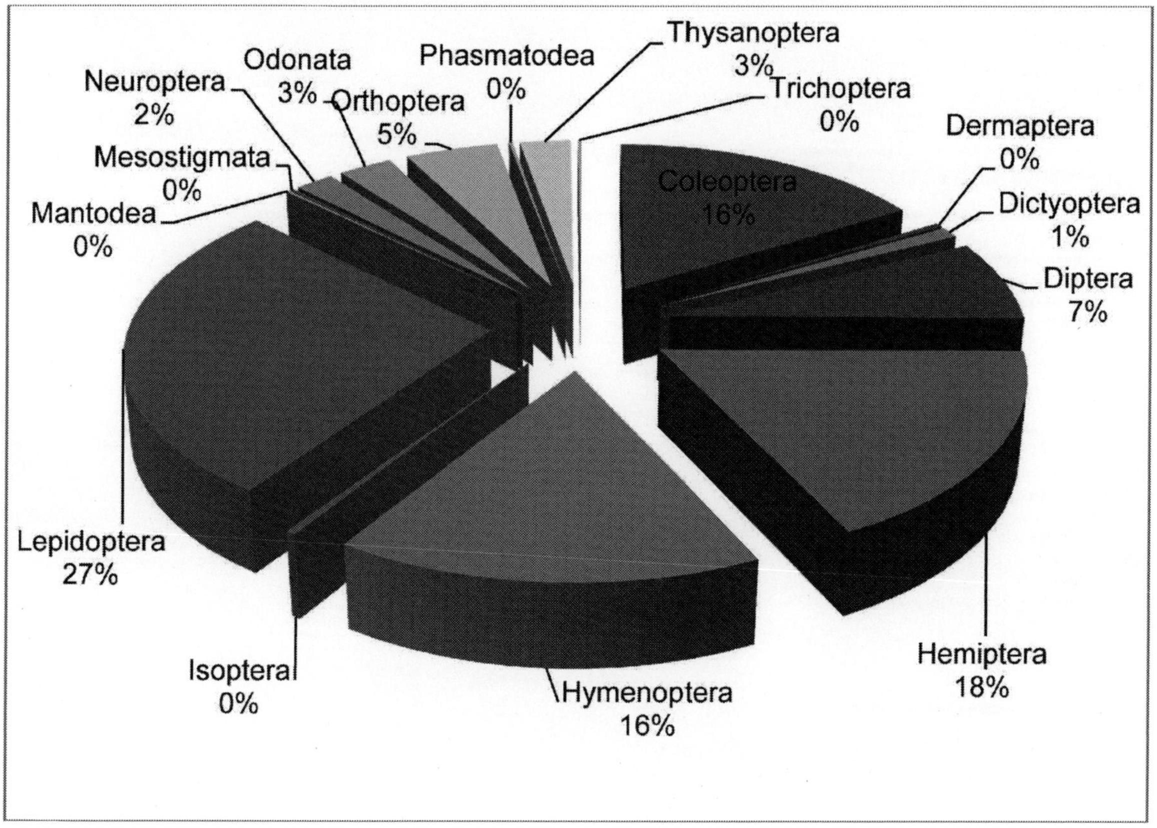

Fig. 20.3 Order wise distribution of insect diversity in JNKVV campus, Jabalpur

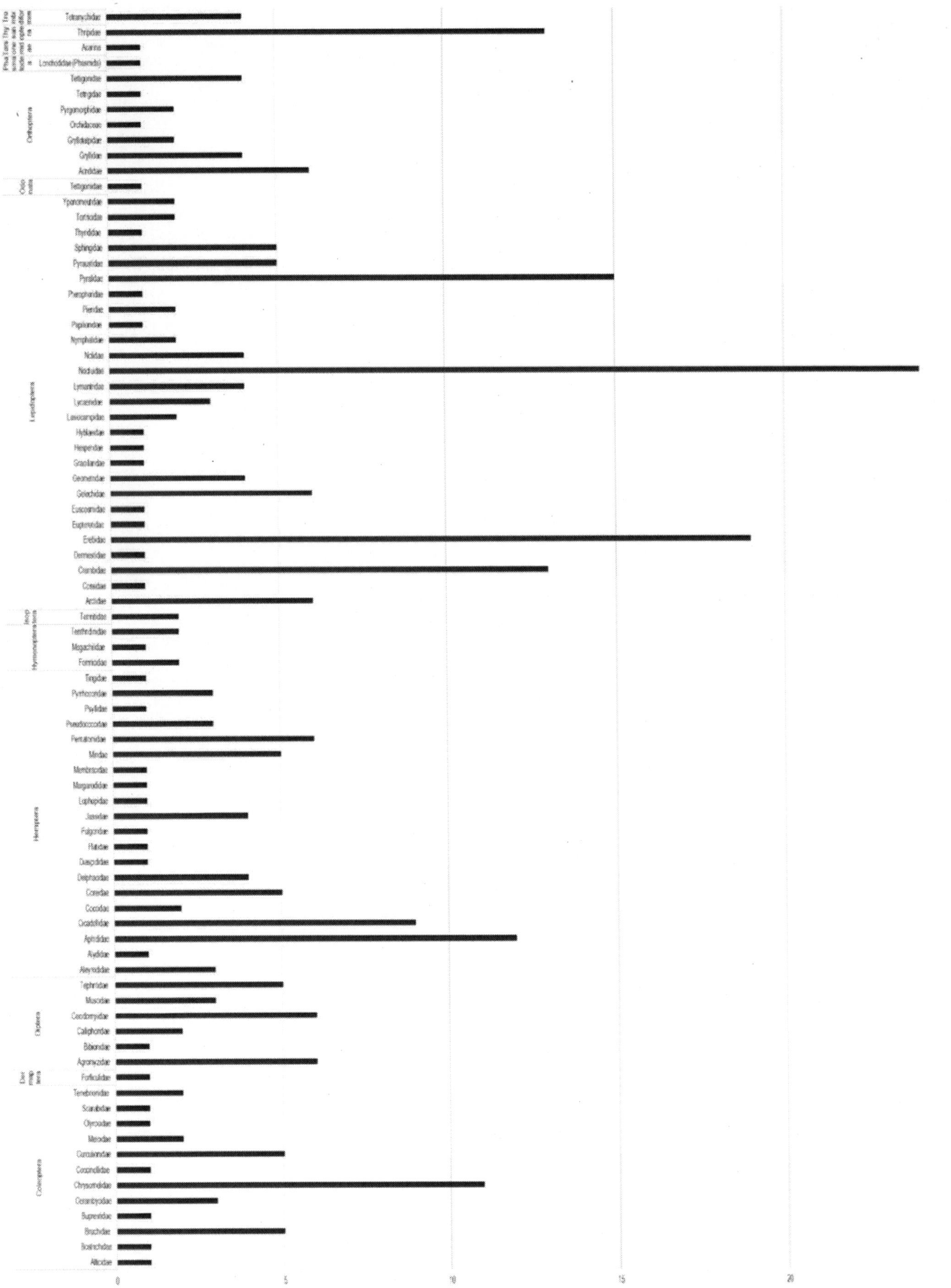

Fig. 20.4 Order and Family wise distribution of Insect diversity

Compilation of the information on insect pollinators revealed that 7 species of pollinators were recorded which represented 3 orders *viz.,* Hymenoptera, Lepidoptera and Diptera and 3 families *viz.,* Apidae, Sphingidae and Syrphidae and included 5,1 and 1 species, respectively (Table 20.3 and Fig. 20.5).

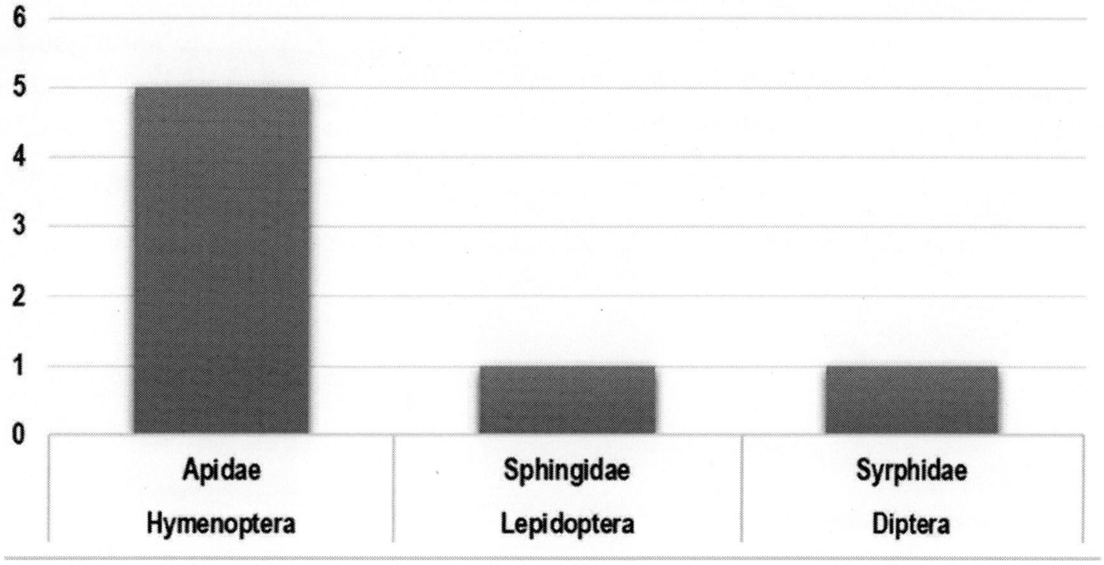

Fig. 20.5 Order and Family wise distribution of Insect pollinator diversity

Compilation of the information on insect predators revealed that 98 species were recorded which represented 13 orders (*viz.*, Coleoptera, Dermaptera, Dictyoptera, Diptera, Hemiptera, Hymenoptera, Lepidoptera, Mantodea, Mesostigmata, Neuroptera, Odonata, Orthoptera and Trichoptera) and 33 families (*viz.*, Aeshnidae, Ascalaphidae, Asilidae, Belostomatidae, Blastobasidae, Carabidae, Ceriogrionidae, Chrysopidae, Cicindellidae, Coccinellidae, Coenagrionidae, Dytisadae, Forficulidae, Formicidae, Hydrophilidae, Hydropsychidae, Hymenopodidae, Laelapidae, Libellulidae, Lygaeidae, Mantidae, Mantispidae, Noctuidae, Pentatomidae, Pyrrhocoridae, Reduvidae, Rhyothemisidae, Scarabaeidae, Scirtidae, Staphylinidae, Syrphidae, Tarachodidae and Tettigoniidae) (Table 20.3. and Fig. 20.6).

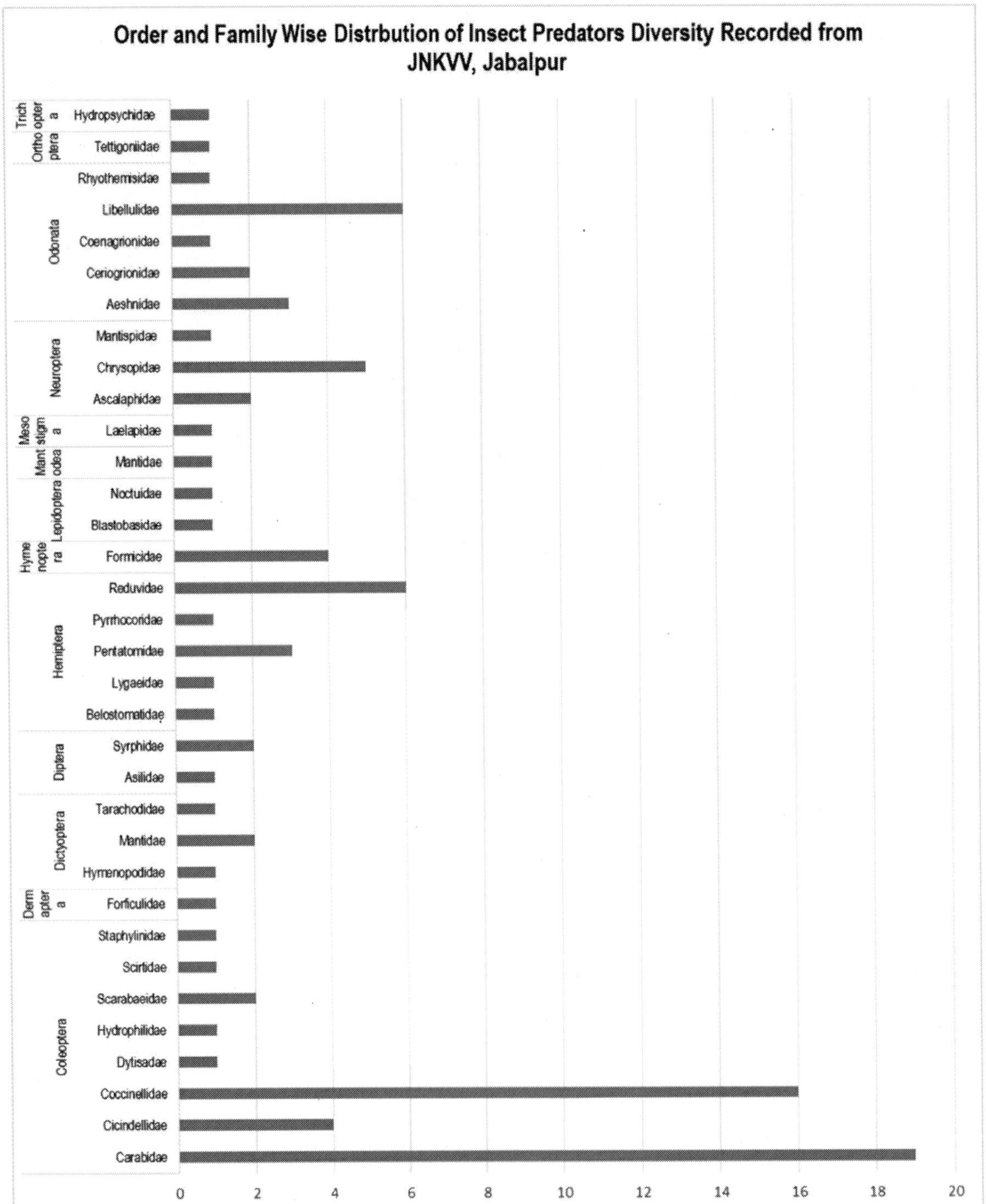

Fig. 20.6 Order and Family wise distribution of Insect Predators diversity

Compilation of the information on insect parasites revealed that 74 species were recorded which represented 5 orders (*viz.,* Diptera, Hemiptera, Hymenoptera, Lepidoptera and Neuroptera) and 20 families (*viz.*, Aphididae, Blastobasidae, Braconidae, Chalcididae, Chrysopidae, Elasmidae, Elasmidae, Encyrtidae, Eucoilidae, Eulophidae, Eupelmidae, Eurytomidae, Formicidae, Hormiini, Ichneumonidae, Pentatomidae, Pteromalidae, Selionidae, Tachinidae and Vespidae) (Table 20.4 and Fig. 20.7).

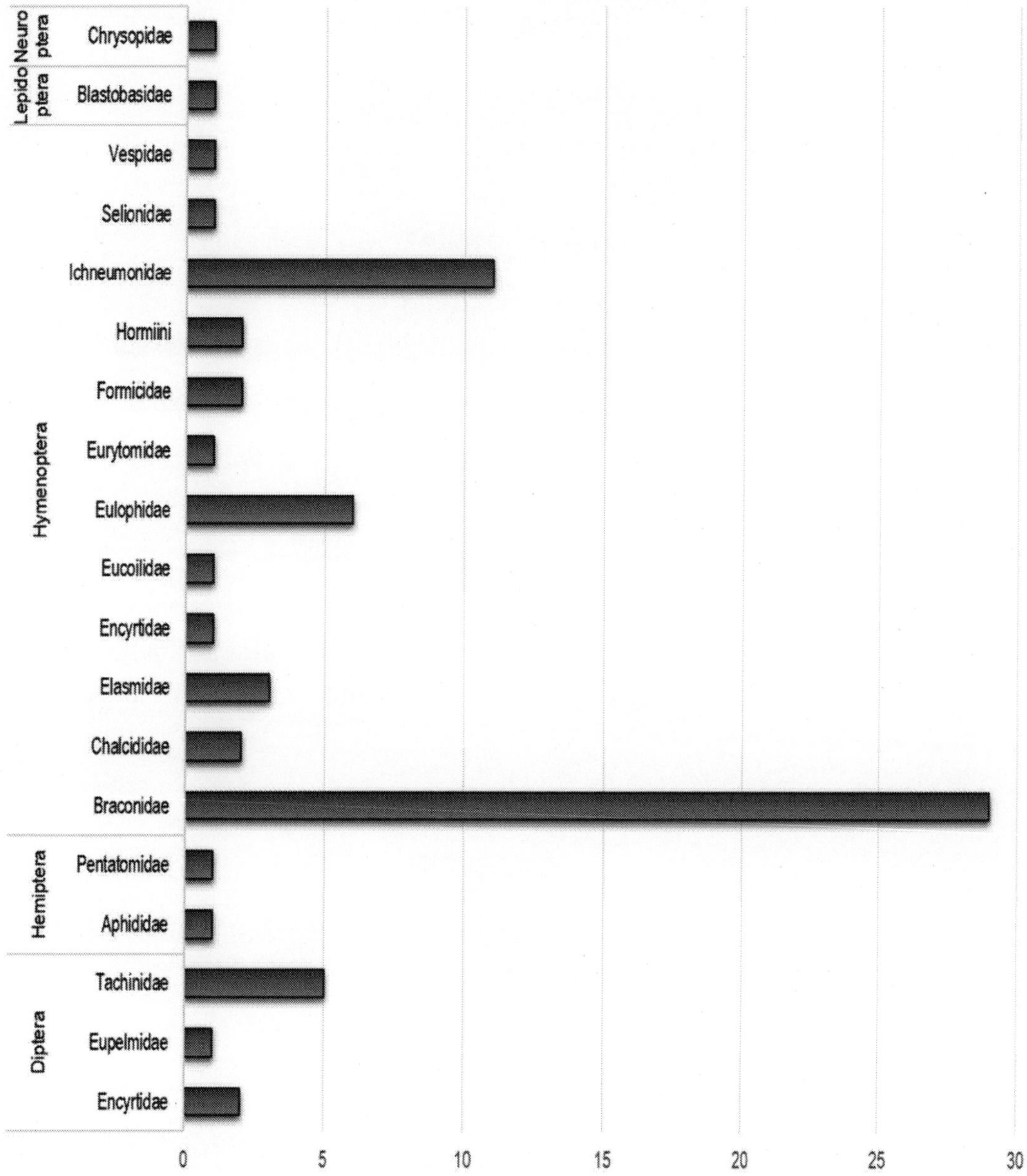

Fig 20.7 Order and Family wise distribution of Insect Parasites diversity

Compilation of the information on predatory spiders revealed that 42 species of spiders were recorded which represented 12 families *viz.,* Araneidae, Clubionidae, Dictynidae, Gnaphosidae, Lycosidae, Oxyopidae, Salticidae, Sparassidae, Tetragnathidae, Therididae, Thomisidae and Uloboridae and included 6,1,1,3,2,6,5,1,5,2,5 and 2 species, respectively (Table 20.3 and Fig. 20.8).

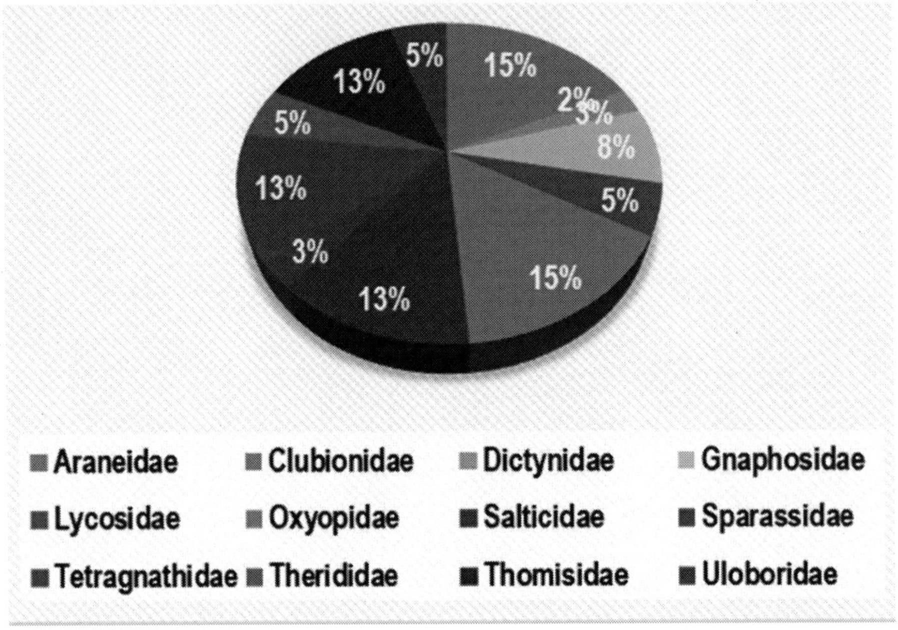

Fig. 20.8 Family wise distribution of Predatory Spider diversity

Compilation of the information weed diversity serving as insect hostrevealed that 10 species of weeds were recorded which represented 3 orders *viz.,* Malvales, Fabales and Aserales and 3 families *viz.,* Malvaceae, Fabaceae and Aseraceae and included 3, 6 and 1 species, respectively (Table 20.5 and Fig.20.9).

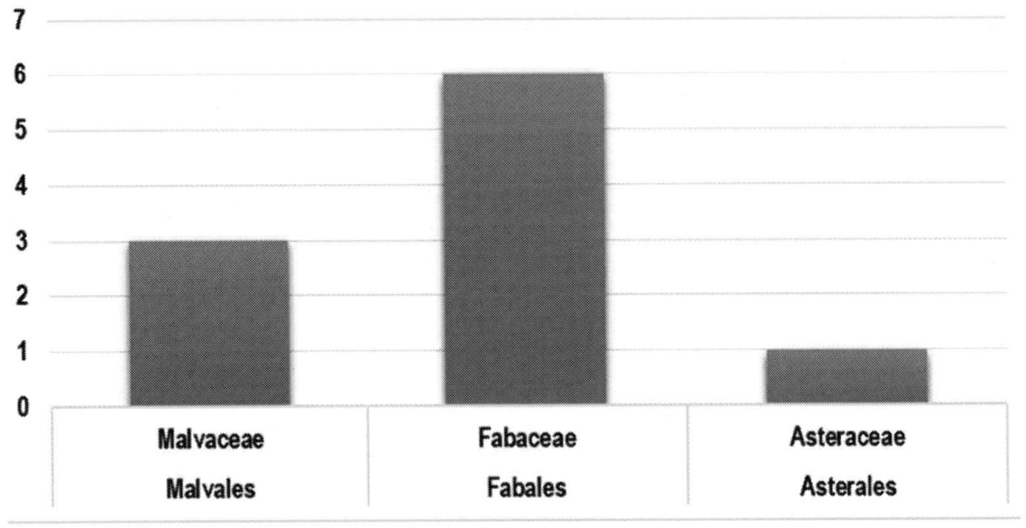

Fig. 20.9 Order and Family wise distribution of weed diversity serving as insect host

Invasive Insect Pests- First Report from JNKVV, Jabalpur

2017- American leaf miner, *Tuta absoluta* (Meyrick) (Lepidoptera: Gelechidae)

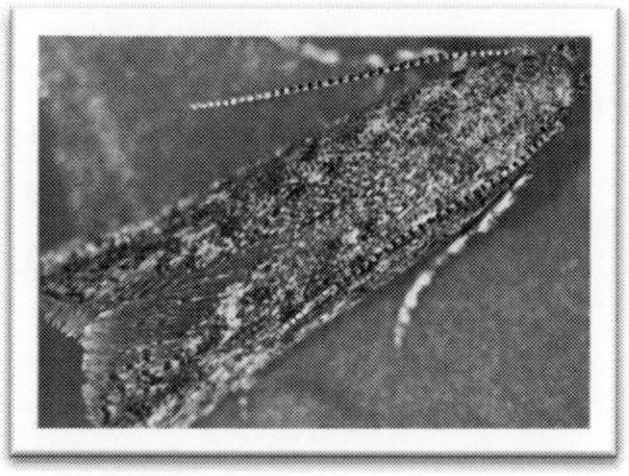

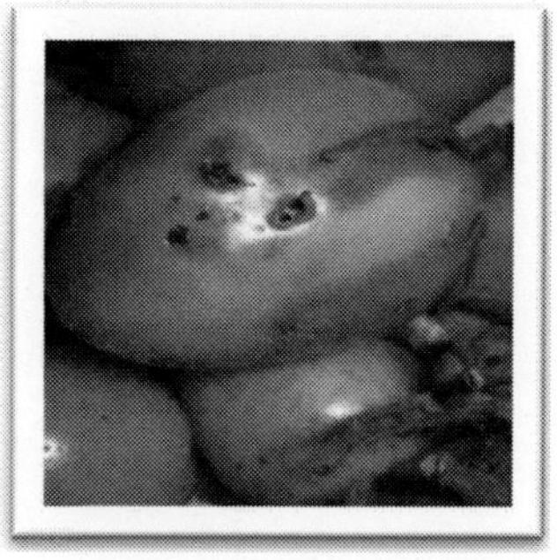

Courtesy: CABI

2019 -Fall Army Worm, *Spodoptera frugiperda* J.E.Smith (Lepidoptera: Noctuidae)

Plate 20.1 Invasive Insect Pests first reported from JNKVV, Jabalpur

Within the JNKVV, Jabalpur campus a complex interdependence of the various organisms was operational. Alike every ecosystem, the plants were the primary producers forming the first trophic level. The herbivorous arthropods constituted the second trophic level. All the natural enemies including the insectivorous birds were the tertiary consumers. Modern concept of ecology calls for the conservation of one and every living organism with the underlying principle that every organism has its own role in the ecosystem. Therefore, rather than complete eradication one must aim towards the sustainable management of the pest species below the economic threshold level. Diversity breeds diversity. Having a diverse array of living organisms allows other organisms to take advantage of the resources provided. If the pest insets are completely eliminated from the ecosystem; gradually, their natural enemies will also meet the same fate. Extinction or elimination of even a single species from the complex assemblage of organism may have serious repercussion on the normal functioning of the ecosystem and disrupt the ecological services. So, biodiversity-friendly practices must be sought after.

CHAPTER 21

Biodiversity Audit of Plant Pathogens and Bio-agents

Vijay Kumar Yadav, Jayant Bhatt, Subhash SS Malempati and Gitanshu Dinkwar
Department of Plant Pathology
College of Agriculture, Jabalpur

So far, 530 microorganism viz., 484 fungi, 18 bacteria, 19 viruses, 8 nematode and one phytoplasma causing diseases on plants are reported from College of Agriculture, Jabalpur apart from 19 mushrooms (Fig. 21.1).

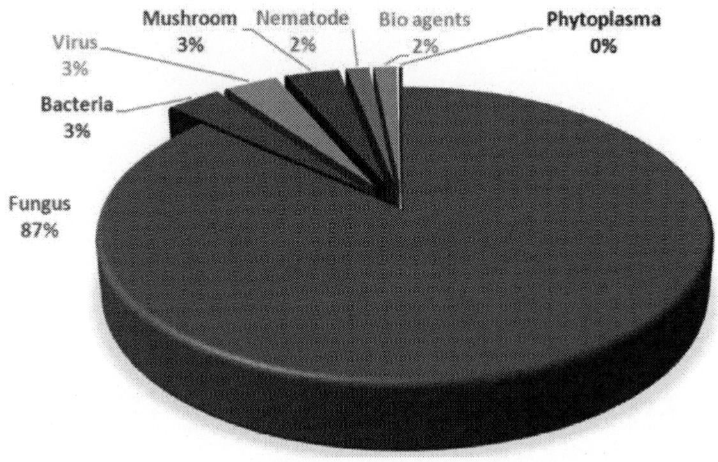

Fig 21.1 Plant pathogenic microorganisms reported from JNKVV campus, Jabalpur

Diseases caused by Fungi

Out of total reported 530 pathogenic microorganism 87 percent are fungi (Fig 1). Out of total 484 fungi studied at College of Agriculture, Jabalpur 101 are on field crops, 85 on post-harvest seed, 77 on forest trees, 43 on weeds, 39 on vegetables, 37 on medicinal plants, 33 on fruit crops, 27 on post-harvest fruits and 25 on ornamental species (Fig. 21.2).

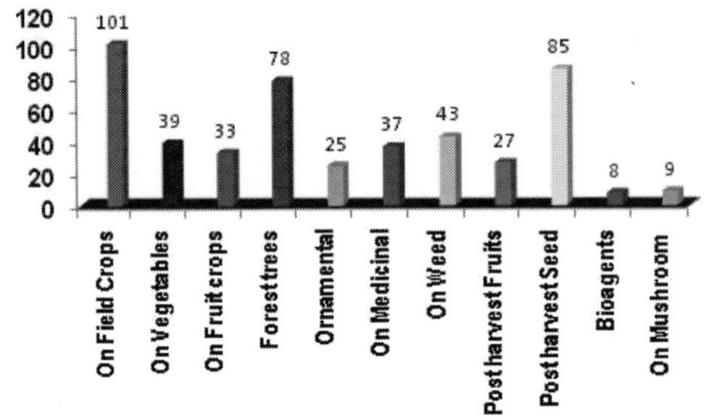

Fig 21.2 Frequency of reported fungal species on various hosts

For fungal diseases, 198 plant species had been studied at the College of Agriculture, Jabalpur that includes 25 field crops; 25 vegetables, 18 fruits, 18 ornamentals, 30 medicinal, 42 forest trees and 40 weeds (Fig. 21.3).

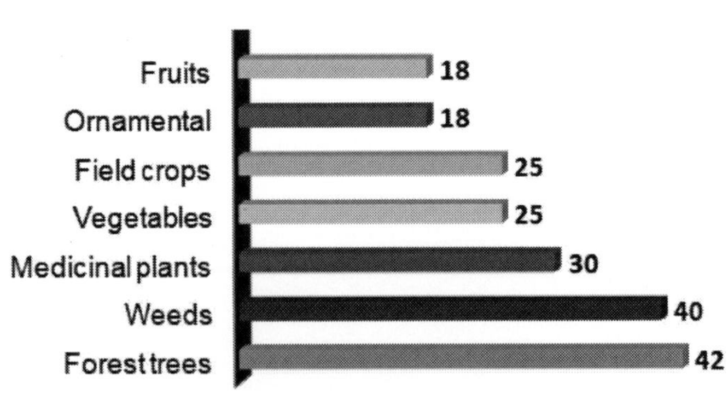

Fig 21.3 Segment wise plant species observed for fungal diseases

The microbial diversity at College of Agriculture, Jabalpur shows prevalence of total 484 fungal species which are associated with field, vegetables, fruit crops, forest trees, ornamentals, medicinal plants, weeds, post-harvest fruits, post-harvest seed, bio-agents and mushrooms respectively. Among the fungal diseases reported from the college, major share was foliar diseases and commonly associated pathogens with this disease were *Alternaria, Colletotrichum* and *Cercospora* they were dominant on the field and vegetables crops, apart from these pathogens *Pestalotiopsis* was also identified on fruit and forest tree diseases (Table 21.1).

Table 21.1 Plant pathogenic fungi reported from College of Agriculture, Jabalpur

Sr	Associated Microorganism	Name of crop	Associated disease
I	**On Field Crops**		
1.	*Alternaria alternata*	Chickpea, soybean, sunflower	Alternaria blight, leaf spot
2.	*Alternaria brassicae*	Black mustard	Alternaria blight
3.	*Alternaria carthami*	Safflower	Leaf spot
4.	*Alternaria sesami*	Sesame	Leaf blight
5.	*Alternaria tenuis*	Wheat	Leaf blight
6.	*Alternaria tenuissima*	Wheat	Leaf spot
7.	*Alternaria* sp.	Soybean	Alternaria blight
8.	*Aristastoma guttulosum*	Soybean	Diffuse yellowing
9.	*Ascochyta rabiei*	Chickpea	Blight
10.	*Ascochyta sorghi*	Sorghum	Rough leaf spot
11.	*Botrytis cinerea*	Chickpea	Botrytis grey mould
12.	*Cercospora arachidicola*	Groundnut	Leaf spot
13.	*Cercospora canescens*	Pigeonpea	Leaf spot
14.	*Cercospora gossypina*	Cotton	Leaf spot
15.	*Cercospora indica*	Pigeonpea	Leaf spot
16.	*Cercospora kikuchii*	Soybean	Leaf blight
17.	*Cercospora lentis*	Lentil	Leaf spot
18.	*Cercospora personata*	Groundnut	Leaf spot
19.	*Cercospora pisi sativae* f. sp. *lathyri-sativae*	Grass pea	Cercospora blight
20.	*Cercospora sesami*	Sesame	Leaf spot
21.	*Cercospora sojina*	Soybean	Leaf spot
22.	*Cercospora sorghi*	Sorghum	Leaf spot
23.	*Cercospora canescens*	Urid and mung bean	Leaf spot
24.	*Cercospora zeae-maydis*	Maize	Grey leaf spot
25.	*Cercosporella* sp.	Pearl millet	Leaf spot
26.	*Cochliobolus carbonum*	Maize	Leaf spots
27.	*Colletotrichum dematium*	Chickpea, soybean	Blight, pod blight, seed rot
28.	*Colletotrichum falcatum*	Sugarcane	Red rot
29.	*Colletotrichum graminicola*	Sorghum	Anthracnose
30.	*Colletotrichum graminicolum*	Sorghum	Anthracnose
31.	Colletotrichum lindemuthianum	Urid and mung bean	Anthracnose
32.	*Colletotrichum sublineolum*	Sorghum	Anthracnose
33.	*Colletotrichum truncatum*	Soybean	Anthracnose
34.	Colletotrichum graminicola	Maize	Anthracnose
35.	*Corynespora cassiicola*	Sesame, soybean	Root rot, target leaf spot
36.	*Curvularia geniculata*	Wheat	Leaf spot
37.	*Curvularia lunata*	Rice, sesame, wheat	Seedling rot, leaf spot, leaf blight

Sr	Associated Microorganism	Name of crop	Associated disease
38.	*Curvularia verruculosa*	Wheat	Leaf spot
39.	*Cylindrocladium scoparium*	Groundnut	Blight
40.	*Diaporthe longicolla*	Soybean	Stem canker
41.	*Diaporthe phaseolorum* var. *caulivora*	Soybean	Stem canker
42.	*Drechslera glycines*	Soybean	Drechslera blight
43.	*Entyloma oryzae*	Rice	Leaf smut
44.	*Erratomyces patelii*	Urid	Leaf spot
45.	*Erysiphe cichoracearum*	Sesame	Powdery mildew
46.	*Erysiphe glycines*	Soybean	Powdery mildew
47.	*Erysiphe polygoni*	Urid and mung bean	Powdery mildew
48.	*Fusarium moniliforme*	Rice, sesame	Foolish seedling disease, Vascular wilt
49.	*Fusarium oxysporum*	Urid bean	Wilt
50.	*Fusarium oxysporum* f. sp. *glycines*	Soybean	Wilt
51.	*Fusarium oxysporum*f. sp. *lini*	Linseed	Wilt
52.	*Fusarium oxysporum* f. sp. *sesami*	Sesame	Wilt
53.	*Fusarium oxysporum* f. sp. *udum*	Pigeonpea	Wilt
54.	*Fusarium oxysporum* f. sp. *vasinfectum*	Cotton	Wilt
55.	*Fusarium oxysporum* f.sp. *ciceris*	Chickpea	Wilt
56.	*Fusarium oxysporum* f.sp. *tracheiphilum*	Sesame	Wilt
57.	*Fusarium solani* f. sp. *glycines*	Soybean	Sudden death, root rot, seed and seedling rot
58.	*Fusarium oxysporum* f. sp. *lentis*	Lentil	Fusarium wilt
59.	*Helminthosporium oryzae*	Rice	Brown spot
60.	*Helminthosporium rostratum*	Wheat	Spot blotch
61.	*Helminthosporium sativum*	Wheat	Spot blotch
62.	*Helminthosporium turcicum*	Maize	Leaf spot
63.	*Leveillula taurica*	Chickpea	Powdery mildew
64.	*Macrophomina phaseolina*	Sesame, soybean, Pigeonpea	Root rot or stem rot or charcoal rot
65.	*Myrothecium roridum*	Soybean	Leaf blight
66.	*Neovossia horrida*	Rice	Kernel smut
67.	*Oidiopsis taurica*	Pigeonpea	Powdery mildew
68.	*Operculella padwickii*	Chickpea	Foot rot
69.	*Phaeodactylum uniseptata*	Soybean	-
70.	*Phakopsora pachyrhizi*	Urid bean, soybean	Rust
71.	*Phoma medicaginis*	Soybean	Leaf spot
72.	*Phoma* sp.	Wheat	Glume blotch
73.	*Phomopsis* sp.	Groundnut	Leaf spot
74.	*Phytophthora nicotianae* var. *parasitica*	French bean	Phytophthora blight
75.	*Phytophthora parasitica* var. *sesami*	Sesame	Stem blight
76.	*Phytophthora vignae*	Cowpea	Stem rot
77.	*Phytophthora drechsleri* f.sp. *cajani*	Pigeonpea	Phytophthora blight
78.	*Plasmopara halstedii*	Sunflower	Downy mildew
79.	*Puccinia arachidis*	Groundnut	Rust
80.	*Puccinia calcitrapae* var. *centaureae*	Safflower	Rust
81.	*Puccinia sorghi*	Maize	Rust

Sr	Associated Microorganism	Name of crop	Associated disease
82.	*Pyricularia oryzae*	Rice	Blast
83.	*Rhizoctonia bataticola*	Urid bean	Root rot
84.	*Rhizoctonia bataticola*	Lentil, linseed	Dry root rot
85.	*Rhizoctonia solani*	Urid bean, Mung, Maize, soybean, rice	Web blight, cob rot, Rhizoctonia aerial blight, seed and seedling rot, sheath blight
86.	*Rhizopus* sp.	Sunflower	Rhizopus rot
87.	*Sclerospora graminicola*	Maize	Downy mildew
88.	*Sclerotinia sclerotiorum*	Chickpea	Sclerotinia stem rot
89.	*Sclerotium rolfsii*	Urid, chickpea, lentil, linseed, soybean, wheat	Collar rot
90.	*Sporisorium cruentum*	Sorghum	Loose smut
91.	*Sporisorium sorghi (=Sphacelotheca sorghi)*	Sorghum	Grain smut
92.	*Sporisorium reilianum (=Sphacelotheca reiliana)*	Sorghum	Head smut
93.	*Stemphylium sarciniformae*	Chickpea	Leaf spot
94.	*Tilletia barclayana*	Rice	Paddy bunt
95.	*Tolyposporium penicillariae*	Pearl millet	Smut
96.	*Uromyces fabae*	Lentil	Rust
97.	*Uromyces setariae-italicae*	Foxtail millet	Rust
98.	*Ustilaginoidea virens*	Rice	False smut
99.	*Ustilago hordei*	Barley	Covered smut
100.	*Ustilago scitaminea*	Sugarcane	Whip smut
101.	*Ustilago tritici*	Wheat	Loose smut
II	**On Vegetable Crops**		
102.	*Alternaria brassicicola*	Cruciferous vegetables	Alternaria leaf spot
103.	*Alternaria cassiae*	Cowpea	Leaf blight
104.	*Alternaria solani*	Potato, tomato	Early blight
105.	*Alternaria* sp.	Pea	Leaf spot
106.	*Alternaria tenuis*	Tomato	Alternaria blotch
107.	*Alternaria alternata*	Brinjal	Alternaria leaf spot
108.	*Cercospora abelmoschi*	Okra	Leaf spot
109.	*Cercospora cruenta*	Beans	Leaf spot
110.	*Cladosporium* sp.	Potato	-
111.	*Colletotrichum capsici*	Chilli	Anthracnose
112.	*Colletotrichum lindemuthianum*	Beans	Anthracnose
113.	*Colletotrichum* sp.	Pumpkin	Anthracnose
114.	*Dendryphiella vinosa*	Cucurbitaceous stem	Saprobe fungus antagonistic
115.	*Erysiphe cichoracearum*	Okra	Powdery mildew
116.	*Erysiphe pisi*	Pea	Powdery mildew
117.	*Fusarium oxysporum f.*sp. *lycopersici*	Tomato	Wilt
118.	*Fusarium oxysporum* f. sp. *melongenae*	Brinjal	Wilt
119.	*Fusarium oxysporum* f. sp. *pisi*	Pea	Wilt
120.	*Peronospora pisi*	Pea	Downy mildew
121.	*Phoma exigua*	Brinjal	Phoma fruit rot
122.	*Phoma* sp.	Chilli	Sunken leaf spot
123.	*Phomopsis vexans*	Brinjal	Phomopsis blight and Fruit rot
124.	*Phyllosticta ipomoea*	Sweet potato	Leaf spot

Sr	Associated Microorganism	Name of crop	Associated disease
125.	*Phytophthora capsici*	Okra	Damping-off
126.	*Phytophthora capsici*	Bottle gourd, brinjal, ridge gourd, sponge gourd, ivy gourd	Phytophthora fruit rot
127.	*Phytophthora cinnamomi*	Pointed gourd	Fruit and vine rot
128.	*Phytophthora citrophthora*	Chilli	Blight and fruit rot
129.	*Phytophthora infestans*	Potato, tomato	Late blight
130.	*Phytophthora nicotianae* var. *nicotianae*	Bell pepper, chilli	Fruit rot and leaf blight
131.	*Phytophthora nicotianae* var. *parasitica*	Lablab bean, tomato	Stem rot or collar rot, Late blight
132.	*Phytophthora parasitica*	Tomato	Buck eye rot
133.	*Phytophthora colocasiae*	Colocasia	Phytophthora blight
134.	*Phytophthora boehmeriae*	Chilli	Fruit rot, blight
135.	*Pythium aphanidermatum*	Tomato, brinjal	Damping off, fruit rot
136.	*Sclerotium rolfsii*	Brinjal, potato, tomato	Collar rot, root rot
137.	*Septoria leguminum*	Bean	Leaf spot
138.	*Septoria lycopersici*	Tomato	Leaf spot
139.	*Stemphylium* sp.	Onion	Tip blight
140.	*Torula herbarum*	Kidney bean	Saprobe on twigs and leaves
III	**On Fruits**		
141.	*Alternaria* sp.	Banana, cashew nut	Leaf blight / spot
142.	*Bartalinia robillardoides*	Citrus, karonda	Leaf spot
143.	*Botryodiplodia* sp.	Papaya	Stem end rot
144.	*Colletotrichum gloeosporioides*	Guava, papaya	Anthracnose
145.	*Colletotrichum papayae*	Papaya	New wilt
146.	*Colletotrichum* sp.	Pomegranate	Anthracnose
147.	*Curvularia lunata*	Pomegranate	Leaf spot
148.	*Curvularia* sp.	Guava	Leaf spot
149.	*Erysiphe alphitoides*	Bael	Powdery mildew
150.	*Fusarium lateritium*	Drumstick tree	Nut gray necrosis
151.	*Fusarium moniliforme*	Banana	Crown rot
152.	*Fusarium oxysporum* f. sp. *psidii*	Guava	Wilt
153.	*Gliocladium* sp.	Pomegranate	Anthracnose
154.	*Gloeosporium artocarpi*	Jack fruit	Leaf spot
155.	*Gloeosporium* sp.	Karonda	Anthracnose
156.	*Helminthosporium* sp.	Pomegranate	Leaf spot
157.	*Macrophoma musae*	Banana	Freckle disease
158.	*Monochaetia jabalpurensis*	Mandarin lime	-
159.	*Pestalotia mangiferae*	Mango	Brown spot
160.	*Pestalotia* sp.	Fig, karonda	Leaf spot
161.	*Pestalotia versicolor*	Mango	Decline
162.	*Pestalotiopsis japonica*	Fig	Leaf spot
163.	*Pestalotiopsis mangiferae*	Mango	Leaf spot
164.	*Pestalotiopsis psidii*	Guava	Scabby fruit canker
165.	*Phyllosticta* sp.	Cashew nut, fig	Leaf spot
166.	*Phyllosticta sulata*	Papaya	Leaf spot
167.	*Phyllostictina artocarpina*	Jackfruit	Leaf spot
168.	*Phytophthora nicotianae* var. *parasitica*	Citron, guava, papaya, Pomegranate	Phytophthora root rot, gummosis and fruit rot
169.	*Pythium aphanidermatum*	Guava, papaya	Root and stem rot

Sr	Associated Microorganism	Name of crop	Associated disease
170.	*Rhizoctonia bataticola*	Guava	Blight
171.	*Rhizoctonia solani*	Guava	Damping off
172.	*Rhizopus* sp.	Guava	Fruit rot
173.	*Verticillium* sp.	Banana	Wilt, Cigar end rot
IV	**On Forest trees:**		

Sr	Microorganism	Common Name	Scientific name	Disease
174.	*Alternaria citri*	Australian black wood	*Acacia melaoxylon*	Leaf spot
175.	*Alternaria gomphrenae*	Climbing aroid	*Pothos scandens*	Leaf spot
176.	*Alternaria infectoria*	Eucalyptus	*Eucalyptus globulus*	Leaf spot
177.	*Alternaria* sp.	Australian black wood Camel foot tree	*Acacia melaoxylon,* *Bauhinia variegata*	Leaf spot
178.	*Alysidium resinae*	Cassia	*Cassia* sp.	Bud blight
179.	*Arthrinium phaeospermum*	Clumping bamboos	*Bambusa* sp.	Bamboo blight
180.	*Ascochyta oleandri*	Axlewood	*Anogeissus latifolia*	Leaf blight
181.	*Ascochyta* sp.	Sandan Fire flame bush	*Ougeinia dalbergioides,* *Woodfordia fructicosa*	Ascochyta blight
182.	*Bartalinia robillardoides*	Sweet acacia Natal plum	*Acacia farnesiana,* *Carissa macrocarpa*	Leaf spot
183.	*Cephaliophora irregularis*	Hiptage	*Hiptage madablata,* *Hiptage benghalensis*	-
184.	*Cercoseptoria balsaminae*	Axlewood	*Anogeissus latifolia*	Leaf spot
185.	*Cercospora justiciidola*	Axlewood	*Anogeissus latifolia*	Leaf spot
186.	*Cercospora neriella*	Crape myrte, Axlewood	*Lagerstroemia parviflora,* *Anogeissus latifolia*	Leaf spot
187.	*Cercospora paramignyae*	Crape myrtle Camel foot tree	*Lagerstroemia parviflora,* *Bauhinia variegata*	Leaf spot
188.	*Cercospora schreberae*	Ceylon tea, Camel foot tree	*Elaeodendron glaucum,* *Bauhinia variegata*	Leaf spot
189.	*Cercospora zinniae*	Golden shower Axlewood	*Cassia fistula,* *Anogeissus latifolia*	Leaf spot
190.	*Chaconia tectonae*	Golden shower Teak	*Cassia fistula,* *Tectona grandis*	Rust
191.	*Ciliochorella mangiferae*	Hiptage	*Hiptage benghaleniss*	-
192.	*Clasterosporium mori*	Mulberry	*Morus alba*	-
193.	*Colletotrichum cylindricum*	Climbing aroid	*Pothos scandens*	Anthracnose
194.	*Colletotrichum gloeosporioides*	Kapok	*Ceiba pentandra*	Die-back
195.	*Colletotrichum pithecolobii*	Manila tamarind	*Pithecolobium dulce*	Anthracnose
196.	*Coniothyrium fuckelii*	Whitebark senna	*Cassia spectabilis*	Stem and graft canker
197.	*Corynespora cassiicola*	Kapok, Hiptage	*Ceiba pentandra,* *Hiptage benghalensis*	Leaf spot/blight
198.	*Cytospora* sp.	Hiptage	*Hiptage benghaleniss*	-
199.	*Dasturella divina*	Calcutta bamboo, Eucalyptus	*Dendrocalamus strictus,* *Eucalyptus globulus*	Rust
200.	*Dendryphiella vinosa*	Malabar embelia	*Embelia robusta*	-
201.	*Dictyoarthrinium sacchari*	Royal poinciana	*Delonix regia*	-
202.	*Dictyosporium intermedium*	Eucalyptus	*Eucalyptus globulus*	-
203.	*Dinemasporium strigosum*	Clumping bamboos	*Bambusa* sp.	On stem

Sr	Microorganism	Common Name	Scientific name	Disease
204.	*Dinemasporium graminum*	Orchid tree	*Bauhinia variegata*	-
205.	*Diplodia sarmentorum*	Siris	*Albizzia lebbeck*	On twigs
206.	*Diplodia* sp.	Axlewood	*Anogeissus latifolia*	Leaf blight
207.	*Diplodia zizyphina*	Jackal jujube	*Ziziphus oenoplia*	On twigs
208.	*Endophragmiella angustispora*	Eucalyptus	*Eucalyptus globulus*	On Bark
209.	*Excipularia narsapurensis*	Sausage tree	*Kigelia pinnata*	On Bark
210.	*Haplosporella beaumontiana*	Siris	*Albizia lebbeck*	-
211.	*Helicosporium elinorage*	Eucalyptus	*Eucalyptus globulus*	Saprobic
212.	*Helminthosporium erythrinae*	Teak, Ceylon tea	*Tectona grandis,* *Erythrina indica*	Leaf blight
213.	*Hermatomyces sphaericum (=Stemphylium sphaericum)*	Eucalyptus	*Eucalyptus globulus*	-
214.	*Microdiplodia purpusila*	Pride of India	*Lagerstroemia flosreginae*	-
215.	*Microsphaeropsis olivacea*	Pride of India	*Lagerstroemia flosreginae*	Endophyte
216.	*Monochaetia jabalpurensis*	Axle wood	*Anogeissus latifolia*	Leaf spot
217.	*Monodictys* sp.	-	*Undetermined wood*	-
218.	*Mycosphaerella populi*	Lombardy poplar	*Populus nigra*	Septoria Leaf spot
219.	*Niesslia exilis*	Pride of India	*Lagerstroemia flosreginae*	Dead leaves
220.	*Oidium* sp.	Squirrel's claws	*Caesalpinia crista*	Powdery mildew
221.	*Paathramaya sundara*	Eucalyptus	*Eucalyptus globulus*	
222.	*Pestalotia palmarum*	Wild date palm	*Phoenix sylvestris*	Leaf spot Petiole blight
223.	*Pestalotia versicolor*	Axlewood	*Anogeissus latifolia*	Leaf spot
224.	*Pestalotiopsis japonica*	Cluster fig	*Ficus glomerata*	Leaf spot
225.	*Pestalotiopsis* sp.	Ceylon tree, Sandan	*Elaeodendron glaucum* *Ougeinia dalbergioides*	Leaf spot
226.	*Pestalotiopsis versicolor*	Eucalyptus	*Eucalyptus globulus*	Leaf spot
227.	*Pestalotiopsis woodfordeae*	Fire flame bush	*Woodfordia fruticosa*	Leaf spot
228.	*Phaeoisaria clematidis*	Royal poinciana	*Delonix regia*	Saprobic on submerged wood
229.	*Phakopsora grewiae*	Falsa	*Grewia asiatica*	Rust
230.	*Phloeospora cassiae*	Golden shower	*Cassia fistula*	Dead leaves
231.	*Phoma eucalyptica*	Eucalyptus	*Eucalyptus globulus*	Blight
232.	*Phoma eucalypticola*	Eucalyptus	*Eucalyptus globulus*	Stem canker
233.	*Phoma minutella*	Siris	*Albizia lebbeck*	Blight
234.	*Phyllachora balansae*	Toona	*Cedrela toona*	Living leaves
235.	*Phyllosticta* sp.	Climbing aroid	*Pothos scandens*	Leaf spot
236.	*Phyllosticta grevilleae*	Southern silky oak	*Grevillea robusta*	Leaf spot
237.	*Phyllosticta sesbaniae*	Indian coral tree	*Erythrina indica*	Leaf spot
238.	*Phyllosticta* sp.	Agati, Fig	*Sesbania grandiflora,* *Ficus glomerata*	Leaf spot
239.	*Phytophthora capsici*	Tree of heaven	*Ailanthus excelsa*	Phytophthora leaf blight
240.	*Phytophthora nicotianae* var. *parasitica*	Sacred fig	*Ficus religiosa*	-
241.	*Pithomyces ellisii*	Eucalyptus	*Eucalyptus globulus*	-

Sr	Microorganism	Common Name	Scientific name	Disease
242.	*Pseudopetrakia kambakkamensis*	Eucalyptus	*Eucalyptus globulus*	with Dead leaves
243.	*Sirosporium ziziphicola*	Indian jujube	*Ziziphus jujuba*	-
244.	*Spegazzinia tessarthra*	Clumping bamboos	*Bambusa sp.*	Leaf blight
245.	*Sphaerophragmium acaciae*	Siris	*Albizia lebbeck*	Rust
246.	*Stemphylium* sp.	Fig, Axlewood	*Ficus glomerata, Anogeissus latifolia*	Leaf blight
247.	*Torula herbarum*	Indian jujube	*Ziziphus jujuba*	Saprobe on twigs and leaves
248.	*Tretopileus sphaerophorus*	Teak	*Tectona grandis*	Endophytic fungi
249.	*Triposporium myrti*	Fire flame bush	*Woodfordia fruticosa*	-
250.	*Zygosporium oscheoides*	Hiptage	*Hiptage benghalensis*	-
V	**On Ornamental Plants**			
251.	*Alternaria gomphrenae*	Globe amaranth	*Gomphrena globosa*	-
252.	*Alternaria* sp.	Night-flowering Jasmine	*Nyctanthes arbor-tristis*	Leaf spot
253.	*Ascochyta oleandri*	Garden balsam, Nerium	*Impatiens balsamina, Nerium oleander*	Ascochyta blight
254.	*Cercospora balsamina*	Garden balsam	*Impatiens balsamina*	Leaf spot
255.	*Cercospora neriella*	Nerium	*Nerium oleander*	Leaf spot
256.	*Cercospora* sp.	Mexican aster	*Cosmos* sp.	Leaf spot
257.	*Cercospora zinniae*	Zinnia	*Zinnia pauciflora*	Leaf spot
258.	*Colletotrichum cylindricum*	Garden balsam	*Impatiens balsamina*	Anthracnose
259.	*Cytospora rosarum*	Rose	*Rosa indica*	Die back
260.	*Diplodia* sp.	Malbar jasmine	*Jasminum malabaricum*	Leaf spot
261.	*Haplosporella rosarum*	Rose	*Rosa indica*	-
262.	*Helminthosporium erythrinae*	Nerium	*Nerium oleander*	Leaf spot
263.	*Monochaetia jabalpurensis*	Nerium	*Nerium oleander*	Leaf spot
264.	*Myrothecium roridum*	Nerium	*Nerium oleander*	Leaf spot
265.	*Paathramaya sundara*	Henna tree	*Lawsonia alba*	-
266.	*Periconia byssoides*	Jasmine	*Nyctanthes arbor-tristis*	Leaf spot
267.	*Pestalotia suffocata*	Rose	*Rosa* sp.	Leaf spot
268.	*Pestalotiopsis japonica*	Jasmine	*Nyctanthes arbor-tristis*	Leaf spot
269.	*Pestalotiopsis woodfordeae*	Geranium	*Pelargonium zonale*	Leaf spot
270.	*Phytophthora citrophthora*	Pinwheel flower, Orange jasmine	*Tabernaemontana divaricata, Murraya exotica*	Leaf blight, Phytophthora root rot
271.	*Phytophthora nicotianae* var. *parasitica*	Paper flower, China rose, Dumb cane	*Bougainvillea spectabilis, Hibiscus rosa-sinensis, Dieffenbachia picta*	Root and stem rot
272.	*Pyrenochaeta* sp.	Horseshoe geranium	*Pelargonium zonale*	Pink root disease
273.	*Pyrenochaeta vexans*	Tiger plant	*Sanchezia nobilis*	Brown rot
274.	*Septoria chrysanthemella*	Chrysanthemum	*Chrysanthemum* sp.	Leaf spot
275.	*Uromyces hobsoni*	Malabar jasmine	*Jasminum malabaricum*	Rust
276.	*Uromyces orientalis*	Heart-leaf indigo	*Indigofera cordifolia*	Rust
VI	**On Medicinal plants**			
277.	*Alternaria alternata*	Chandrasur	*Lepidium sativum*	Leaf spot
278.	*Alternaria rephani*	Sarpagandha	*Rauwolfia serpentina*	Leaf spot
279.	*Alternaria sp.*	Ashwagandha	*Withania somnifera*	Leaf blight

Sr	Microorganism	Common Name	Scientific name	Disease
280.	*Biharia smilacis*	Sarsaparilla	*Smilax macrophylla*	-
281.	*Cercoseptoria balsaminiana*	Justicia	*Justicia simplex*	Leaf spot
282.	*Cercospora agarwalii*	Chastetree	*Vitex negundo*	Leaf spot
283.	*Cercospora hyalospora*	Country mallow	*Sida cordifolia*	Leaf spot
284.	*Cercospora justiciidola*	Justicia	*Justicia simplex*	Leaf spot
285.	*Cercospora rauwolfia*	Sarpagandha	*Rauwolfia serpentina*	Leaf spot
286.	*Cercospora viachueli*	Bush grape	*Vitis trifolia*	Leaf spot
287.	*Cercospora withaniae*	Ashwagandha	*Withania somnifera*	Leaf spot
288.	*Cerebella* sp.	Vetiver	*Vetiveria zizanioides*	-
289.	*Colletotrichum piperis*	Betel vine	*Piper betel*	Anthracnose
290.	*Corynespora cassiicola*	Sarpagandha	*Rauwolfia serpentina*	Target leaf spot
291.	Curvularia andropogonis	Lemongrass	*Cymbopogon flexuosus*	Leaf blight
292.	*Curvularia lunata*	Lemongrass	*Cymbopogon citratus*	Leaf spot
293.	*Curvularia* sp.	Curry leaf tree	*Murraya koenigii*	Endophytic fungus
294.	*Drechslera spicifera*	Gudmar	*Gymnema sylvestre*	Leaf spot
295.	*Erysiphe biocellata*	Tulsi	*Ocimum sanctum*	Powdery mildew
296.	*Fusarium solani*	Alovera	*Aloe barbadensis*	Leaf spot and root rot
297.	*Oidium* sp.	Babchi, Sarpagandha	*Psoralea corylifolia* *Rauwolfia serpentina*	Powdery mildew
298.	*Pestalotiopsis carbonacea*	Chinese chaste tree, Indian sarsaparilla	*Vitex negundo, Hemidesmus Indicus*	Leaf spot
299.	*Pestalotiopsis* sp.	Baheda	*Terminalia bellerica*	Leaf spot
300.	*Pestalotiopsis woodfordeae*	Chinese chastetree	*Vitex negundo*	Leaf spot
301.	*Phoma glomerata*	Harra, Saragandha	*Terminalia chebula,* *Rauwolfia serpentina*	Phoma leaf blight
302.	*Phyllosticta ipomoeae*	Baheda	*Terminalia bellerica*	Leaf spot
303.	*Phyllostictina murrayae*	Curry leaf tree	*Murraya koenigii*	Leaf spot
304.	*Phytophthora capsici*	Betel vine, Long pepper, Bryophyllum	*Piper betel,* *Piper longum,* *Kalanchoe mortagei*	Foot and leaf rot
305.	*Phytophthora infestans*	Ashwagandha	*Withania somnifera*	Leaf spot
306.	*Phytophthora nicotianae* var. *nicotianae*	Musk mallow	*Abelmoschus moschatus*	Leaf blight
307.	*Phytophthora nicotianae* var. *parasitica*	Periwinkle	*Catharanthus roseus*	Leaf rot
308.	*Ramularia aromatici*	Sweet flag	*Acorus calamus*	Leaf spot
309.	*Ravenelia emblicae*	Indian gooseberry	*Emblica officinalis*	Rust
310.	*Ravenelia ornate*	Rosary pea	*Abrus precatorius*	Rust
311.	*Sclerotinia sclerotiorum*	Ashwagandha	*Withania somnifera*	Sclerotinia stem rot
312.	*Sclerotium rolfsii*	Kenaf, Safed musli	*Hibiscus cannabinus,* *Chlorophytum borivilianum*	Collar rot
313.	*Ustilago vetiveriae*	Vetiver	*Vetiveria zizanioides*	Smut
VII	On Weeds			
314.	*Alternaria sp.*	Rough cocklebur, Swamp morning glory	*Xanthium strumarium,* *Ipomea reptans*	Leaf spot
315.	*Alternaria zinniae*	Chickweed	*Ageratum conyzoides*	Leaf spot
316.	*Arthrinium sacchari*	Kans grass	*Saccharum spontaneum*	Wilt

Sr	Microorganism	Common Name	Scientific name	Disease
317.	*Bartalinia robillardoides*	Bermuda grass	*Cynodon dactylon*	-
318.	*Cercospora avicennae*	Indian mallow	*Abutilon indicum*	Leaf spot
319.	*Cercospora sp.*	Field milk thistle	*Sonchus arvensis*	Leaf spot
320.	*Cintractia axicola*	Tall fringe rush	*Fimbristylis diphylla*	Smut
321.	*Cintractia clintonii*	Annual fimbry	*Fimbristylis annua*	Smut
322.	*Cintractia minor*	Annual sedge	*Cyperus compressus*	Smut
323.	*Corynespora cassiicola*	Painted euphorbia	*Euphorbia geniculata*	Target leaf spot
324.	*Deightoniella jabalpurensis*	Painted euphorbia	*Euphorbia geniculate*	-
325.	*Dictyoarthrinium sacchari*	Broadleaf cattail	*Typha latifolia*	-
326.	*Melanopsichium pennsylvanicum*	Common marsh buckwheat	*Polygonum glabrum*	-
327.	*Myrothecium roridum*	Giant milkweed	*Calotropis gigantea*	Leaf spot
328.	*Nodulisporium gregarium*	Kans grass	*Saccharum spontaneum*	Endophytic fungi
329.	*Periconia byssoides*	Wild senna	*Cassia tora*	Leaf spot
330.	*Pestalotiopsis sp.*	Wild senna	*Cassia tora*	Leaf spot
331.	*Phytophthora cyperi*	Purple nutsedge	*Cyperus rotundas*	-
332.	*Puccinia chloridis*	Finger grass	*Chloris incomplete*	Rust
333.	*Puccinia hookeri*	Beard grass	Andropogon foveolatus	Rust
334.	*Puccinia lateritia*	False buttonweed	*Spermacoce stricta*	Rust
335.	*Puccinia purpurea*	Johnsan grass	*Sorghum halepense*	Rust
336.	*Puccinia pusilla*	Pitted beard grass	*Amphilophis pertusa*	Rust
337.	*Puccinia romagnoliana*	Umbrella sedge, Yefen, Cyprus	*Cyprus diffusus, Fuirena umbellate, Cyperus iria*	Rust
338.	*Puccinia sp.*	Indian goose grass	*Eleusine indica*	Rust
339.	*Pyrenochaeta tandonii*	Wild indigo	*Tephrosia purpurea*	Pink root disease
340.	*Sphacelotheca annulata*	Marvel grass	*Dichanthium annulatum*	Smut
341.	*Sphacelotheca apludae-aristatae*	Perennial grass	*Apluda aristata*	Smut
342.	*Sporisorium bursum (=Sphacelotheca bursa)*	Kangaroo grass	*Themeda quadrivalvis*	Smut
343.	*Sporisorium cornutum*	Sedge	*Ophiurus corymbosus*	Smut
344.	*Sporisorium inayati (=Sphacelotheca inayati)*	Musal grass	*Iseilema laxum*	Smut
345.	*Stauronema sacchari*	Kans grass	*Saccharum spontaneum*	-
346.	*Stemphylium sp.*	Wild senna	*Cassia tora*	Leaf blight
347.	*Synchytrium cookei*	Buffalo clover	*Alysicarpus* sp.	-
348.	*Tilletia dactyloctenii*	Egyptian crow foot grass	*Dactyloctenium aegyptium*	Smut
349.	*Tilletia sp.*	Saramolla grass	*Ischaemum rugosum*	Smut
350.	*Tilletia sumati*	Adlay millet	*Coix lachryma-jobi*	Smut
351.	*Tilletia vittata*	Burmann's basket grass	*Oplismenii burmanii*	Smut
352.	*Ustilaginoidea oplismenii*	Burmann's basket grass	*Oplismenus burmanii*	-
353.	*Ustilago coicis*	Adlay millet	*Coix lacryma-jobi*	Smut
354.	*Ustilago cynodontis*	Bermuda grass	*Cynodon dactylon*	Smut
355.	*Ustilago neglecta*	Yellow foxtail	*Setaria glauca*	Smut
356.	*Ustilago trichophora*	Deccan grass	*Echinochloa colona*	Smut

VIII	**Post-harvest fungal diseases on fruits**

Fruits are perishable products, if they are infected with fungi after harvest or during transit and storage; they easily lose their keeping quality within lesser time. In nature, the cause of most of the post-harvest diseases of fruits due to presence of latent infection of pathogen in the fruits; they mostly do not cause infection until the fruit is ripened. Among all these fungal associations *Aspergillus, Penicillium* and *Rhizopus* are non-pathogenicde graders and they are commonly found associated with post-harvest fruit rot.

Sr	Fungus	Crop	Disease
357.	*Alternaria alternata*	Apple	Alternaria blotch
358.	*Alternaria citri*	Mandarin lime	Black rot
359.	*Alternaria tenuis*	Grapes	Fruit rot
360.	*Aspergillus flavus*	Banana	Fruit rot
361.	*Aspergillus niger*	Apple, Banana, Pomegranate	Fruit rot
362.	*Aspergillus* sp.	Guava, Orange	Fruit spoilage
363.	*Botryodiplodia* sp.	Apple, Orange	Fruit decay
364.	*Botryodiplodia theobromae*	Banana	Finger rot
365.	*Botrytis cinerea*	Banana	Gray mold
366.	*Ceratocystis major*	Apple	Black rot
367.	*Coniella noviae-zealanidiae*	Pomegranate	Fruit rot
368.	*Curvularia lunata*	Banana	Fruit rot
369.	*Deightoniella torulosa*	Banana	Fruit spot (speckle)
370.	*Fusarium* sp.	Papaya	Fruit rot
371.	*Gloeosporium musarum*	Banana	Anthracnose
372.	*Haplosporella* sp.	Apple	-
373.	*Lasiodiplodia theobromae*	Apple	Fruit rot
374.	*Monilia* sp.	Pomegranate	Fruit rot
375.	*Penicillium digitatum*	Papaya	Fruit rot
376.	*Penicillium expansum*	Apple	Blue mold
377.	*Penicillium italicum*	Orange	Blue mold
378.	*Penicillium* sp.	Banana	Fruit rot
379.	*Phytophthora capsici*	Watermelon	Phytophthora fruit rot
380.	*Rhizopus arrhizus*	Pomegranate	Fruit rot
381.	*Rhizopus nigricans*	Apple, Lime	Fruit rot
382.	*Rhizopus* sp.	Banana	Soft rot
383.	*Sclerotinia sclerotiorum*	Banana	Fruit rot

IX	**Post-harvest fungal diseases on seeds**

Seed health testing carried out at the University to know the health status of the seeds after harvest as well as storage, much fungal diversity has been reported in association with the crop seeds. All the fungal associations on seeds are recorded using standard methods of detection of seed borne pathogens recommended by ISTA, the maximum number of species of *Fusarium*(13) were recovered during the investigation followed by *Alternaria, Curvularia*(7) and *Aspergillus, Colletotrichum*(5) (Table 21.2).

Dr MN Khare
Served
4.1.1956 to 30.6.1994
In all 173 fungi have been reported by the group of Plant Pathologist from College of Agriculture, Jabalpur under his leadership

Table 21.2 Plant pathogenic fungi reported on seed from College of Agriculture, Jabalpur

Sr	Microorganism	Associated crop
384.	*Alternaria alternata*	Soybean, mung, jatropha
385.	*Alternaria citri*	Jatropha
386.	*Alternaria longissima*	Rice
387.	*Alternaria padwickii*	Rice
388.	*Alternaria porri*	Onion
389.	*Alternaria* sp.	Urid, brinjal, chilli, cowpea, mung, lentil, okra, pea, Pigeonpea, radish, tomato, wheat
390.	*Alternaria alternata*	Sorghum (Grain mold)
391.	*Aspergillus clavatus*	Jatropha
392.	*Aspergillus flavus*	Urid, cowpea, mung, jatropha, lentil, pea, pigeonpea, rice (grain discolouration), soybean
393.	*Aspergillus niger*	Urid, mung, cowpea, chickpea, lentil, rice, pea, pigeonpea, sorghum
394.	*Aspergillus ochraceus*	Chickpea, lentil, pea, mung
395.	*Aspergillus* sp.	Urid, brinjal, cabbage, chickpea, chilli, cowpea, mung, okra, onion, pea, pigeonpea, radish, sesame, tomato, wheat
396.	*Botrytis cinerea*	Soybean, urid
397.	*Cephalosporium* sp.	Chickpea, rice, urid, cowpea
398.	*Cephalothecium* sp.	Wheat
399.	*Cercospora kikuchii*	Soybean (Purple seed stain)
400.	*Chaetomium globosum*	Soybean, jatropha
401.	*Chaetomium* sp.	Urid, chickpea, cowpea, lentil, rice, sesame, wheat
402.	*Choanephora* sp.	Urid
403.	*Cladosporium tenuissium*	Mung
404.	*Colletotrichum dematium*	Jatropha
405.	*Colletotrichum gloeosporioides*	Jatropha
406.	*Colletotrichum graminicola*	Urid
407.	*Colletotrichum* sp.	Soybean
408.	*Colletotrichum truncatum*	Urid
409.	*Curvularia intermedia*	Urid
410.	*Curvularia lunata*	Jatropha, rice, mung, urid, sorghum
411.	*Curvularia pallescens*	Soybean
412.	*Curvularia* sp.	Urid, brinjal, chickpea, chilli, cowpea, mung, lentil, pea, pigeonpea, radish, tomato, wheat
413.	*Curvularia trifoli*	Urid
414.	*Curvularia verruculosa*	Sesame, cowpea
415.	*Curvularia pallescens*	Rice (Grain discolouration)
416.	*Drechslera haveliensis*	Mung
417.	*Drechslera oryzae*	Rice (Grain discolouration)
418.	*Epicoccum* sp.	Urid
419.	*Eurotium* sp.	Urid, chickpea, lentil, pea, wheat
420.	*Fusarium moniliforme*	Rice
421.	*Fusarium semitectum*	Rice
422.	*Fusarium* sp.	Wheat
423.	*Fusarium equiseti*	Rice, sesame, mung, cowpea
424.	*Fusarium fusarioides*	Sesame, cowpea
425.	*Fusarium graminiarum*	Jatropha
426.	*Fusarium moniliforme*	Soybean, urid, mung, sorghum(grain mould)
427.	*Fusarium oxysporum*	Jatropha, mung

Sr	Microorganism	Associated crop
428.	*Fusarium oxysporum* f. sp. *tracheiphilum*	Cowpea
429.	*Fusarium peniseti*	Urid
430.	*Fusarium semitectum*	Mung
431.	*Fusarium solani*	Soybean, urid, mung
432.	*Fusarium* sp.	Chickpea, cowpea, mung, lentil, pea, pigeonpea
433.	*Geotrichum candidum*	Jatropha
434.	*Helminthosporium bambus*	Jatropha
435.	*Helminthosporium* sp.	Urid, chickpea, cowpea, mung, lentil, pea, pigeonpea, soybean, wheat
436.	*Macrophomina phaseolina*	Rice, sesame, urid, mung, cowpea
437.	*Macrospore levispora*	Jatropha
438.	*Memnospora* sp.	Rice
439.	*Mucor* sp.	Cabbage, soybean
440.	*Myrothecium roridum*	Soybean, urid, mung
441.	*Myrothecium verrucaria*	Urid
442.	*Nigrospora* sp.	Chickpea, rice, wheat, urid
443.	*Oedocephalum* sp.	Chickpea, urid
444.	*Oospora* sp.	Tomato, brinjal
445.	*Penicillium frequentans*	Jatropha
446.	*Penicillium* sp.	Urid, cowpea, mung, onion, pigeonpea, rice, sesame, soybean, wheat
447.	*Pestalotia* sp.	Mung
448.	*Phoma bakeriana*	Sesame, cowpea
449.	*Phoma sorghina*	Sorghum Grain mold
450.	*Phoma* sp.	Rice (grain discoloration), urid, mung
451.	*Pithomyces* sp.	Rice
452.	*Pleospora infectoria*	Sesame, mung, cowpea
453.	*Pythium aphanidermatum*	Wheat
454.	*Pythium graminicolum*	Wheat
455.	*Rhizoctonia bataticola*	Soybean
456.	*Rhizoctonia solani*	Jatropha, Wheat
457.	*Rhizoctonia* sp.	Urid, chickpea, chilli, cowpea, mung, lentil, okra, pea, Pigeonpea, radish
458.	*Rhizopus* sp.	Urid, brinjal, cabbage, chickpea, chilli, cowpea, mung, jatropha, lentil, okra, onion, pea, pigeonpea, radish, rice, soybean, tomato, wheat
459.	*Sclerotium rolfsii*	Wheat
460.	*Sclerotium* sp.	Brinjal
461.	*Stachybotrys* sp.	Onion, rice
462.	*Syncephalastrum racemosum*	Sesame, cowpea
463.	*Trichoconiella padwickii* (=*Trichoconis padwickii)*	Rice Grain discoloration
464.	*Trichoderma* sp.	Chickpea, soybean
465.	*Trichothecium* sp.	Urid, chickpea
466.	*Trichothecium* sp.	Mung
467.	*Ulocladium chartarum*	Cowpea

Bio-agents

All the antagonistic fungal genera (05) were isolated from soil, their efficacy against plant pathogens and insects is well documented by several researchers. Among all these fungal genera *Trichoderma* is extensively used as potential bio-control agent against soil borne plant pathogens by several researchers of JNKVV. Whereas, *Beauveria bassiana, Verticillium lecanii, Metarhizium anisopliae* were used for the management of insect pest (Table 21.3).

Table 21.3 Bio-agents against plant pathogens/Insects reported from College of Agriculture, Jabalpur

X			
Sr	**Microorganism**	**Habitat**	**Effective against**
468.	*Beauveria bassiana*	Soil	Insect
469.	*Chaetomium* sp.	Rhizospheric fungi	Plant pathogens
470.	*Lecanicillium lecanii (=Verticillium lecanii)*	Soil	Insect
471.	*Metarhizium anisopliae*	Soil	Insect
472.	*Trichoderma hamatum*	Rhizospheric fungi	Plant pathogens
473.	*Trichoderma harzianum*	Rhizospheric fungi	Plant pathogens
474.	*Trichoderma longibrachiatum*	Rhizospheric fungi	Plant pathogens
475.	*Trichoderma viride*	Rhizospheric fungi	Plant pathogens

Mushroom associated Microflora:

The mushrooms are cultivated under high humid conditions on agricultural wastes having lignin and cellulose, others saprobes also compete with the mushroom for space and nutrition during spawn run and fruiting period, 9 fungal genera causing various diseases on mushroom are given in the table. Among all these fungal genera *Trichoderma* is a major constraint in mushroom production and lowering the yield substantially by hyper parasitizing the mushroom mycelium, while, *Coprinus* is a common weed on mushroom beds which arrest the growth of spawn and development of fruiting bodies (Table 21.4).

Table 21.4 Mushroom associated Microflora reported from College of Agriculture, Jabalpur

XI	**Associated microorganism**	**Associated disease**
Sr	**On Mushrooms**	
476.	*Chaetomium* sp.	Olive green Mould
477.	*Cladobotryum* sp.	Cobweb
478.	*Coprinus* sp.	Inky cap
479.	*Fusarium* sp.	Fusarial Rot
480.	*Mucor* sp.	White Mould
481.	*Rhizopus* sp.	Black Mould
482.	*Sclerotium* sp.	Sclerotium rot
483.	*Trichoderma* sp.	Green mould
484.	*Verticillium* sp.	Dry bubble

Percentage distribution of microflora in district of Jabalpur

Data collected from different sources and converted into percentage share of each fungal genera in the diversity of microflora in Jabalpur district exhibits that, out of total 485 fungal species of different genera recorded on different hosts (Fig 4), *Alternaria* species accounts 8% followed by *Fusarium* (7%), *Aspergillus, Phytophthora* (6%), *Cercospora, Curvularia* (5%), *Colletotrichum* (4%), *Rhizoctonia, Rhizopus, Helminthosporium* (3%), *Sclerotium, Phoma, Puccinia, Pestalotiopsis,*

Penicillium (2%) and *Phyllosticta* was 1%. The remaining genera are included in the others whose frequency was less than ten, they merged together and they occupied 39% share of these fungi. All the fungal associations on host in the nature were finally grouped into different categories as pathogenic, saprophytic, bioagents and endophytes (Fig. 21.4).

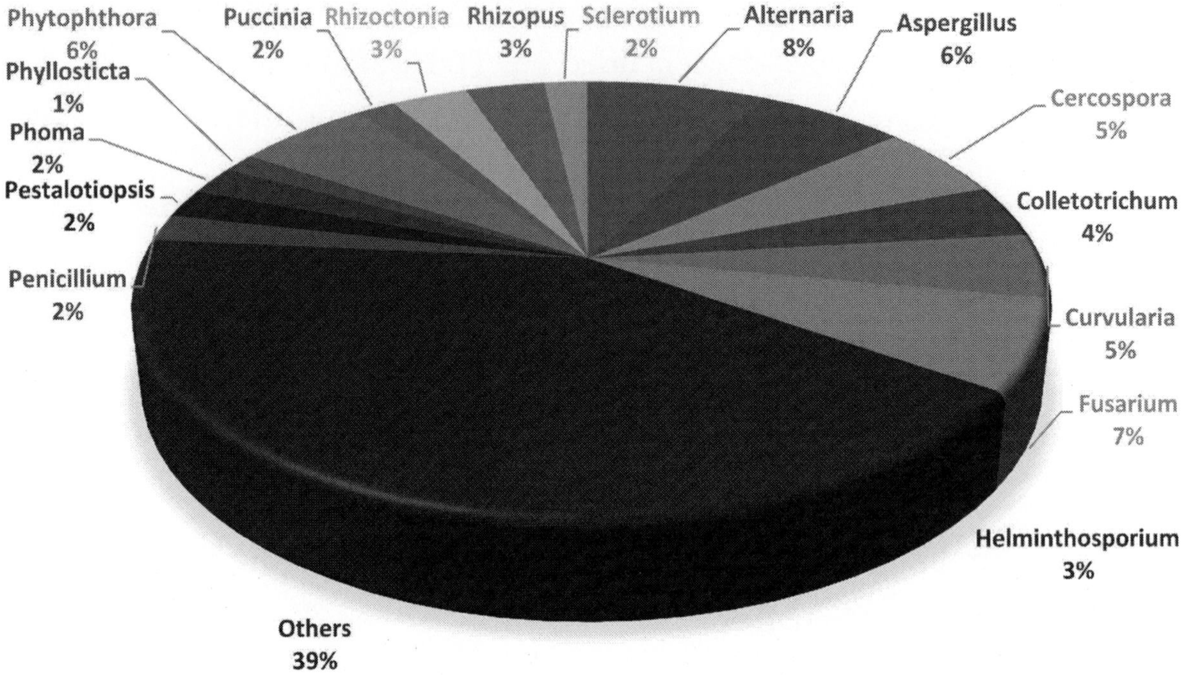

Fig 21.4 Reported Diversity of Phytopathogenic Microflora at JNKVV Campus Jabalpur

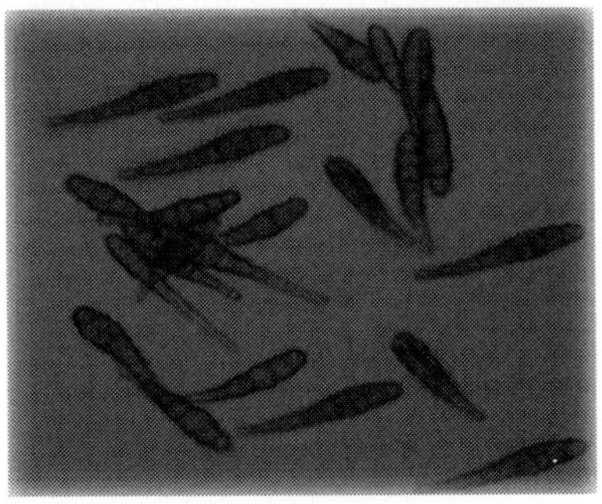

At College of Agriculture, Jabalpur maximum frequency of *Alternaria* is reported among all the fungal genera

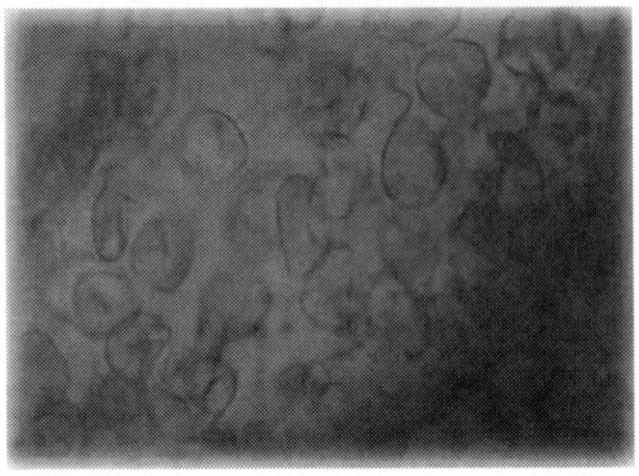

Phakopsora pachyrhizi cause soybean rust an epidemic in Madhya Pradesh

2 Bacterial Diseases:

Out of 24 species / pathovar of bacteria representing five genera studied at the College of Agriculture, Jabalpur, 19 cause disease on 16 plant species (8 vegetables; 6 field crops; 1 forest tree, and one mushroom); three are endophytic and two are bio-control agents (Fig. 21.5).

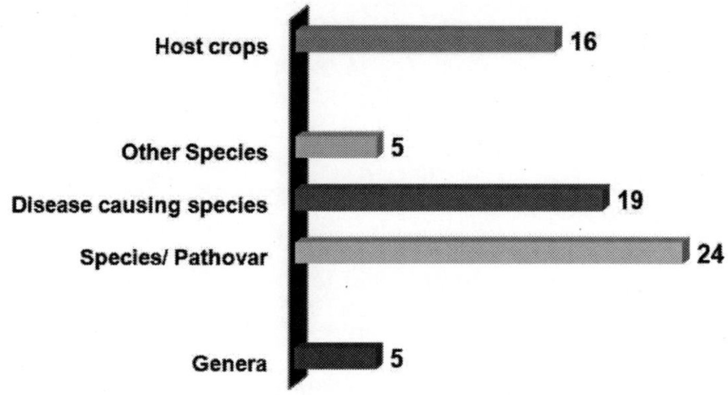

Fig 21.5 Data on bacterial diseases studied at College of Agriculture, Jabalpur

58% studied bacterial diseases at the College of Agriculture, Jabalpur are from genera *Xanthomonas* and 26% by *Pseudomonas* (Fig. 21.6).

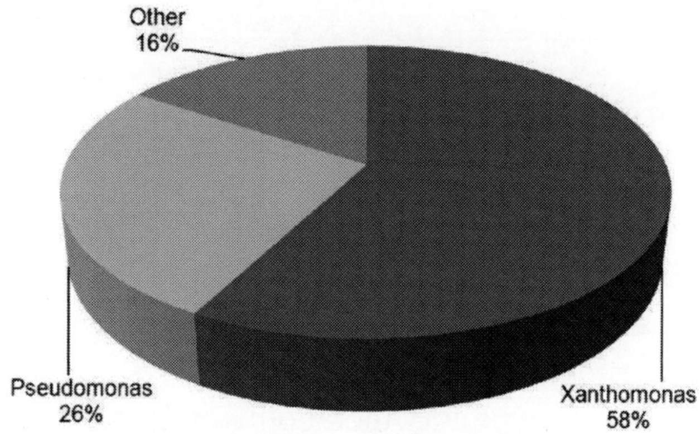

Fig 21.6 Percentage of studied bacterial diseases caused by Genera

Bacterial diseases is a matter of concern after the fungal diseases, in which the genus associated with plant diseases most often belongs to *Xanthomonas*, followed by *Pseudomonas, Erwinia* and *Streptomyces*. Among all these bacterial diseases leaf spot or blight are common symptoms found associated with the crop. While soft rot of many vegetables under field condition and even after harvest and during storage is a problem of vegetable growing areas and the genus *Erwinia* was isolated as one of the cause during the investigation (Table 21.5).

Table 21.5 Plant pathogenic bacteria reported from College of Agriculture, Jabalpur

Sr	Microorganism	Season	Crop	Disease
1.	*Erwinia carotovora*	Kharif	Vegetables Maize	Bacterial soft-rot of vegetables Stalk rots
2.	*Pseudomonas tolasii*	Rabi	Button mushroom	Bacterial Blotch
3.	*Pseudomonas fluorescens*	Rabi	Button mushroom	Bacterial spot
4.	*Pseudomonas* sp.	-	Vegetables	Bacterial soft-rot of vegetables
5.	*Pseudomonas syringae* pv. *sesame*	Rabi	Sesame	Bacterial leaf spot
6.	*Pseudomonas syringae pv. glycinea*	Kharif	Soybean	Bacterial pustule, Bacterial blight
7.	*Ralstonia solanacearum* (=*Pseudomonas solanacearum*)	Rabi	Tomato, Brinjal	Bacterial wilt
8.	*Streptomyces* sp.	-	Jatropha	-
9.	*Xanthomonas campestris* pv. *phaseoli*	Kharif	Soybean	Common blight

Sr	Microorganism	Season	Crop	Disease
10.	*Xanthomonas axonopodis* pv. *phaseoli*	Rabi	Cowpea	Leaf blight
11.	*Xanthomonas axonopodis* pv. *glycines*	Kharif	Soybean	Bacterial pustule
12.	*Xanthomonas campestris* pv. *betlicola*	-	Betel vine	Leaf spot, leaf blight and stem infection
13.	*Xanthomonas oryzae* pv. *oryzae* (=*Xanthomonas campestris* pv. *oryzae*)	Kharif	Rice	Bacterial leaf blight
14.	*Xanthomonas campestris* pv. *campestris*	Rabi	Cruciferous vegetables	Black rot
15.	*Xanthomonas oryzae* pv. *oryzicola*	Kharif	Rice	Bacterial leaf streak
16.	*Xanthomonas phaseoli*	Rabi	Urid, Mung	Bacterial blight
17.	*Xanthomonas phaseoli* var. *fuscans*	Rabi	Beans	Bacterial blight
18.	*Xanthomonas sesami*	Rabi	Sesame	Bacterial leaf spot
19.	*Xanthomonas vesicatoria*	Rabi	Tomato	Bacterial leaf spot

2. b. Bacteria as Bio-Control Agent / Endophytes:

Bacillus subtilis and *Pseudomonas fluorescens* isolated from the soil which are potential bioagent. Between these two *P. fluorescens* are the most exploited bacteria for biological control of soil borne and foliar plant pathogens because of their catabolic versatility, excellent root colonizing abilities and their capacity to produce a wide range of metabolites. Rest of the three bacterial genera isolated from aerial part of the plant and recognized as endophytes. These endophytes have important roles in host plant growth either by secondary metabolite or nutrient assimilation or by preventing induction of plant disease symptoms by different pathogens (Table 21.6).

Table 21.6 Bacteria as Bio-Control Agent reported form College of Agriculture, Jabalpur

Sr	Bacteria	Nature	Habitat
20.	*Bacillus* sp.	Endophytic bacteria	Leaf and stem
21.	*Bacillus subtilis*	Bio-control agent	Soil
22.	*Pseudomonas fluorescens*	Bio -control agent	Soil
23.	*Pseudomonas* sp.	Endophytic bacteria	Leaf and stem
24.	*Xanthomonas* sp.	Endophytic bacteria	Leaf and stem

3 Viral diseases

Total 19 viruses have been reported with disease in 12 crops. Out of 19 diseases 10 are mosaic, 3 are necrosis, 3 are bud blight, two are leaf curl and one wilt (Table 21.7).

Table 21.7 Plant pathogenic viruses reported from College of Agriculture, Jabalpur

Sr	Microorganism	Crop	Disease
1.	Dasheen mosaic virus	Colocasia	Mosaic
2.	Enation Leaf Curl Virus	Okra	Enation leaf curl
3.	Gemini virus	Frenchbean	Yellow mosaic
4.	Mung bean yellow mosaic virus (MYMV)	Soybean, Mung, Urid	Mung bean yellow mosaic
5.	Bhendi yellow vein mosaic virus (BYVMV)	Okra	Yellow Vein Mosaic
6.	Papaya leaf curl virus (PaLCuV)	Papaya	Leaf curl
7.	Papaya ringspot virus (PRSV)	Papaya	Mosaic
8.	Pigeonpea Yellow Mosaic virus	Pigeonpea	Yellow mosaic
9.	Pigeonpea sterility mosaic virus (PPSMV)	Pigeonpea	Sterility mosaic

Sr	Microorganism	Crop	Disease
10.	Potato leaf roll virus	Potato	Net necrosis of potato
11.	Potato virus X	Potato	Mosaic
12.	Potato virus Y	Potato	Mosaic
13.	Soybean mosaic virus (SMV)	Soybean	Mosaic
14.	Spotted wilt virus	Groundnut Soybean	Bud necrosis Spotted wilt
15.	Tobacco ringspot virus (TRSV)	Soybean	Bud blight
16.	Tobacco streak virus	Soybean	Bud blight
17.	Tomato leaf curl virus (TLCV)	Tomato	Leaf curl
18.	Tomato mosaic virus (ToMV)	Tomato	Mosaic
19.	Tomato spotted wilt virus- Bud necrosis virus	Potato	Stem necrosis

4 Phytoplasmic diseases

Only one phytoplasma causing disease on soybean, sesame and brinjal has been reported from the college (Table 21.8).

Table 21.8 Phytoplasma reported from College of Agriculture, Jabalpur

	Causal Agent	Crop	Disease
1	Phytoplasma	Soybean, Sesame, Brinjal	Phyllody, Little leaf of brinjal

5. Plant Disease caused by nematode

Total eight species of nematodes have been reported associated with disease in 10 crops. Out of 10 crops nine are of Kharif season. Among all the nematodes *Meloidogyne* spp. is reported on maximum number of crops (Table 21.9).

Table 21.9 Plant pathogenic nematode reported form College of Agriculture, Jabalpur

Sr	Microorganism	Season	Crop	Disease
1.	*Aphelenchoides besseyi*	Kharif	Rice Proso millet	White tip, Foliar disease
2.	*Heterodera cajani*	Kharif	Pigeonpea, Urid	Pearly root disease
3.	*Hirschmanniella oryzae*	Kharif	Rice	Mentek
4.	*Meloidogyne graminicola*	Rabi	Wheat	Root knot
5.	*Meloidogyne incognita*	Kharif	Soybean, Tomato, bean, Urid	Root knot
6.	*Meloidogyne javanica*	Kharif	Mung	Root knot
7.	*Pratylenchus thornei*	Kharif- Rabi	Soybean, Chickpea	Root lesion
8.	*Rotylenchulus reniformis*	Kharif	Soybean	Reniform

The frequency of microorganisms on different group of hosts (Fig 7) expressed maximum 39% on pulses (10% on urid bean, 8% on mung bean, 6% on chickpea, 6 % on cowpea, 5% on pigeonpea and 4% on lentil); followed by 19% in cereals (10% in rice, 6% in wheat and 3% in sorghum); 17% on oilseeds(12% on soybean and 5% on sesame); 13% on vegetables; 8% on fruits and 7% on trees (Fig. 21.7).

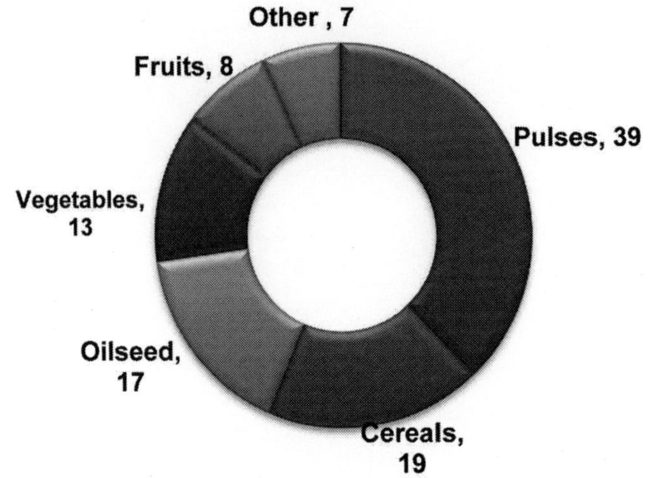

Fig 21.7 Percent distribution of pathogen on different groups of hosts

Number of microbes associated with host

The graphical presentation of number of microorganisms associated with different hosts (Fig 8) clearly depicts that maximum number of microbes association was found with soybean (52) followed by Urid bean (43), Rice (36), Mung bean (33), Chickpea (27), wheat, Cowpea (25), Sesame (23), Pigeon pea (20), Tomato (19), Jatropha (17), Lentil (16), Pea, Brinjal (15), Sorghum, Eucalyptus (13), Chilli (11), Pomegranate, Guava and Banana (10) (Fig.21.8).

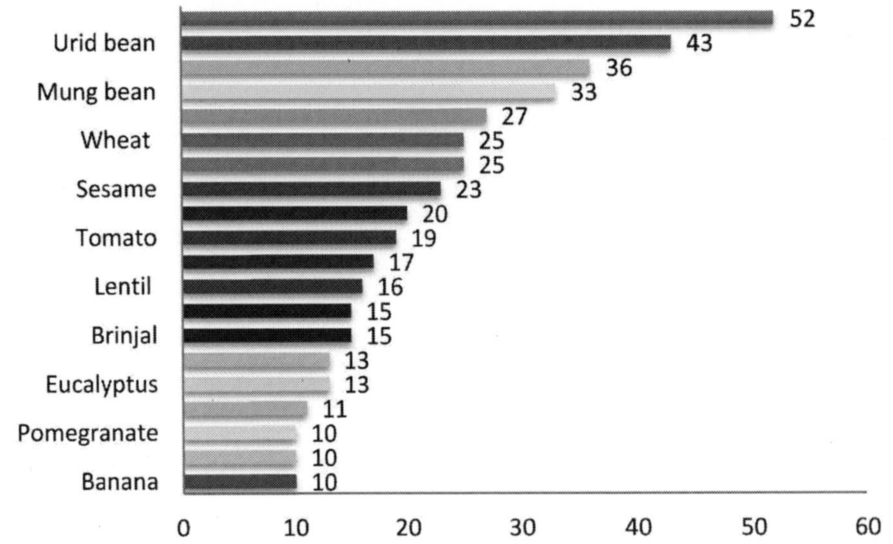

Fig 21.8 Number of microbes associated with host observed at JNKVV campus Jabalpur

New Pathogens/ Diseases Reported:

Dr ND Sharma (served from June 5, 1972 to 31 December 2000) Professor, Department of Plant Pathology, College of Agriculture, Jabalpur discovered four genera of fungi, four second species of monotypic genera, 69 nova species of fungi 20 new diseases and nine Fungal Pathogens of Weeds (Table 21.9).

- Discovered four Genera of fungi Included in 8th ed. of **"Dictionary of fungi"** CAB. UK, 1995
- Discovered one Genus based on collection as Type Species

- Discovered four Second Species of monotypic genera, earlier only one species known in the world.
- Discovered 67 new Species of fungi to Science
- Discovered 46 Species of fungi new to Indian Peninsula

- Discovered 20 New diseases to the World/Indian Peninsula/State of Rice, chickpea, Lentil, Groundnut, soybean, Little millet, Custard apple, Pomegranate, Falsa, Karonda, Chrysanthemum, Gurmar, Obnoxious weed: Imperata and of several other weeds
- Identified nine Fungal Pathogens of Weeds potential for evaluation under biological control of obnoxious weeds.
- One of powdery mildew of India I[st] entry in Dictionary of Fungi on Powdery Mildew from India

Table 21.9 New plant pathogenic fungi reported by Dr ND Sharma

Associated Microorganism	Host	Disease
Aristastoma quadrilosus	Soybean	Yellow spot
Cercospora thirumalachari	Pigeonpea	Leaf spot
Colletotrichum dematium	Chickpea	Blight
Conielle novi	Pomegranate	Fruit rot
Corynespora cassiicola	Kapok	Kapok blight
Eballistra oryzae	Rice	Leaf smut
Erysiphe sp.	Soybean	Powdery mildew
Macalpinomyces sharmae	Kutki	Smut
Memnoniella echinata	Ginger	Black rot
Oidium grewiae	Falsa	Powdery mildew
Phakopsora pachyrhizi	Soybean	Rust
Phytophthora boehmeriae	Chilli	Phytophthora blight
Pirostoma sp.	Cogon grass (weed)	-
Pithomyces chartarum	Rice	Glume blotch
Pyrochaeton sp.	Citrus, Mango	Sooty mould
Septoria sp.	Eucalyptus	Leaf spot
Tilletia narsimhanii	Kutki	Smut
Tilletia polypogonis	Polyllpogon (weed)	Smut
Trichurus spiralis	Ginger	Gray rot
Triposporium sp.	Citrus, Mango	Sooty mould
Verticillium sp.	Ginger	Red rot

New discovered taxa of fungi:

- *Botryotrichum longibrachiatum* var. *macrospora*
- *Cormichaelis lignicela*
- *Discellouyeas calumi*
- *Phragmotenium radicum*
- *Pseudopesilospora cendiseks*
- *Pycnidiocrxicllea casii*
- *Pycnothera cordiae*

Two fungal pathogenic species dedicated to the name of District Jabalpur
Monochaetia jabalpurensis
Deightoniella jabalpurensis

Fungal pathogenic species in the name of
Dr. N. D. Sharma
Macalpinomyces sharmae

CHAPTER 22

Energy Audit

AK Rai and Bharati Dass
Directorate of Instrumentation
JNKVV, Jabalpur

Greenhouse gases are the primary contributors to climate change and global warming. These gases are capable of accumulating infrared light, which leads to the retention and trapping of heat in the atmosphere. As a result, the heat emitted by the surface of the planet will continue rising. The use of fossil fuels to generate power can potentially cause global warming. Carbon dioxide would be released into the atmosphere as a result of the burning of fossil fuels. Carbon dioxide and some other greenhouse gases behave as an insulating blanket, allowing solar energy to enter but blocking it from reflecting into space. Once solar radiation heat is trapped on the Earth's surface, it causes the problem of global warming. As a result, using energy-efficient equipment, such as an air conditioner and refrigerator, helps to save energy and also reduce harmful emissions, thus protecting the environment.

Generation of solar energy

The University is generating solar energy through two approaches, first for Electricity Regulatory Commission of the Government of Madhya Pradesh from January 2017 and second through Deployment of Grid Connected Solar Rooftop Power Plant under RESCO model since March 2021. Out of total 696 KW solar power generated each year at campus Jabalpur since installment, 300 KW solar power is being used by the University and 396 KW is being given to National Grid.

Electricity Regulatory Commission of the Government of Madhya Pradesh: Government of Madhya Pradesh made mandatory for distribution companies to install module and procure a certain percentage of its total power procurement through renewable energy sources on zero lease rental bases. The approved firm installed the MP Government project for 396 KWp grid-connected Rooftop Solar PV project on buildings. For promotion of renewable energy, the University has allotted 4410 m^2 rooftop area to install 1764 modules for generation of 396 KW solar electricity in following buildings on January 2017

College of Agricultural Engineering, Jabalpur: A 217 KW unit consisting of 835 modules was installed on 2400 sq-m area at roof top of second and third floor connected to 11 KV grid using 440/11000V transformer (Plate 22.1).

Plate 22.1 A 217 KW unit module on College of Agricultural Engineering, Jabalpur

College of Agriculture, Jabalpur: A 40 KW unit consisting of 150 modules was installed on 880 sq-m area at roof top of the building of the college; a 115 KW unit consisting of 689 modules on 850 sq-m area at roof top of the Department of Food Science and Technology and a 24 KW unit consisting of 90 modules on 280 sq-m area at roof top of the building of Seed Technology centre all are connected to 33 KV grid using 440/33000 V transformer (Plate 22.2).

Plate 22.2 A 40 KW unit module on College of Agriculture, Jabalpur

Table 22.1 Generation of solar energy by MP Electricity Regulatory Commission on zero lease rentals at campus Jabalpur

Location	College of Agricultural Engineering, Jabalpur	College of Agriculture, Jabalpur	Food Science and Technology	Seed Technology Centre	Total
Year of establishment					
Area (m²)	2400	880	850	280	**4410**
Capacity (KWp)	217	040	115	024	**0396**
Modules	835	150	689	90	**1764**
Size of module		1960x992x40 mm			
Module Capacity	325W	325 W	325W	325W	
Connected to Grid	11 KV	33KV	33KV	33KV	
Vale in Rupees	3000000KWhx100 Rs5=Rs15000000/-				

The University is contributing, **396 KW** solar power to the National Grid towards the Nation's drive for Green and Clean Energy w.e.f. January 2017. 3000000 KWh solar energy is contributed in nation grid from Jan 2017 costing Rs 15000000/- (Fifteen million) per year (Table 22.1 and Fig. 22.3).

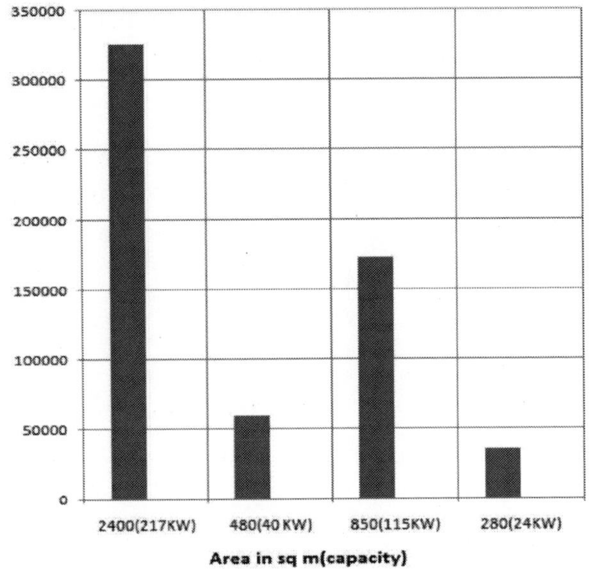

Fig 22.3 Solar energy generated in Kwh per year

Deployment of Grid Connected Solar Rooftop Power Plant under RESCO model at JNKVV, campus, Jabalpur:

Government of M.P. has nominated M.P. *Urja Vikas Nigam* (MPUVN) as State Nodal Agency of the Central and State Government for implementation of Deployment of Grid Connected Solar Rooftop Power Plant under RESCO model. On difficulty faced by the institutes in arranging the capital investment for these systems, Government deployed a private firm for development of solar power project, generation and sale of solar energy under RESCO model with net metering. The power producer has been notified by Nodal Agency for design, engineering, supply, installation, erection, testing and commissioning along with comprehensive Operation and Maintenance of the Grid Connected Rooftop Solar Project for sale of Solar Power for a period of 25 Operational Years and has been awarded 300 KW of capacity. In this model, Institute does not have to incur upfront capital investment for the Solar Rooftop Power Plant, but will still get benefit through purchase of Solar power at a discovered tariff i.e. approximately two rupees per KWh from March 2021 (Plate 22.1).

From March to Dec 2021, the University received solar energy of 4,20,473 KWh under this programme with discovered tariff @ Rs 2/- per KWh.

Plate 22.1 Solar Power system installed in the University

Consumption of energy of the campus

The University campus at Jabalpur has four feeders for supply of electricity with total consumption of 2529924 units per year (Table 22.2).

Table 22.2 Electricity feeder at the University campus Jabalpur

Sr.	Location of feeders	Units / year (KWh)
1	Registrar Feeder	983544 (500 KVA)
2	EE Feeder	1045800 (250 KVA)
3	Director Farms-1	362160 (150 KVA)
4	Director Farms-2 (Tank area)	138420 (100 KVA)
	Total	**2529924 (1000 KVA)**

Renewal energy based appliances: In five out of 14 hostels in headquarters, 3300 lit of water is heated through seven solar water heaters of 10 KW capacity of each. Apart from this only one Solar Water Pump is under operation at KVK, Jabalpur mainly for demonstration purpose (Table 22.3).

Table 22.3 Renewal energy based appliances under operation at Jabalpur campus

	Source	Location	Number	Volume lit	Capacity
1	Solar water heater	Hostel			
		Girls Tribal	01	500	10KW
		Awanti Bai	01	500	10KW
		International Hostel	01	300	10KW
		Kalpana Chawla	02	500	10KW
		Maharani Laxmi Bai	02	500	10KW
2	Solar Water Pump	KVK Jabalpur	01		5 HP
3	Gasifier				
	Wood Gasifier	FRes. Lab College of Agril Engg	01		35 KW
	Rice Husk Gasifier	(FRes. Lab College of Agril Engg.)	01		3.75KW

Use of Energy Efficient Appliances (EEA) and Technologies in the Campus:

In all 896 number of EEA are operating in the Jabalpur campus that reduces 78502 units of energy/ month. It includes 947 LED lights that reduce 17424 unit apart from star rated air conditioner and refrigerator. This is a saving of Rs 7,85,020/- per month (Table 22.4).

Table 22.4 Details of Energy Efficient Appliances under operation at Jabalpur campus

Sr.	Energy Efficient Appliances	Rating	Number	Reduction in energy (in units / month)
1	**Star rated Air Conditioners** (Score is proportional to rating) If all AC s are of 5 star rating then annual energy saving will be 83520 in kWh	1	041	0
		2	061	5917
		3	149	26075
		4	005	1190
		5	042	12096
2	**Star rated refrigerators** (Score is proportional to rating) If all Refrigerators s are of 5 star rating then annual energy saving will be 11340 in kWh	1	021	0
		2	003	192
		3	015	1725
		4	010	1560
		5	007	1323

Sr.	Energy Efficient Appliances	Rating	Number	Reduction in energy (in units / month)
3	**LED lights / bulbs**		947	17424
4	**Solar water heater and Pump**		008	11000
5	**Sensor based automatic lights**		Nil	Nil
6	**Any others (Mobile app, sub meter etc.)**		Nil	Nil
	Total		**896**	**78502**

Star rated appliances installed at campus Jabalpur

In all 298 air conditioners and 56 refrigerators are installed in various offices/ colleges/lab etc at campus Jabalpur. Out of these electric appliances 46% are three, 18% one, 18% two, 14% five and 4% four star ratings. Out of 298 installed air-conditioners 50% are three star and 14% five, whereas 34% are of 1 and 2 star rating. 57% refrigerator has > or = to three whereas 43% has <3 rating. If all AC and refrigerator will be of 5 star rating then annual energy saving will be 94860 KWh, amounting Rs 9,48,600/- per year(Table 22.5 and 22.6 and Fig. 22.4).

Table 22.5 Number and Star rating of air conditioner and refrigerator installed in various offices at campus Jabalpur

Location	Type	Air Conditioner					Refrigerator				
		Star					Star				
		1	2	3	4	5	1	2	3	4	5
Administrative Building											
Vice Chancellor Office				4		1		1			
Office of the Registrar				1		1					
Office of the Comptroller				1	1						
Directorates											
Dean Faculty Agriculture		1		1		1	1				
Directorate of Research		4	2	1					1		
Directorate of Instruction		2	1	1					1		
Directorate of Extension		1		9						1	
Communication Centre		1				3					
KVK			4	1				1			
ATIC		1					1				
Gyan Vani		1				1					
Directorate of Farms		2				1	1				
Offices											
DSW			4	1					1		
Executive Engineer						1					
IPRO											
College of Agriculture											
Office of the Dean	Split window		5	2			1			1	
	Ceiling CA 3 ph			2							
Sardar Patel Auditorium	Wall AC 3 phase			8							
	Tower AC			2							
	Green room			2							
Swami Vivekananda hall			4								
Departments											
Agricultural Economics				8			1				
Agronomy				5					1		
Entomology							2				

Location	Type	Air Conditioner					Refrigerator				
		Star					Star				
		1	2	3	4	5	1	2	3	4	5
Extension Education			7						1		
Food Science and Technology		1	-	9	-	2			1		
Forestry		3	1	2		4	1				
Horticulture							1			1	
Plant Breeding and Genetics	Split/window		6	1			2	1	2		
Plant Pathology	Split/window	1	5								
Plant Physiology	Split/ window		7				1		2		
Soil Sci and Agril Chemistry	Split/window			7					1	2	
PC Unit Sesame and Niger			5						2		
Establishment of the College in separate buildings											
Seed Technology Centre			4	3							1
ELP on Biotechnology											
ABM			2	5		2					
Bio-control Research Centre			3		2						
Plant Residual Lab											
AERC				2							
Micro Res & Production Centre				10		3					5
Biotechnology Centre		7		5		2	6		1		
College of Agricultural Engineering											
Dean Chamber							1				
Prof. S.V. Arya Auditorium					2	2					
College Meeting Hall						2					
Video Conferencing Hall						2					
Department											
Agro-met						1					
Directorate of Instrumentation		1		3			1				
Farm Machinery and Power			1	5						2	
Post Harvest Technology				3						1	
Soil Water Engineering						7					
Laboratories											
Water Resources Lab						1					
ARIS Lab UG		2									
R.S. & GIS Lab				7							
Central Library				4							
Guesthouses											
Krishak Bhawan				6			1				
Narbada				9							
Tapti				14							
Wanganga				5		4					1
Seed Breeding Unit		**13**				**1**			**1**		
Total		**41**	**61**	**149**	**5**	**42**	**21**	**3**	**15**	**10**	**7**
Grand total						**354**					

Overall Star Rating:

Table 22.6 Number of star rated electric appliances installed in various offices at campus Jabalpur

Star rating	Number of items
One star	62
Two star	64
Three star	164
Four star	15
Five star	49
Total	354

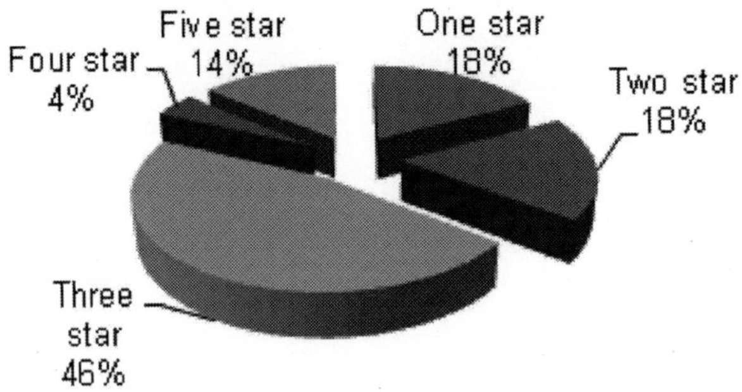

Fig 22. 4 Percentage of star rated electric appliances installed in various offices at campus Jabalpur

CHAPTER 23

Biodiversity Audit of Trees

GK Vani and SB Nahatkar
Department of Agricultural Economics and Farm management
College of Agriculture, Jabalpur

In all 19422 trees of 239 species representing 59 families are present on 543.54 ha land divided into seven locations at Jabalpur campus of the University (Table 23.1). The maximum number of trees i.e. 6404 is present at Dusty acre and Agro-forestry block representing 32.97% of the total trees available at the headquarters. Whereas, the representation of the maximum number of species (186) and families (45) are from Administrative, College and Hostel block of 34.18 ha land.

Table 23.1 Distribution of trees at JNKVV Campus Jabalpur

Location	Area (ha)	Species	Family	Trees	Trees (%)
		Number			
Administrative, College and Hostel	34.18	186	45	4321	22.25
Dusty Acre and Agro Forestry	72.62	30	08	6404	32.97
Fruit Orchard and Forestry	28.70	57	14	3896	20.06
Instructional Farm and Nursery	56.85	76	19	1061	05.46
Research Farm	142.24	23	08	451	02.32
Residential area	25.43	42	10	2665	13.72
Vegetable Res and Soybean Seed Farm	183.52	30	10	624	03.21
Total	**543.54**	**239**	**59**	**19422**	**100.00**

At the Jabalpur campus of the University, tree density is 36 trees per ha with the highest 136 trees/ha among the seven locations at Fruit Orchard and Forestry Block of 28.70 ha (Table 23.2). Density of species (5.44/ha) and family (1.32/ha) of trees are the highest in Administrative, College and Hostel block of the campus mainly due to presence of rich botanical garden in this block.

Table 23.2 Density of treesin different blocks at the Jabalpur campus of the University

Location	Density (Number/ Hectare)		
	Tree	Species	Family
Administrative, College and Hostel	126	5.44	1.32
Dusty Acre and Agro Forestry	88	0.41	0.11
Fruit Orchard and Forestry	136	1.99	0.49
Instructional Farm and Nursery	19	1.34	0.33
Research Farm	03	0.16	0.06
Residential area	105	1.65	0.39
Vegetable Res and Soybean Seed Farm	03	0.16	0.05
Average	**36**	**0.44**	**0.11**

In the main campus, 20 trees are available on each student, whereas it is 13.87 on each resident and 7.71 on per person i.e., all the staff, student and family members of the staff members residing in the residential area (Table 23.3).

Table 23.3 Ratio of trees with number of human being at Jabalpur campus of the University

Location	Number of trees per		
	Student	Resident	Person
Administrative, College and Hostel	4	3.09	1.71
Dusty Acre and Agro Forestry	7	4.57	2.54
Fruit Orchard and Forestry	4	2.78	1.55
Instructional Farm and Nursery	1	0.76	0.42
Research Farm	0.1	0.32	0.18
Residential area	3	01.9	1.06
Vegetable Res and Soybean Seed Farm	1	0.45	0.25
Total Result	**20**	**13.87**	**7.71**

A single mature tree can absorb 48 lbs. carbon dioxide/year and releases enough oxygen back into the atmosphere to fulfill the requirement of twohuman beings. https://wiki.nurserylive.com/t/how-many-trees-are-needed-to-produce-enough-oxygen-for-one-person/3430 stated that at least 1 mature tree or 4 to 5 small trees are needed to make enough oxygen for 1 person. According to this quote the headquarters of the University is self-sufficient to provide enough oxygen for students and staff.

Tree diversity estimated using Composite Entropy Index (CEI) with two components, number of trees/ species and their density shows that CEI increases with the decrease in density and rises with the number of trees/species (Table 23.4). Both the components of index are bounded by zero and one to range the value of CEI between zero and one. It is converted in to index by multiplication of CEI with 100 to range between 0 to 100. The highest value of 75.09 per cent of CEI at Administrative, College and Hostel block indicates higher tree diversification as compared to Research Farm (56.74%). The overall 66.97 per cent CEI at the headquarters indicates moderate tree diversification in the campus as high percent CEI indicates high diversity.

Table 23.4 Tree diversity based on Composite Entropy Index at the Jabalpur campus of the University

Location	Species (Number)	Trees (%)	Composite Entropy Index(%)
Administrative, College and Hostel	186	22.25	75.09
Dusty Acre and Agro Forestry	30	32.97	57.17
Fruit Orchard and Forestry	57	20.06	73.62
Instructional Farm and Nursery	76	05.46	70.01
Research Farm	23	02.32	56.74
Residential area	42	13.72	64.10
Vegetable Res and Soybean Seed Farm	30	03.21	72.99
Total Result	**239**	**100.00%**	**66.97**

CHAPTER 24

Climate Audit of Jabalpur Campus (1968-2022)

Anay Rawat
Department of Agronomy
College of Agriculture, Jabalpur

Climate:

Jabalpur (latitude 23.1815° N, longitude 79.9864°E and 1348 feet above MSL) has a humid subtropical climate, with mild winter and hot summer. About 94 per cent of rainfall is received during monsoon season and 5 percent during winter season. The annual average rainfall is 1394mm. May is the hottest month, with an average temperature exceeding 40 °C. January is the coldest month, with an average daily temperature near 15°C.

Annual temperature and rainfall

The annual maximum temperature varied from 41.6 to 48.3°C with a mean of 42.7°C at Jabalpur in last 42 years the lowest annual maximum temperature (41.6°C) was recorded during 2021 and the highest annual maximum temperature was recorded (48.3°C) during 1984 (Table 24.1).

The annual minimum temperature lowest value (0.5°C) was recorded during 1986 and 2011 and the highest minimum temperature (6.0°C) recorded during 1982. The annual mean of minimum temperature is 3.0°C (Table 24.1).

The highest annual rainfall of 2541.5 mm was recorded during the year 2013 at Jabalpur. The lowest annual rainfall of 577.2 mm was recorded in the year 1981 (Table 24.1). The annual average number of rainy days was 62 days at Jabalpur (Table 24.2).

Seasonal rainfall and rainy days

The average monsoon and winter rainfall are 1264 mm and 56 mm respectively at Jabalpur. The highest monsoon rainfall of 2331.4 mm was recorded during 2013 and the lowest monsoon rainfall of 467.2 mm during 1979. The highest winter rainfall of 193.9 mm was received during 1997 and the lowest rainfall of 0.0 mm were recorded during 1991 at Jabalpur (Table 24.2 and Fig. 24.2).

The average monsoon and winter season rainy days are 52 and 5 day respectively. The highest monsoon rainy days of 74 days were recorded during the year 2003 and the lowest number of rainy days (26 days) was observed during 1979. The highest number of winter season rainy days (12 days) was recorded during 1983, 1995 and 1997 while the lowest or nil rainy days was recorded in 1991 at Jabalpur.

Average Annual and Seasonal temperature

The highest value of average annual maximum temperature (33.5°C) was recorded during the year 2022 and the lowest (29.9°C) was during 1971. The average summer maximum temperature was observed to be 39.3°C during year 1980 and average summer minimum temperature was observed to be 32.3°C during year 1968. Annual maximum winter temperature (19.6°C) was recorded during the year 1977 and the lowest (17°C) during 2017. Average winter maximum temperature (12.1°C) was recorded during the year 1997 and Average winter minimum temperature (7.6°C) was recorded during the year 1996 (Table 24.3 and Fig. 24.3).

Table 24.1 Yearly Maximum, Minimum Temperature and Rainfall at Jabalpur

Year	Maximum Temp. (°C)	Minimum Temp. (°C)	Annual RF (mm)
1980	44.6	3.4	1641.0
1981	45.4	3.4	577.2
1982	42.6	6.0	1456.5
1983	44.2	3.5	1397.9
1984	48.3	3.3	1532.2
1985	44.5	5.7	1732.0
1986	44.4	0.5	1314.1
1987	44.2	3.9	1358.0
1988	46.7	5.3	1074.1
1989	45.7	2.0	744.5
1990	42.9	5.6	1824.0
1991	45.0	1.5	1370.0
1992	44.5	2.9	1065.8
1993	45.2	3.9	1486.2
1994	46.3	4.8	2083.2
1995	46.0	3.3	1417.2
1996	44.9	2.5	918.8
1997	42.5	4.0	1483.5
1998	45.3	4.0	943.3
1999	44.0	2.9	1806.8
2000	43.0	4.4	1109.0
2001	44.7	2.9	1509.3
2002	45.2	3.4	1274.8
2003	46.0	2.2	1747.4
2004	45.0	4.5	1224.3
2005	44.8	3.5	1853.4
2006	45.5	3.8	1038.0
2007	44.4	2.7	1188.5
2008	43.2	1.5	1398.7
2009	44.7	4.9	1492.0
2010	45.7	2.5	1643.2
2011	44.2	0.5	1857.9
2012	45.0	2.5	1353.5
2013	45.3	0.9	2541.5
2014	45.9	1.9	1090.0
2015	45.2	1.2	1201.1
2016	44.6	1.2	1729.3
2017	43.7	1.2	1060.9
2018	44.0	1.6	1170.1
2019	45.5	2.2	1706.9
2020	44.0	2.7	1307.5
2021	41.6	1.2	842.8
2022	43.6		

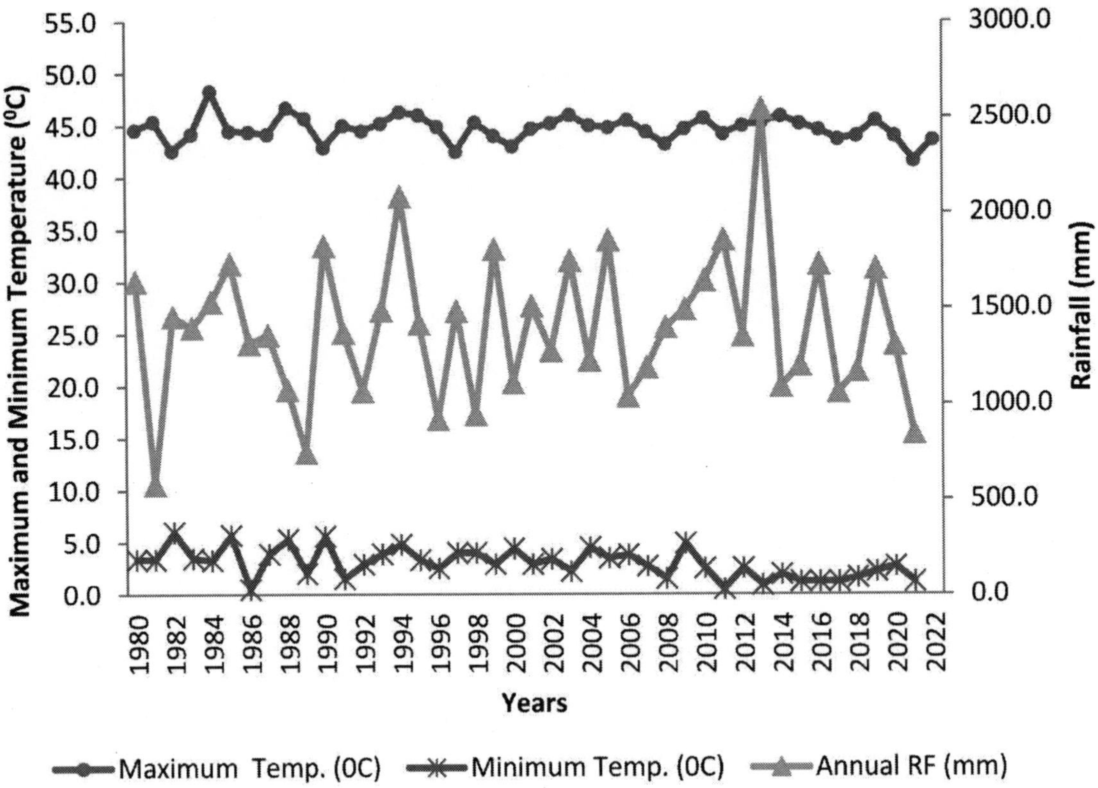

Fig 24.1 Yearly Lowest and Highest Maximum, Minimum Temperature and Rainfall at Jabalpur

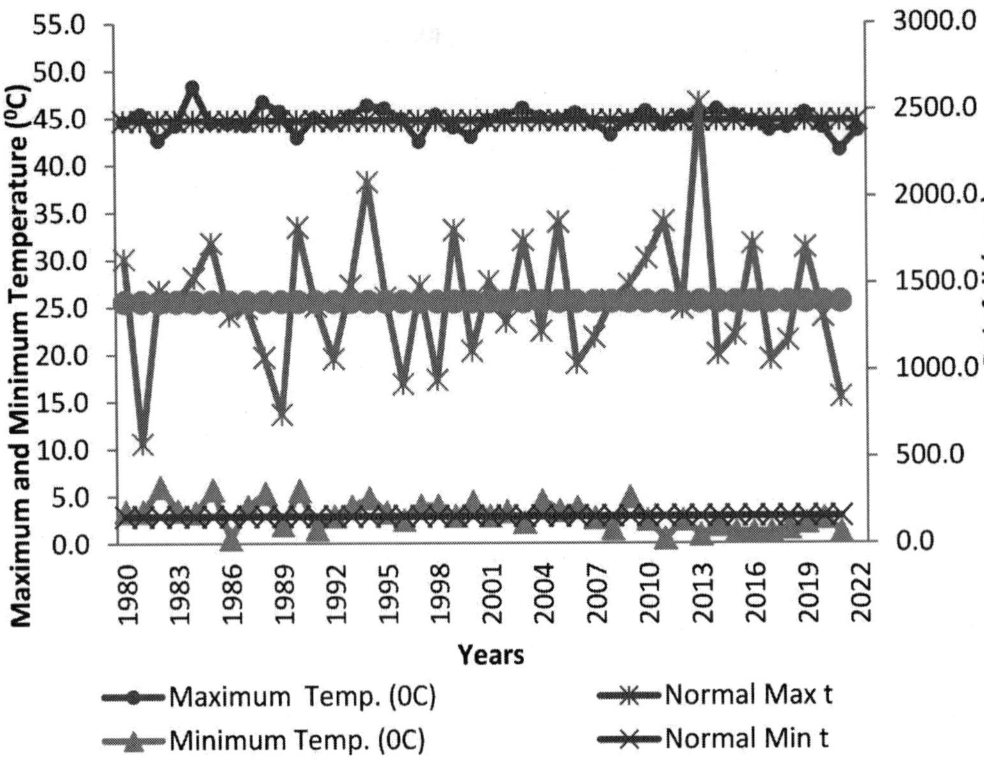

Fig 24.2 Yearly Lowest and Highest Maximum, Minimum Temperature and Rainfall at Jabalpur

Table 24.2 Yearly and Seasonal Rainfall and Rainy days

	Rainfall (mm)			Rainy days		
Year	Annual	Monsoon	Winter	Annual	Monsoon	Winter
1968	1045.0	952.0	10.9	54	43	1
1969	1261.4	1210.3	48.1	57	52	5
1970	1145.3	1058.5	11.8	66	58	2
1971	1723.3	1573.1	10.9	76	63	2
1972	1454.6	1344.4	101.6	43	34	6
1973	1427.9	1268.7	40.1	74	61	3
1974	934.4	835.4	9.7	43	37	2
1975	1518.0	1343.3	8.4	71	57	1
1976	1024.2	981.7	7.4	58	50	2
1977	2092.4	1729.0	133.3	69	53	11
1978	1507.9	1250.1	156.7	76	60	9
1979	620.2	467.2	12.2	40	26	2
1980	1641.0	1551.0	78.4	60	56	2
1981	577.2	491.2	160.0	46	39	11
1982	1456.5	1273.9	53.9	65	50	3
1983	1397.9	1271.1	82.6	59	52	12
1984	1532.2	1443.2	135.8	59	48	9
1985	1732.0	1470.6	118.1	74	56	9
1986	1314.1	1097.1	153.9	83	65	10
1987	1358.0	1123.4	12.2	61	45	2
1988	1074.1	1011.8	4.3	56	50	1
1989	744.5	685.0	73.6	50	46	3
1990	1824.0	1708.9	24.2	76	66	2
1991	1370.0	1334.6	0.0	51	47	0
1992	1065.8	1050.2	21.5	51	47	2
1993	1486.2	1428.9	8.8	65	60	1
1994	2083.2	2052.0	49.0	64	60	5
1995	1417.2	1170.8	158.1	61	44	12
1996	918.8	658.7	25.4	62	44	1
1997	1483.5	1084.8	193.9	86	59	12
1998	943.3	793.8	58.2	57	44	6
1999	1806.8	1670.9	5.8	78	66	1
2000	1109.0	1059.5	21.9	52	47	2
2001	1509.3	1350.0	43.7	82	66	4
2002	1274.8	1171.6	43.2	51	42	4
2003	1747.4	1613.1	66.0	83	74	3
2004	1224.3	1106.4	85.5	51	44	6
2005	1853.4	1710.2	7.2	71	59	1
2006	1038.0	886.6	83.7	58	45	4
2007	1188.5	1048.2	13.0	58	50	3
2008	1398.7	1366.5	37.8	46	41	3
2009	1492.0	1313.7	45.0	48	36	4
2010	1643.2	1532.0	13.2	51	44	2
2011	1857.9	1829.2	49.4	65	62	4
2012	1353.5	1298.6	63.4	49	44	5
2013	2541.5	2331.4	100.8	80	65	8
2014	1090.0	916.6	84.1	66	51	10
2015	1201.1	966.8	12.2	64	43	2

Year	Rainfall (mm)			Rainy days		
	Annual	Monsoon	Winter	Annual	Monsoon	Winter
2016	1729.3	1603.6	16.6	80	68	2
2017	1060.9	1007.1	18.0	59	50	3
2018	1170.1	1115.1	10.7	53	47	1
2019	1706.9	1624.8	55.2	62	53	6
2020	1307.5	1212.9	15.8	56	50	1
2021	842.8	652.4	51.3	55	44	7

Table 24.3 Yearly and Seasonal Maximum and Minimum Temperature

Year	Annual Max T	Summer Max T	Annual Mini t	Winter Min T
1968	30.2	32.3	17.8	9.1
1969	31.7	37.4	18.7	10.2
1970	31.0	37.4	18.3	9.1
1971	29.9	36.8	18.0	8.4
1972	31.2	37.0	18.6	11.8
1973	31.1	38.5	19.3	9.2
1974	31.2	38.0	18.6	9.8
1975	30.8	37.7	18.7	9.4
1976	31.4	36.9	19.3	10.5
1977	31.1	37.4	19.6	11.2
1978	30.4	36.7	18.8	10.5
1979	32.6	37.1	18.8	10.6
1980	32.3	39.3	19.1	10.2
1981	31.7	37.6	18.3	11.0
1982	31.3	36.2	18.7	10.3
1983	31.3	36.9	18.1	10.0
1984	31.3	38.7	17.9	9.2
1985	31.6	38.5	18.2	10.1
1986	31.6	37.3	18.2	10.3
1987	31.9	37.0	18.5	9.8
1988	32.3	38.8	18.7	8.2
1989	31.9	37.5	17.7	10.0
1990	30.9	36.3	18.6	10.4
1991	31.8	38.2	18.5	8.8
1992	32.2	38.0	18.1	9.1
1993	31.7	37.3	18.3	9.6
1994	31.5	38.3	18.2	9.2
1995	31.6	35.9	18.9	11.1
1996	32.0	38.1	18.5	7.6
1997	30.7	36.3	18.5	12.1
1998	31.1	35.8	19.1	9.3
1999	31.3	38.5	18.4	8.8
2000	31.4	36.6	18.0	8.7
2001	31.3	36.9	18.2	9.5
2002	32.3	38.3	18.9	10.4
2003	31.5	37.8	18.8	9.7
2004	31.6	38.5	18.6	10.3
2005	31.3	37.1	18.3	9.2
2006	31.6	36.2	18.9	10.5
2007	31.6	37.0	18.2	7.8

Year	Annual Max T	Summer Max T	Annual Mini t	Winter Min T
2008	31.8	37.8	16.6	9.8
2009	32.9	38.6	17.9	9.5
2010	33.0	39.9	18.9	8.4
2011	31.6	37.3	16.9	8.6
2012	32.1	37.7	17.5	9.0
2013	30.7	37.2	17.4	9.8
2014	31.5	36.3	18.0	9.1
2015	31.6	35.9	18.3	8.9
2016	32.0	38.0	17.6	8.3
2017	32.0	37.9	17.0	9.8
2018	32.1	38.0	18.0	8.4
2019	31.3	36.9	17.5	9.2
2020	31.3	36.4	18.2	9.6
2021	31.2	36.3	17.7	8.9
2022	33.5	38.2		

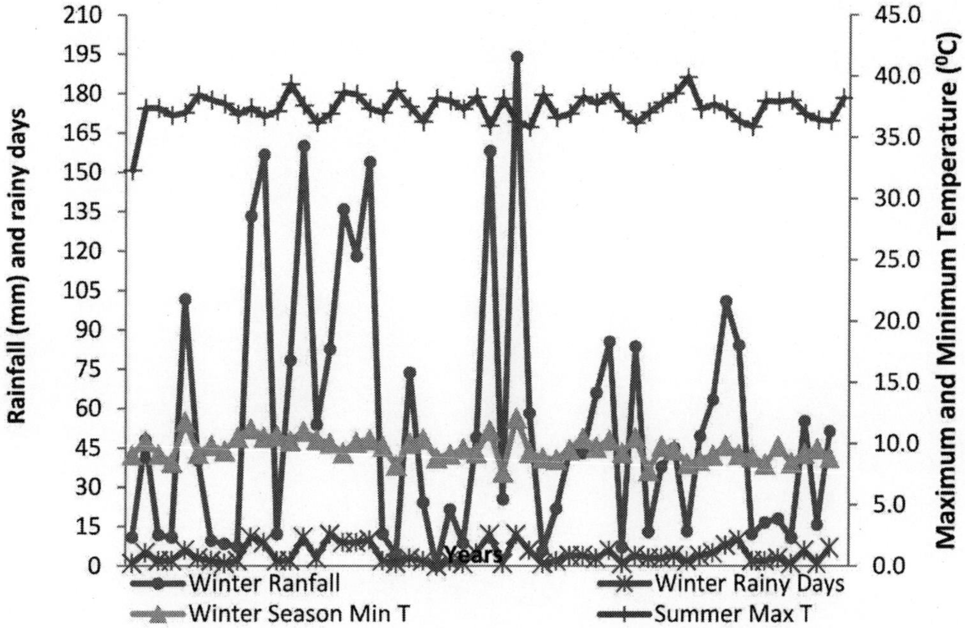

Fig 24.3 Season wise Maximum, Minimum Temperature and Rainfall and Rainy Days at Jabalpur

CHAPTER 25

Quantification of Carbon Sequestration Potential of Tree Species

Mudaliyar RSRK, Supriya Debnath, Gyanendra Tiwari and [1]S.D. Upadhyaya

Department of Plant Physiology,
College of Agriculture, Jabalpur 482004, MP

[1]Dean Faculty of Agriculture
Medi-Caps University, Indore

Urban Green spaces provide various ecosystem services such as improving air quality (Singh et al., 2018), buffering noise pollution, biodiversity conservation, microclimate regulation, stabilization of soil, groundwater recharge, prevention of soil erosion, and carbon sequestration (Shah & Gavali, 2017). In addition, these upgrade urban living standards by facilitating people's health and well-being by alleviating stress and enabling relaxation. Such areas also deliver various cultural services like spiritual and religious, recreation, ecotourism, and aesthetics (Chang et al., 2017). The maximum benefit of these spaces largely depends on the judicious selection of an appropriate and diverse mix of tree species and their proper management in the urban areas (Bhalla & Bhattacharya, 2015; Singh et al., 2017). Trees are considered to be the major sinks or sponges of carbon. The carbon assimilated by trees is retained for a longer duration with little leakage into the atmosphere. Annual carbon sequestration rates largely depend on the tree size at maturity, life span, and growth rates (Nowak, 2002). After the trees die, the biomass either enters the food chain or the soil as soil carbon (Suryavanshi et al., 2014). University campuses provide large areas for urban tree plantations that can be a potential solution for climate change mitigation. Knowing how much carbon an urban green space can sequester is helpful because it can help an institution or organization offset its emissions and values its green spaces. Therefore an effort was made to estimate the carbon sequestration, oxygen release and air pollutant removal potential of trees in JNKVV, Jabalpur campus. Based on detailed information about the tree that has been gathered in the biodiversity audit survey, the contribution of the JNKVV Jabalpur campus to carbon sequestration was calculated. For making the observation and analysis convenient, the campus is divided into seven blocks *viz.*, Administrative College and hostel, Dusty acre and Agroforestry, fruit orchard and forestry, Instructional Farm and Nursery, Research farm, Residential area, Vegetable Research and Soybean seed farm block for studying the carbon pool. For the calculation of carbon sequestration, the total tree samples available were 18262.

Steps involved in calculation of carbon-sequestration potential is as follows

A. Tree height and girth at breast height

A non-destructive method of biomass estimation was used to measure the tree height and girth at breast height of individual trees of the campus. Species level identification of the trees was obtained through visual observations and the doubtful samples were collected and stored in herbarium for later identification by the taxonomist.

B. Estimation of above ground biomass

First you have to calculate the above ground biomass as followed:

$W_{\text{above ground}} = 0.25D^2H$ (for trees when, D<11)

$W_{\text{above ground}} = 0.15D^2H$ (for trees when, D>11) where,

D= Diameter of tree trunk in inches,

H= Height of the tree in feet.

C. Determination of total green biomass of the tree

The root system weight is 20% of the above ground weight. Therefore, to determine the total green weight of the tree, multiply the above ground weight by 1.2.

$$W_{\text{total green biomass}} = 1.2 \times W_{\text{above ground}}$$

D. Determine the dry weight of the trees

The average tree is 72.5% dry matter and 27.5% moisture. Therefore, to determine the dry weight of the tree, multiply the total green weight of the tree by 72.5%.

$$W_{\text{dry-weight}} = 0.725 \times W_{\text{total green weight}}$$

E. Estimation of carbon content

Generally, for any plant species, 50% of its biomass is its carbon content (IPCC, 2006).

$$W_{\text{carbon}} = 0.5 \times \text{total biomass}$$

F. Estimation of Carbon-dioxide sequestration in the tree

CO_2 has one molecule of carbon and two molecules of oxygen. The atomic weight of carbon is 12 and the atomic weight of oxygen is 16. The weight of CO2 in trees is determined by the ratio of CO_2 to C is 44/12 = 3.67. Therefore, to determine the weight of carbon dioxide sequestered in the tree, multiply the weight of carbon in the tree by 3.67.

$$W_{\text{carbon-dioxide}} = 3.67 * W_{\text{carbon}}$$

Status of total carbon stocks, carbon-sequestration potential of trees

A total of two-thirty-three different species of trees were enumerated on the campus (Table 25.1). The most dominant species on the campus is *Psidium guajava*, with a total of 2900 trees. The second and the third most common tree species are *Gmelina arborea* and *Tectona grandis*, with their tree number equal to 1417 and 1847, respectively. The above-ground biomass (AGB) and total biomass of all the campus trees are equivalent to 1,57,371.40 Kg and 1,88,845.70 Kg, respectively. The total carbon stock of the campus trees is equal to 68,456.64 Kg. The total carbon sequestered by all the trees in a year is 250982.40 Kg or 250.98 tons. *Mangifera indica* stood first in total carbon dioxide sequestration with 50173.82 kg, followed by *Psidium guajava* and *Tectona grandis* with 20810.34 Kg and 18401.64 Kg, respectively. In other words, on average, carbon sequestered by an individual tree on the campus is 13.74 kg. The average diameter at breast height is 12.91 inches. Nowak et al. (2007) revealed that for trees on land uses with a park-like structure (e.g., parks, cemeteries, golf courses), average dbh growth was 0.61 cm/year (0.24 in/year). On that basis, the average age of the tree was 19.67 years. On that basis, the calculation of CO2 sequestered and O2 released per year was calculated. The total carbon dioxide sequestered by campus trees is 12759.66 kg per year.

Table 25.1 Total carbon stock, CO2 sequestered and molecular O2 released by the tree species at the University headquarters

Tree species	Family	Number	Above ground biomass	Total Biomass	Total Dry biomass	Total Carbon stock	CO_2 sequestration	Molecular O_2 released
			Kg					
Acacia nilotica	Fabaceae	123	473.31	567.98	411.78	205.89	754.86	2012.96
Acacia auriculiformis	Mimosaceae	24	316.15	379.38	275.05	137.53	504.21	1344.56
Acacia catechu	Fabaceae	7	26.94	32.32	23.43	11.72	42.96	114.56
Acacia conncinna	Mimosaceae	10	44.99	53.99	39.15	19.57	71.76	191.36
Acacia mangium	Mimosaceae	2	17.93	21.51	15.60	7.80	28.59	76.24
Acacia senegal	Fabaceae	66	354.94	425.93	308.80	154.40	566.07	1509.52
Acalypha hispida	Euphorbiaceae	4	52.69	63.23	45.84	22.92	84.04	224.09
Acarus sapota	Sapotaceae	65	312.66	375.19	272.01	136.01	498.64	1329.70
Adansonia digitata	Bombacaceae	3	8.66	10.39	7.53	3.77	13.81	36.82
Adenanthera parvania	Fabaceae	2	18.44	22.13	16.04	8.02	29.41	78.43
Aegel marmelos	Rutaceae	186	894.68	1073.61	778.37	389.18	1426.87	3804.98
Albizia lebbeck	Mimosaceae	26	183.80	220.56	159.91	79.95	293.14	781.70
Albizia procera	Mimosaceae	39	308.25	369.90	268.18	134.09	491.61	1310.95
Alstonia scholaris	Apocynaceae	56	927.32	1112.78	806.76	403.38	1478.92	3943.79
Annona muricata	Annonaceae	25	108.44	130.13	94.35	47.17	172.95	461.20
Annona reticulata	Annonaceae	8	113.82	136.59	99.02	49.51	181.53	484.07
Annona squamosa	Annonaceae	226	980.32	1176.39	852.88	426.44	1563.46	4169.23
Anogeissus latifolia	Combretaceae	1	4.34	5.21	3.77	1.89	6.92	18.45
Anthocephalus cadamba	Rubiaceae	6	91.27	109.52	79.40	39.70	145.55	388.14
Areca catechu	Arecaceae	1	2.12	2.55	1.85	0.92	3.38	9.02
Artabotrys hexapetalus	Annonaceae	5	21.68	26.01	18.86	9.43	34.57	92.20
Artocarpus heterophyllus	Moraceae	78	559.30	671.16	486.59	243.29	891.99	2378.64
Artocarpus lacucha	Moraceae	19	136.24	163.49	118.53	59.26	217.28	579.41
Azadirachta indica	Meliaceae	412	3256.36	3907.63	2833.03	1416.52	5193.38	13849.01
Bambusa tulda	Poaceae	100	4810.09	5772.11	4184.78	2092.39	7671.33	20456.89
Bassia latifolia	Sapotaceae	19	125.14	150.17	108.87	54.44	199.58	532.22
Bauhinia purpurea	Caesalpiniaceae	7	46.11	55.33	40.11	20.06	73.53	196.08
Bauhinia vahli	Leguminaceae	1	3.23	3.87	2.81	1.40	5.15	13.72
Bauhinia variegata	Caesalpiniaceae	74	238.78	286.53	207.74	103.87	380.81	1015.49
Bersera spp.	Burseraceae	1	8.96	10.76	7.80	3.90	14.29	38.12
Bixa orellana	Bixaceae	7	21	25.20	18.27	9.13	33.49	89.30
Bombax ceiba	Malvaceae	17	134.36	161.24	116.90	58.45	214.29	571.44
Bombax malabaricum	Bombacaceae	5	39.52	47.42	34.38	17.19	63.03	168.07
Borassus flabellifer	Arecaceae	9	51.95	62.34	45.20	22.60	82.85	220.93
Boswellia serrata	Burseraceae	2	11.54	13.85	10.04	5.02	18.41	49.10
Bougainvillea species	Nyctaginaceae	14	50.39	60.47	43.84	21.92	80.37	214.32
Buchanania lanzan	Anacardiaceae	4	8.43	10.12	7.33	3.67	13.45	35.86
Butea frondosa	Leguminaceae	6	19.36	23.23	16.84	8.42	30.88	82.34
Butea monosperma	Fabaceae	57	306.54	367.84	266.69	133.34	488.88	1303.68
Caesalpinia crista	Caesalpiniaceae	3	13.50	16.20	11.74	5.87	21.53	57.41
Calendrina sp.	Mimosaceae	5	9.82	11.78	8.54	4.27	15.66	41.76
Calliandra surinamensis	Caesalpiniaceae	53	764.80	917.77	665.38	332.69	1219.74	3252.64
Callistemon lanceolatus	Myrtaceae	8	71.70	86.05	62.38	31.19	114.36	304.95
Callistemon rigidus	Myrtaceae	2	17.93	21.51	15.60	7.80	28.59	76.24
Careya arbarea	Lecythidaceae	52	367.61	441.13	319.82	159.91	586.27	1563.40
Carissa carandas	Apocynaceae	152	360.34	432.41	313.50	156.75	574.68	1532.49

Tree species	Family	Number	Above ground biomass	Total Biomass	Total Dry biomass	Total Carbon stock	CO_2 sequestration	Molecular O_2 released
			Kg					
Cascabela thevetia	Apocynaceae	18	28.57	34.28	24.86	12.43	45.57	121.51
Cassia fistula	Fabaceae	30	28.57	34.28	24.86	12.43	45.57	121.51
Cassia siamea	Caesalpiniaceae	16	86.05	103.25	74.86	37.43	137.23	365.94
Casuarina equisetifolia	Casuarinaceae	1	6.41	7.70	5.58	2.79	10.23	27.28
Ceiba pentandra	Malvaceae	22	173.88	208.66	151.28	75.64	277.32	739.51
Celastrus paniculatus	Celastraceae	5	65.86	79.04	57.30	28.65	105.04	280.12
Cestrum nocturnam	Solanaceae	14	62.99	75.59	54.80	27.40	100.46	267.90
Cestrum nocturnam	Solanaceae	16	71.99	86.39	62.63	31.32	114.81	306.17
Cicca acida	Phyllanthaceae	28	134.68	161.62	117.17	58.59	214.80	572.79
Cinnamomum camphora	Lauraceae	2	12.91	15.49	11.23	5.61	20.58	54.89
Cinnamomum tamala	Lauraceae	5	24.05	28.86	20.92	10.46	38.36	102.28
Cinnamomum zeylanicum	Lauraceae	1	3	3.60	2.61	1.30	4.78	12.76
Citrus aurantiifolia	Rutaceae	179	497.10	596.52	432.48	216.24	792.80	2114.13
Citrus grandis	Rutaceae	1	2.78	3.33	2.42	1.21	4.43	11.81
Citrus jambhiri	Rutaceae	1	2.78	3.33	2.42	1.21	4.43	11.81
Citrus limetta	Rutaceae	1	2.78	3.33	2.42	1.21	4.43	11.81
Citrus limonia	Rutaceae	12	33.33	39.99	28.99	14.50	53.15	141.73
Citrus paradisi	Rutaceae	1	2.78	3.33	2.42	1.21	4.43	11.81
Citrus sinensis	Rutaceae	49	136.08	163.29	118.39	59.19	217.02	578.73
Cleistanthus collinus	Phyllanthaceae	1	4.81	5.77	4.18	2.09	7.67	20.46
Cocos nucifera	Arecaceae	9	22.82	27.38	19.85	9.93	36.39	97.04
Codiaeum variegatum	Euphorbiaceae	23	58.31	69.97	50.73	25.36	92.99	247.98
Coffea arabica	Rubiaceae	2	1.28	1.54	1.12	0.56	2.05	5.46
Copernicia alba	Arecaceae	2	5.07	6.08	4.41	2.21	8.09	21.56
Cordia dichotoma	Ehretiaceae	4	28.68	34.42	24.95	12.48	45.74	121.98
Cordia macleoidi	Boraginaceae	8	57.36	68.84	49.91	24.95	91.49	243.96
Cordia myxa	Boraginaceae	7	15.87	19.05	13.81	6.90	25.31	67.50
Crataeva nurvala	Cappraceae	2	4.50	5.40	3.91	1.96	7.18	19.14
Crataeva religia	Capparaceae	1	4.50	5.40	3.91	1.96	7.18	19.14
Dalbergia latifolia	Fabaceae	2	0	0	0	0	0	0
Dalbergia sissoo	Caesalpiniaceae	315	2874.85	3449.82	2501.12	1250.56	4584.93	12226.48
Delonix regia	Caesalpiniaceae	108	824.67	989.60	717.46	358.73	1315.21	3507.22
Dendrocalamus strictus	Poaceae	153	11149.01	13378.81	9699.64	4849.82	17780.89	47415.72
Dillenia indica	Dilleniaceae	3	16.13	19.36	14.04	7.02	25.73	68.61
Diospyros melanoxylon	Ebenaceae	2	17.93	21.51	15.60	7.80	28.59	76.24
Diospyros montana	Ebenaceae	5	19.24	23.09	16.74	8.37	30.69	81.83
Elaeis guineensis	Arecaceae	44	170.37	204.44	148.22	74.11	271.71	724.57
Elaeocarpus angustifolius	Elaeocarpaceae	4	11.54	13.85	10.04	5.02	18.41	49.10
Emblica officinalis	Euphorbiaceae	359	1036.09	1243.31	901.40	450.70	1652.41	4406.41
Epipremnum aureum	Arecaceae	1	3.85	4.62	3.35	1.67	6.14	16.37
Erythria variegata	Fabaceae	6	23.09	27.71	20.09	10.04	36.82	98.19
Erythrina indica	Fabaceae	1	6.41	7.70	5.58	2.79	10.23	27.28
Erythrina orientalis	Fabaceae	3	19.24	23.09	16.74	8.37	30.69	81.83
Eucalyptus globulus	Myrtaceae	422	4447.20	5336.64	3869.06	1934.53	7092.57	18913.53
Eucalyptus citriodora	Myrtaceae	116	1222.45	1466.94	1063.53	531.77	1949.62	5198.98
Eucalyptus gigantea	Myrtaceae	9	94.85	113.81	82.52	41.26	151.26	403.37
Eucalyptus sp.	Myrtaceae	1016	10707	12848.40	9315.09	4657.55	17075.96	45535.89

Tree species	Family	Number	Above ground biomass	Total Biomass	Total Dry biomass	Total Carbon stock	CO_2 sequestration	Molecular O_2 released
			Kg					
Euginea uniflora	Myrtaceae	1	4.71	5.66	4.10	2.05	7.52	20.04
Euphorbia pulcherrima	Euphorbiaceae	2	15.71	18.85	13.67	6.83	25.05	66.81
Feronia elephandum	Rutaceae	1	6.41	7.70	5.58	2.79	10.23	27.28
Ficus banjamiana	Moraceae	109	524.30	629.16	456.14	228.07	836.18	2229.80
Ficus benghalensis	Moraceae	35	250.97	301.16	218.34	109.17	400.25	1067.34
Ficus carica	Moraceae	1	4.81	5.77	4.18	2.09	7.67	20.46
Ficus elastica	Moraceae	3	21.51	25.81	18.71	9.36	34.31	91.49
Ficus glomerata	Moraceae	16	161.62	193.94	140.61	70.30	257.76	687.35
Ficus racemosa	Moraceae	23	147.51	177.01	128.33	64.17	235.25	627.34
Ficus religiosa	Moraceae	83	488.96	586.76	425.40	212.70	779.82	2079.52
Ficus retusa	Moraceae	1	7.17	8.60	6.24	3.12	11.44	30.50
Ficus virens	Moraceae	1	4.50	5.40	3.91	1.96	7.18	19.14
Gardenia gummifera	Rubiaceae	4	7.26	8.71	6.31	3.16	11.57	30.86
Gliricidia sepium	Fabaceae	5	32.07	38.48	27.90	13.95	51.14	136.38
Glycosmis pentaphylla	Rutaceae	5	13.89	16.66	12.08	6.04	22.15	59.05
Gmelina arborea	Lamiaceae	1417	8347.74	10017.29	7262.54	3631.27	13313.32	35502.18
Gossypium arboreum	Malvaceae	2	8.33	10	7.25	3.62	13.29	35.43
Grevillea robusta	Proteaceae	9	80.67	96.80	70.18	35.09	128.65	343.07
Grewia asiatica	Proteaceae	38	24.37	29.25	21.20	10.60	38.87	103.65
Haloptelea integrifolia	Ulmaceae	4	31.62	37.94	27.51	13.75	50.42	134.46
Helicteres isora	Malvaceae	1	1.96	2.36	1.71	0.85	3.13	8.35
Hevea brasiliensis	Euphorbiaceae	1	1.96	2.36	1.71	0.85	3.13	8.35
Hibiscus rosa-sinensis	Malvaceae	32	62.84	75.41	54.67	27.33	100.22	267.25
Hyophorbe lagenicaulis	Arecaceae	168	1187.65	1425.18	1033.26	516.63	1894.12	5050.98
Inga dulcis	Caesalpiniaceae	2	25.45	30.54	22.14	11.07	40.59	108.25
Ixora coccinea	Euphorbiaceae	4	19.24	23.09	16.74	8.37	30.69	81.83
Jacaranda mimosifolia	Bignomiaceae	2	21.30	25.55	18.53	9.26	33.96	90.57
Jasminum sambac	Oleaceae	29	28.47	34.17	24.77	12.39	45.41	121.10
Juniperus communis	Cupressaceae	9	37.49	44.99	32.62	16.31	59.79	159.45
Justicia adhatoda	Acanthaceae	300	221.76	266.11	192.93	96.46	353.67	943.11
Kigelia Pinnata	Bignomiaceae	23	312.92	375.50	272.24	136.12	499.05	1330.80
Lagerstroemia parviflora	Lythraceae	2	15.81	18.97	13.75	6.88	25.21	67.23
Lagerstroemia flosregini	Lythraceae	4	15.21	18.25	13.23	6.62	24.26	64.69
Lagerstroemia indica	Lythraceae	22	83.66	100.39	72.78	36.39	133.42	355.80
Lagerstroemia parviflora	Lythraceae	7	55.33	66.39	48.13	24.07	88.24	235.30
Lagerstroemia speciosa	Lythraceae	57	450.52	540.62	391.95	195.97	718.50	1916
Lantana camara	Verbenaceae	3	4.71	5.66	4.10	2.05	7.52	20.04
Lawsonia inermis	Lythraceae	15	44.99	53.99	39.15	19.57	71.76	191.36
Legerstroemia indica	Lythraceae	6	47.42	56.91	41.26	20.63	75.63	201.68
Leucaena leucocephala	Mimosaceae	509	4023.03	4827.64	3500.04	1750.02	6416.09	17109.57
Leucas aspera	Fabaceae	1	13.17	15.81	11.46	5.73	21.01	56.02
Limonia acidissima	Rutaceae	15	96.20	115.44	83.70	41.85	153.43	409.14
Litchi chinensis	Sapindaceae	21	123.71	148.46	107.63	53.82	197.30	526.14
Litoria umarensis	Pelodryadiaceae	1	0	0	0	0	0	0
Litsea glutinosa	Lauraceae	3	21.21	25.45	18.45	9.23	33.82	90.20
Livistona chinensis	Arecaceae	23	162.60	195.11	141.46	70.73	259.31	691.50
Livistona rotundifolia	Arecaceae	1	7.07	8.48	6.15	3.08	11.27	30.07

Tree species	Family	Number	Above ground biomass	Total Biomass	Total Dry biomass	Total Carbon stock	CO_2 sequestration	Molecular O_2 released
			Kg					
Madhuca indica	Sapotaceae	14	92.21	110.65	80.22	40.11	147.06	392.16
Madhuca latifolia	Sapotaceae	1	6.59	7.90	5.73	2.87	10.50	28.01
Madhuca longifolia	Sapotaceae	4	26.35	31.62	22.92	11.46	42.02	112.05
Magnolia champa	Magnoliaceae	6	38.48	46.18	33.48	16.74	61.37	163.66
Mallotus philippensis	Euphorbiaceae	15	96.20	115.44	83.70	41.85	153.43	409.14
Mangifera indica	Anacardiaceae	834	31460.09	37752.10	27370.27	13685.14	50173.82	133796.85
Manilkara hexandra	Sapotaceae	24	189.69	227.63	165.03	82.52	302.53	806.74
Melaleuca citrina	Myrtaceae	2	0	0	0	0	0	0
Melia azedarach	Meliaceae	36	232.32	278.79	202.12	101.06	370.52	988.05
Millettia pinnata	Fabaceae	4	15.21	18.25	13.23	6.62	24.26	64.69
Mimusops elengi	Sapotaceae	10	47.13	56.55	41	20.50	75.16	200.44
Mitragyna parviflora	Rubiaceae	6	63.23	75.88	55.01	27.51	100.84	268.91
Moringa oleifera	Moringaceae	72	516.27	619.53	449.16	224.58	823.37	2195.66
Morus alba	Moraceae	22	98.99	118.78	86.12	43.06	157.87	420.98
Murraya exotica	Rutaceae	6	9.43	11.31	8.20	4.10	15.03	40.09
Murraya koenigii	Rutaceae	1	2.27	2.72	1.97	0.99	3.62	9.64
Murraya panniculata	Rutaceae	504	1142.82	1371.38	994.25	497.13	1822.62	4860.31
Musa paradisiaca	Musaceae	14	62.99	75.59	54.80	27.40	100.46	267.90
Mussaenda frondosa	Rubiaceae	3	4.71	5.66	4.10	2.05	7.52	20.04
Mussaenda rosa	Rubiaceae	2	3.14	3.77	2.73	1.37	5.01	13.36
Neolamarckia cadamba	Rubiaceae	43	462.49	554.99	402.37	201.19	737.61	1966.95
Nerium alba	Apocynaceae	2	3.15	3.78	2.74	1.37	5.02	13.38
Nerium indicum	Apocynaceae	6	4.43	5.32	3.86	1.93	7.07	18.86
Nerium oleander	Apocynaceae	15	23.60	28.32	20.53	10.27	37.64	100.37
Nyctanthes arbor-trisitis	Oleaceae	4	7.85	9.43	6.83	3.42	12.53	33.41
Oroxylum indicum	Bignomiaceae	144	774.41	929.29	673.74	336.87	1235.06	3293.50
Parkia biglandulosa	Mimosaceae	7	111.67	134.01	97.16	48.58	178.10	474.93
Parkinsonia aculeata	Fabaceae	2	15.81	18.97	13.75	6.88	25.21	67.23
Peltaphorum pterocarpum	Caesalpiniaceae	42	331.96	398.35	288.80	144.40	529.42	1411.79
Phoenix dactylifera	Arecaceae	46	296.86	356.23	258.27	129.13	473.44	1262.51
Phyllanthus emblica	Euphorbiaceae	183	1968.29	2361.95	1712.41	856.21	3139.11	8370.97
Phyllantun acidus	Euphorbiaceae	5	32.07	38.48	27.90	13.95	51.14	136.38
Pimenta dioica	Myrtaceae	3	5.89	7.07	5.13	2.56	9.40	25.05
Pinus roxburghii	Pinaceae	1	8.02	9.62	6.97	3.49	12.79	34.09
Pinus syivestris	Pinaceae	2	16.03	19.24	13.95	6.97	25.57	68.19
Pithecellobium dulce	Mimosaceae	146	656.92	788.30	571.52	285.76	1047.68	2793.81
Plumeria alba	Apocynaceae	8	25.40	30.48	22.09	11.05	40.50	108.01
Plumeria indica	Apocynaceae	29	92.06	110.47	80.09	40.05	146.82	391.52
Plumeria rubra	Apocynaceae	5	15.87	19.05	13.81	6.90	25.31	67.50
Polyalthea longifolia	Annonaceae	831	10813.93	12976.71	9408.12	4704.06	17246.49	45990.64
Pongamia pinnata	Fabaceae	894	4807.80	5769.36	4182.78	2091.39	7667.67	20447.12
Prosopis cineraria	Mimosaceae	12	71.99	86.39	62.63	31.32	114.81	306.17
Prunus amygdalus	Rosaceae	6	53.78	64.53	46.79	23.39	85.77	228.71
Prunus dulcis	Rosaceae	2	0	0	0	0	0	0
Psidium cattleianum	Myrtaceae	1	1.96	2.36	1.71	0.85	3.13	8.35
Psidium guajava	Myrtaceae	2900	13048.35	15658.02	11352.06	5676.03	20810.04	55493.43
Pterocarpus marsupium	Fabaceae	13	91.90	110.28	79.95	39.98	146.57	390.85

Tree species	Family	Number	Above ground biomass	Total Biomass	Total Dry biomass	Total Carbon stock	CO_2 sequestration	Molecular O_2 released
			Kg					
Pterospermum acerifolium	Malvaceae	19	200.23	240.28	174.20	87.10	319.33	851.56
Punica granatum	Punicaceae	115	344.96	413.95	300.11	150.06	550.15	1467.07
Putranjiva roxburghii	Annonaceae	6	78.08	93.69	67.93	33.96	124.52	332.06
Putranjiva roxburghii	Putaranjivaceace	22	144.90	173.88	126.07	63.03	231.10	616.26
Roystonea regia	Arecaceae	2	14.14	16.97	12.30	6.15	22.55	60.13
Russelia juncea	Scrophulariaceae	2	0					
Salmalis malaberica	Malvaceae	7	37.64	45.17	32.75	16.38	60.04	160.10
Santalum album	Sanatalaceae	13	93.22	111.86	81.10	40.55	148.66	396.44
Sapindus mukorossi	Sapindaceae	3	19.24	23.09	16.74	8.37	30.69	81.83
Saraca asoca	Detarioideae	174	1115.94	1339.13	970.87	485.43	1779.75	4746
Saraca indica	Detarioideae	4	19.24	23.09	16.74	8.37	30.69	81.83
Schleichera oleosa	Sapindaceae	7	44.89	53.87	39.06	19.53	71.60	190.93
Semecarpus anacardium	Anacardiaceae	7	0	0	0	0	0	0
Senna siamea	Fabaceae	2	6.45	7.74	5.61	2.81	10.29	27.45
Sesbania augsata	Arecaceae	10	96.20	115.44	83.70	41.85	153.43	409.14
Spathodea companulata	Bignomiaceae	4	28.28	33.93	24.60	12.30	45.10	120.26
Sphaeranthus indicus	Rubiaceae	1	0	0	0	0	0	0
Spondias cythera	Anacardiaceae	1	2.11	2.53	1.83	0.92	3.36	8.96
Sterculia alata	Sterculiaceae	2	15.81	18.97	13.75	6.88	25.21	67.23
Sterculia foetida	Sterculiaceae	1	7.90	9.48	6.88	3.44	12.61	33.61
Sterculia guttata	Sterculiaceae	4	18.85	22.62	16.40	8.20	30.07	80.17
Sterculia urens	Malvaceae	8	15.71	18.85	13.67	6.83	25.05	66.81
Stereospermum suavealens	Bignomiaceae	2	14.14	16.97	12.30	6.15	22.55	60.13
Strychnos nux-vomica	Strychnaceae	4	28.68	34.42	24.95	12.48	45.74	121.98
Swietenia mahagoni	Meliaceae	1	7.90	9.48	6.88	3.44	12.61	33.61
Syzygium aromaticum	Myrtaceae	1	0.64	0.77	0.56	0.28	1.02	2.73
Syzygium cumini	Myrtaceae	112	885.22	1062.27	770.15	385.07	1411.79	3764.78
Syzygium fruiticosum	Myrtaceae	1	7.17	8.60	6.24	3.12	11.44	30.50
Syzygium jambos	Myrtaceae	2	14.34	17.21	12.48	6.24	22.87	60.99
Syzygium saliafelium	Myrtaceae	1	7.17	8.60	6.24	3.12	11.44	30.50
Tabebuia argentea	Bignomiaceae	8	30.78	36.94	26.78	13.39	49.10	130.92
Tabebuia pentaphylla	Bignomiaceae	50	192.40	230.88	167.39	83.70	306.85	818.28
Tabebuia rosea	Bignomiaceae	38	268.64	322.36	233.71	116.86	428.43	1142.48
Tabernaemontana divaricata	Apocynaceae	4	8.48	10.18	7.38	3.69	13.53	36.08
Tamarindus indica	Fabaceae	334	2639.87	3167.84	2296.68	1148.34	4210.17	11227.11
Tecoma stans	Bignomiaceae	6	15.21	18.25	13.23	6.62	24.26	64.69
Tectona grandis	Verbenaceae	1847	12165.25	14598.30	10583.77	5291.88	19401.64	51737.69
Terminalia arjuna	Combretaceae	285	1877.15	2252.58	1633.12	816.56	2993.76	7983.35
Terminalia bellirica	Combretaceae	60	474.23	569.07	412.58	206.29	756.32	2016.85
Terminalia catapa	Combretaceae	6	47.42	56.91	41.26	20.63	75.63	201.68
Theobroma cacao	Sterculiaceae	2	6	7.20	5.22	2.61	9.57	25.51
Thespesia populnea	Malvaceae	56	212.95	255.54	185.27	92.63	339.62	905.66
Thevetia alba	Apocynaceae	1	2.41	2.89	2.09	1.05	3.84	10.23
Thevetia peruviana	Apocynaceae	23	55.93	67.11	48.66	24.33	89.20	237.85
Thuja occidentalis	Cupressaceae	29	65.76	78.91	57.21	28.60	104.87	279.66
Total		**18262**	**157371.4**	**188845.7**	**136913.1**	**68456.64**	**250982.4**	**669286.2**

Oxygen release, oxygen demand offset, and monetary values of carbon sequestration and air pollution removal of total trees

The net amount of oxygen a tree produces during a year is directly related to the amount of carbon sequestered by the tree, which is tied to the accumulation of tree biomass. The total O_2 released by the campus trees is 6,69,286.20 kg, calculated as per the formulae given by Nowak et al. 2007. The annual oxygen produced by campus trees is 34025.74 kg per year (Table 25.2). Data on the average adult human oxygen consumption rate of 0.84 kg/day (1.85 lb/day) (Perry and Le Van, 2003) was used to estimate how much human oxygen consumption would be offset by campus trees' oxygen production annually. In order to estimate how much human oxygen consumption is offset by the carbon stock of the University, annual oxygen production was divided by average annual oxygen consumption per person. This net oxygen production offsets the oxygen consumption of 4.40% of the total population of the University campus.

Oxygen production is one of many environmental benefits trees produce, and campus trees produce a significant amount of oxygen. Thus, although campus trees produce significant amounts of oxygen, it is not a significant ecological benefit given the global nature of oxygen and the sheer volume of oxygen in the atmosphere. The relative impacts of carbon sequestration and air pollution removal are much more significant than oxygen production.

In the University campus, tree carbon sequestration was estimated at 250982.4 Kg with a monetary value of USD 8282.42. Air pollution removal by trees on the University campus was estimated at 80962.07 tons with a $39.23 million value which was calculated as per the formulae given by Nowak et al. (2007) (Table 25.3). Oxygen production by Campus tree is estimated at 669826.34 (67 million tons), but the monetary value to society is negligible. The oxygen production value of campus trees is insignificant due to a large amount of oxygen within the atmosphere (approximately 21% of the atmosphere's volume is oxygen). The oxygen content of the atmosphere remains essentially constant, with the oxygen consumed by all animals, bacteria, and respiration processes roughly balanced by the oxygen released by land and sea plants during photosynthesis.

Table 25.2 Oxygen production and population oxygen demand offset by trees at JNKVV, Jabalpur University Campus

Population	Oxygen consumed by total population per day (kg)	Oxygen consumed by total population per year (Kg)	Oxygen produced by trees at campus (Kg)	Oxygen Offset (%)	Monetary values of carbon sequestration by campus tress (Dollars)
2520	2116.8	772635	34025.74	4.40	8282.42

Table 25.3 Air pollution removal by trees at JNKVV Campus

Air pollution removal by the tree campus in the University (tons)	80962.071
Monetary value of air pollution removal (million $)	39.23

Suggestions

The average number of trees needed to offset the annual oxygen consumption of one adult is two trees. However, currently, the University campus has seven trees per person. Hence, the university has the lead of five trees/ person. Considering the university's current population density of 4.63 persons/ ha, 32 trees/ ha is needed to offset 100% oxygen consumption demand. However, the current tree density is 34 trees /ha. Hence, University tree numbers and tree density are enough to offset the oxygen consumption demand of 100% of the population.

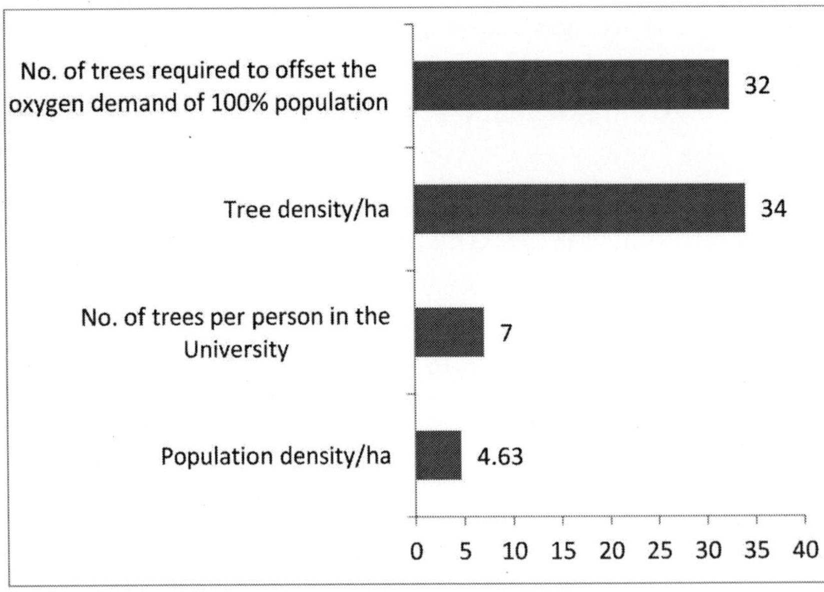

Fig 25.1 Population density, tree density, no. of trees/person and no. of trees/ha required to offset the oxygen demand of 100% population.

Tree mortality is a significant factor affecting change in the urban ecosystem. Medium- to low-density residential areas exhibited low mortality rates of 2.2% (Nowak, 2004). Considering the University's current tree population of 18262 and with a 2.2% mortality rate per year, 401 trees will get dead and need to replenish through a plantation. Hence, plantation of one tree per hectare is needed, which help maintain the present forest area of 12% in the University. The tree species recommended for plantation are *Alstonia scholaris, Annona squamosa, Saraca ashoka, Dalbergia sissoo, Embilica officinalis, Mangifera indica, Gmelina arborea, Psidium guajava, Mamphal, Acacia catechu, Acacia nilotica* as these trees are found suitable for the agro-forestry system of the University (Plate 25.1).

Plate 25.1 *Dalbergia sissoo*

Saraca ashoka

Reference

Bhalla P and Bhattacharya P. 2015. Urban Biodiversity and Green Spaces in Delhi: A Case Study of New Settlement and Lutyens' Delhi. Journal of Human Ecology 51(1, 2): 83-96.

Canu DM, Ghermandib A, Nunes PALD, Lazzari P, Cossarini G and Solidoro C. 2015. Estimating the carbon sequestration ecosystem services in the Mediterranean Sea: An ecological economics approach. Global Environmental change 32: 87-95.

Chang J, Qu Z, Xu R, Pan K, Xu B, Min Y, Ren Y, Yang G and Ge Y. 2017. Assessing the ecosystem services provided by urban green spaces along urban center-edge gradients. Nature Scientific reports 7(1): 11226.

Fransen, B. 2019. EcoMatcherhttps://www.ecomatcher.com/author/bashk/.

IPCC 2006. IPCC Guidelines for National Greenhouse Gas Inventories, Prepared by the National Greenhouse Gas Inventories Programme, Eggleston HS, Buendia L, Miwa K, Ngara T and Tanabe K (eds) Published: IGES, Japan.

Nowak DJ & Crane DE. 2002. Carbon storage and sequestration by urban trees in the USA. Environmental Pollution 116(3): 381-389.

Nowak DJ, Hoehn R and Crane DE. 2007. Oxygen Production by Urban Trees in the United States. Arboriculture & Urban Forestry 33(3): 220-226

Nowak DJ, Crane DE and Stevens JC. 2006a. Air pollution removal by urban trees and shrubs in the United States. Urban Forestry and Urban Greening 4: 115–123.

Perry J and Levan MD. 2003. Air Purification in Closed Environments: Overview of Spacecraft Systems. US Army Natrick Soldier Center. http://nsc.natick.army.mil/jocotas/ColPro_Papers/Perry-LeVan.pdf (accessed 6/06).

Richa Sharma, Lolita Pradhan, Maya Kumari and Prodyut Bhattacharya.2020. Assessment of Carbon Sequestration Potential of Tree Species in Amity University Campus Noida. Environmental Science. Proceeding. 3: 52-58.

Singh S, Bhattacharya P & Gupta NC. 2018. Dust particles characterization and innate resistance for *Thevetia peruviana* in different land-use pattern of urban area. International Journal of Environmental Science and Technology 15(5): 1061-1072.

Suryavanshi MN, Patel AR, Kale TS and Patil PR. 2014. Carbon Sequestration Potential of tree Species in the Environment of North Maharashtra University Campus, Jalgaon (MS) India. Bioscience Discovery 5(2): 175-179.

CHAPTER 26

Ethnic Diversity Audit

Amit Sharma and Vivek Barde
JNKVV, Jabalpur

The University had representation of 16 states and two countries in its faculty and staff upto 1993 with use of 16 languages but in 2022 the representation of the states is reduced to 10 and language 11 (Table 26.1).

Table 26.1 Ethnic diversity in JNKVV campus, Jabalpur

Country	
Upto 1993	**In 2022**
2	**2**
India	**India**
Nepal	**Nepal**
Representation of States	
Up to 1993	**In 2022**
16	**10**
Andhra Pradesh	Bihar
Bihar	Haryana
Gujarat	Jammu Kashmir
Haryana	Kerala
Himachal Pradesh	Madhya Pradesh
Jammu Kashmir	Maharashtra
Karnataka	Punjab
Kerala	Rajasthan
Madhya Pradesh	Uttar Pradesh
Maharashtra	Uttarkhand
Orissa	
Punjab	
Rajasthan	
Tamil Nadu	
Uttar Pradesh	
West Bengal	

CHAPTER 27

Overall diversity of flora and fauna

Total existing diversity in Jabalpur campus of the University

	Name	Family	Purpose
1.	*Acacia auriculiformis*	Mimosaceae	Tree
2.	*Acacia caesia*	Mimosaceae	Tree
3.	*Acacia catechu*	Mimosaceae	Tree
4.	*Acacia conncinna*	Mimosaceae	Tree
5.	*Acacia mangium*	Mimosaceae	Tree
6.	*Acacia nilotica*	Fabaceace	Tree
7.	*Acacia senegal*	Fabaceace	Tree
8.	*Acalypha hispida*	Euphorbiaceae	Tree
9.	*Acarus sapota*	Sapotaceae	Tree
10.	*Adansonia Digitata*	Bombacaceae	Tree
11.	*Adenanthera parvania*	Fabaceace	Tree
12.	*Aegel marmelos*	Rutaceae	Tree
13.	*Albizia lebbeck*	Mimosaceae	Tree
14.	*Albizia procera*	Mimosaceae	Tree
15.	*Alstonia scholaris*	Apocynaceae	Tree
16.	*Annona muricata*	Annonaceae	Tree
17.	*Annona reticulata*	Annonaceae	Tree
18.	*Annona squamosa*	Annonaceae	Tree
19.	*Anogeissus latifolia*	Combretaceae	Tree
20.	*Anthocephalus cadamba*	Rubiaceae	Tree
21.	*Areca catechu*	Arecaceae	Tree
22.	*Artabotrys hexapetalus*	Annonaceae	Tree
23.	*Artocarpus heterophyllus*	Moraceae	Tree
24.	*Artocarpus lacucha*	Moraceae	Tree
25.	*Azadirachta indica*	Meliaceae	Tree
26.	*Bambusa tulda*	Poaceae	Tree
27.	*Bambusa vulgaris*	Poaceae	Tree
28.	*Bauhinia purpurea*	Caesalpiniaceae	Tree
29.	*Bauhinia vahli*	Leguminosae	Tree
30.	*Bauhinia variegata*	Caesalpiniaceae	Tree
31.	*Bersera spp.*	Burseraceae	Tree
32.	*Bixa orellana*	Bixaceae	Tree
33.	*Bombax ceiba*	Malvaceae	Tree
34.	*Bombax mlabaricum*	Bombacaceae	Tree
35.	*Borassus flabellifer*	Arecaceae	Tree
36.	*Boswellia serrata*	Burseraceae	Tree
37.	*Bougainvillea species*	Nyctaginaceae	Tree
38.	*Buchanania lanzan*	Anacardiaceae	Tree
39.	*Butea frondosa*	Leguminosae	Tree
40.	*Butea monosperma*	Fabaceace	Tree
41.	*Caesalpinia crista*	Caesalpiniaceae	Tree
42.	*Calendrina sp.*	Mimosaceae	Tree
43.	*Calliandra surinamensis*	Caesalpiniaceae	Tree
44.	*Callistemon lanceolatus*	Myrtaceae	Tree
45.	*Callistemon rigidus*	Myrtaceae	Tree
46.	*Careya arbarea*	Lecythidaceae	Tree
47.	*Carissa carandas*	Apocynaceae	Tree
48.	*Cascabela thevetia*	Apocynaceae	Tree
49.	*Cassia fistula*	Caesalpiniaceae	Tree

	Name	Family	Purpose
50.	*Cassia siamea*	Caesalpiniaceae	Tree
51.	*Casuarina equisetifolia*	Casuarinaceae	Tree
52.	*Ceiba pentandra*	Malvaceae	Tree
53.	*Celastrus paniculatus*	Celastraceae	Tree
54.	*Cestrum nocturnam*	Solanaceae	Tree
55.	*Cicca acida*	Phyllanthaceae	Tree
56.	*Cinnamomum camphora*	Lauraceae	Tree
57.	*Cinnamomum tamala*	Lauraceae	Tree
58.	*Cinnamomum zeylanicum*	Lauraceae	Tree
59.	*Citrus aurantifolia*	Rutaceae	Tree
60.	*Citrus grandis*	Rutaceae	Tree
61.	*Citrus jambhiri*	Rutaceae	Tree
62.	*Citrus limetta*	Rutaceae	Tree
63.	*Citrus limonia*	Rutaceae	Tree
64.	*Citrus paradisi*	Rutaceae	Tree
65.	*Citrus sinensis*	Rutaceae	Tree
66.	*Cleistanthus collinus*	Phyllanthaceae	Tree
67.	*Cocos nucifera*	Arecaceae	Tree
68.	*Codiaeum variegatum*	Euphorbiaceae	Tree
69.	*Coffea arabica*	Rubiaceae	Tree
70.	*Copernicia alba*	Arecaceae	Tree
71.	*Cordia dichotoma*	Ehretiaceae	Tree
72.	*Cordia macleoidi*	Boraginaceae	Tree
73.	*Cordia myxa*	Boraginaceae	Tree
74.	*Crataeva nurvala*	Cappraceae	Tree
75.	*Crateva religia*	Capparaceae	Tree
76.	*Dalbergia latifolia*	Fabaceae	Tree
77.	*Dalbergia sisoo*	Caesalpiniaceae	Tree
78.	*Delonix regia*	Caesalpiniaceae	Tree
79.	*Dendrocalamus strictus*	Poaceae	Tree
80.	*Dillenia indica*	Magnoliaceae	Tree
81.	*Diospyros melanoxylon*	Ebenaceae	Tree
82.	*Diospyros montana*	Ebenaceae	Tree
83.	*Elaeis guineensis*	Arecaceae	Tree
84.	*Elaeocarpus angustifolius*	Elaeocarpaceae	Tree
85.	*Emblica officinalis*	Euphorbiaceae	Tree
86.	*Epipremnum aureum*	Arecaceae	Tree
87.	*Erythria variegata*	Fabaceae	Tree
88.	*Erythrina indica*	Fabaceae	Tree
89.	*Erythrina orientalis*	Fabaceae	Tree
90.	*Erythrina suberosa*	Fabaceae	Tree
91.	*Eucalyptus globulus*	Myrtaceae	Tree
92.	*Eucalyptus citriodora*	Myrtaceae	Tree
93.	*Eucalyptus gigantea*	Myrtaceae	Tree
94.	*Eucalyptus sp.*	Myrtaceae	Tree
95.	*Euginea uniflora*	Myrtaceae	Tree
96.	*Euphorbia pulcherrima*	Euphorbiaceae	Tree
97.	*Feronia elephandum*	Rutaceae	Tree
98.	*Ficus banjamiana*	Moraceae	Tree
99.	*Ficus benghalensis*	Moraceae	Tree

	Name	Family	Purpose
100.	*Ficus carica*	Moraceae	Tree
101.	*Ficus elastica*	Moraceae	Tree
102.	*Ficus glomerata*	Moraceae	Tree
103.	*Ficus racemosa*	Moraceae	Tree
104.	*Ficus religiosa*	Moraceae	Tree
105.	*Ficus retusa*	Moraceae	Tree
106.	*Ficus virens*	Moraceae	Tree
107.	*Gardenia gummifera*	Rubiaceae	Tree
108.	*Gliricidia sepium*	Fabaceae	Tree
109.	*Glycosmis pentaphylla*	Rutaceae	Tree
110.	*Gmelina arborea*	Lamiaceae	Tree
111.	*Gossypium arborecum*	Malvaceae	Tree
112.	*Grevillea robusta*	Proteaceae	Tree
113.	*Grewia asiatica*	Proteaceae	Tree
114.	*Haloptelea integrifolia*	Ulmaceae	Tree
115.	*Helicteres isora*	Malvaceae	Tree
116.	*Hevea brariliensis*	Euphorbiaceae	Tree
117.	*Hibiscus rosa-sinensis*	Malvaceae	Tree
118.	*Hyophorbe lagenicaulis*	Arecaceae	Tree
119.	*Inga dulcis*	Caesalpiniaceae	Tree
120.	*Ixora coccinea*	Euphorbiaceae	Tree
121.	*Jacaranda mimosifolia*	Bignomiaceae	Tree
122.	*Jasminum sambac*	Oleaceae	Tree
123.	*Juniperus communis*	Cupressaceae	Tree
124.	*Justicia adhatoda*	Acanthaceae	Tree
125.	*Kigelia africana*	Bignomiaceae	Tree
126.	*Kigelia pinnata*	Bignomiaceae	Tree
127.	*Lagerstroemia parviflora*	Lythraceae	Tree
128.	*Lagerstroemia flosregini*	Lythraceae□	Tree
129.	*Lagerstroemia indica*	Lythraceae□	Tree
130.	*Lagerstroemia speciosa*	Lythraceae□	Tree
131.	*Lantana camara*	Verbenaceae	Tree
132.	*Lawsonia inermis*	Lythraceae	Tree
133.	*Legerstroemia indica*	Lythraceae	Tree
134.	*Leucaena leucocephala*	Mimosaceae	Tree
135.	*Leucas aspera*	Fabaceae	Tree
136.	*Limonia acidissima*	Rutaceae	Tree
137.	*Litchi chinensis*	Sapindaceae	Tree
138.	*Litoria umarensis*	Pelodryadiaceae	Tree
139.	*Litsea glutinosa*	Lauraceae	Tree
140.	*Livistona chinensis*	Arecaceae	Tree
141.	*Livistona rotundifolia*	Arecaceae	Tree
142.	*Madhuca indica*	Sapotaceae	Tree
143.	*Madhuca latifolia*	Sapotaceae	Tree
144.	*Madhuca longifolia*	Sapotaceae	Tree
145.	*Magnolia champa*	Magnoliaceae	Tree
146.	*Mallotus philippensis*	Euphorbiaceae	Tree
147.	*Mangifera indica*	Anacardiaceae	Tree
148.	*Manilkara hexandra*	Sapotaceae	Tree
149.	*Melaleuca citrina*	Myrtaceae	Tree

	Name	Family	Purpose
150.	*Melia azedarach*	Meliaceae	Tree
151.	*Millettia pinnata*	Fabaceae	Tree
152.	*Mimusops elengi*	Sapotaceae	Tree
153.	*Mitragyna parviflora*	Rubiaceae	Tree
154.	*Moringa oleifera*	Moringaceae	Tree
155.	*Morus alba*	Moraceae	Tree
156.	*Murraya exotica*	Rutaceae	Tree
157.	*Murraya koenigii*	Rutaceae	Tree
158.	*Murraya paniculata*	Rutaceae	Tree
159.	*Musa paradisiaca*	Musaceae	Tree
160.	*Mussaenda frondosa*	Rubiaceae	Tree
161.	*Mussaenda rosa*	Rubiaceae	Tree
162.	*Neolamarckia cadamba*	Rubiaceae	Tree
163.	*Nerium alba*	Apocynaceae	Tree
164.	*Nerium indicum*	Apocynaceae	Tree
165.	*Nerium oleander*	Apocynaceae	Tree
166.	*Nyctanthes arbor-trisitis*	Oleaceae	Tree
167.	*Oroxylum indicum*	Bignomiaceae	Tree
168.	*Parkia biglandulosa*	Mimosaceae	Tree
169.	*Parkinsonia aculeata*	Fabaceae	Tree
170.	*Peltaphorum pterocarpum*	Caesalpiniaceae	Tree
171.	*Phoenix dactylifera*	Arecaceae	Tree
172.	*Phyllanthus acidus*	Euphorbiaceae	Tree
173.	*Phyllanthus emblica*	Euphorbiaceae	Tree
174.	*Pimenta dioica*	Myrtaceae	Tree
175.	*Pinus roxburghii*	Pinaceae	Tree
176.	*Pinus sp.*	Pinaceae	Tree
177.	*Pinus syivestris*	Pinaceae	Tree
178.	*Pithecellobium dulce*	Mimosaceae	Tree
179.	*Plumeria alba*	Apocynaceae	Tree
180.	*Plumeria indica*	Apocynaceae	Tree
181.	*Plumeria rubra*	Apocynaceae	Tree
182.	*Polyalthea longifolia*	Annonaceae	Tree
183.	*Pongamia pinnata*	Fabaceae	Tree
184.	*Prosopis cineraria*	Mimosaceae	Tree
185.	*Prunus amygdalus*	Rosaceae	Tree
186.	*Prunus dulcis*	Rosaceae	Tree
187.	*Psidium cattleianum*	Myrtaceae	Tree
188.	*Psidium guajava*	Myrtaceae	Tree
189.	*Pterocarpus marsupium*	Fabaceae	Tree
190.	*Pterospermum acerifolium*	Malvaceae	Tree
191.	*Punica granatum*	Punicaceae	Tree
192.	*Putranjiva roxburghii*	Annonaceae	Tree
193.	*Roystonea regia*	Arecaceae	Tree
194.	*Russelia juncea*	Scrophulariaceae	Tree
195.	*Salmalia malabarica*	Bombacaceae	Tree
196.	*Santalum album*	Sanatalaceae	Tree
197.	*Sapindus mukorossi*	Sapindaceae	Tree
198.	*Saraca asoca*	Detarioideae	Tree
199.	*Saraca indica*	Detarioideae	Tree

	Name	Family	Purpose
200.	*Schleichera oleosa*	Sapindaceae	Tree
201.	*Semecarpus anacardium*	Anacardiaceae	Tree
202.	*Senna siamea*	Fabaceae	Tree
203.	*Sesbania augsata*	Arecaceae	Tree
204.	*Spathodea companulata*	Bignomiaceae	Tree
205.	*Sphaeranthus indicus*	Rubiaceae	Tree
206.	*Spondias cythera*	Anacardiaceae	Tree
207.	*Sterculia alata*	Sterculiaceae	Tree
208.	*Sterculia foetida*	Sterculiaceae	Tree
209.	Sterculia *guttata*	Sterculiaceae	Tree
210.	*Sterculia urens*	Malvaceae	Tree
211.	*Stereospermum suavealens*	Bignomiaceae	Tree
212.	*Strychnos nux-vomica*	Strychnaceae	Tree
213.	*Swietenia mahagoni*	Meliaceae	Tree
214.	*Syzygium aromaticum*	Myrtaceae	Tree
215.	*Syzygium cumini*	Myrtaceae	Tree
216.	*Syzygium fr uiticosum*	Myrtaceae	Tree
217.	*Syzygium jambos*	Myrtaceae	Tree
218.	*Syzygium saliafelium*	Myrtaceae	Tree
219.	*Tabebuia argentea*	Bignomiaceae	Tree
220.	*Tabebuia entaphuylla*	Bignomiaceae	Tree
221.	*Tabebuia pentaphylla*	Bignomiaceae	Tree
222.	*Tabebuia rosea*	Bignomiaceae	Tree
223.	*Tabernaemontana divaricata*	Apocynaceae	Tree
224.	*Tamarindus indica*	Caesalpiniaceae	Tree
225.	*Tecoma stans*	Bignomiaceae	Tree
226.	*Tectona grandis*	Verbenaceae	Tree
227.	*Terminalia arjuna*	Combretaceae	Tree
228.	*Terminalia bellirica*	Combretaceae	Tree
229.	*Terminalia catapa*	Combretaceae	Tree
230.	*Theobroma cacao*	Sterculiaceae	Tree
231.	*Thespesia populnea*	Malvaceae	Tree
232.	*Thevetia alba*	Apocynaceae	Tree
233.	*Thevetia peruviana*	Apocynaceae	Tree
234.	*Thuja occidentalis*	Cupressaceae	Tree
235.	*Triadica sebifera*	Euphorbiaceae	Tree
236.	*Vachellia nilotica*	Fabaceae	Tree
237.	*Vitex negundo*	Verbenaceae	Tree
238.	*Vitis vinifera*	Vitaceae	Tree
239.	*Ziziphus jujuba*	Rhamnaceae	Tree
240.	*Ziziphus mauritiana*	Rhamnaceae	Tree
241.	*Ziziyphus xylopyra*	Rhamnaceae	Tree
242.	Rasutula		Tree
243.	*Acalypha hispida*	Euphorbiaceae	Shrub
244.	*Bixa orellana L*	Bixaceae	Shrub
245.	*Bougainvillea species*	Nyctaginaceae	Shrub
246.	*Carissa carndus L*	Apocynaceae	Shrub
247.	*Cestrum nocturnam*	Solanaceae	Shrub
248.	*Cestrum norcturnum*	Solanaceae	Shrub
249.	*Codiaeum variegatum*	Euphorbiaceae	Shrub

	Name	Family	Purpose
250.	*Coffea arabica*	Rubiaceae	Shrub
251.	*Euphorbia pulcherrima*	Euphorbiaceae	Shrub
252.	*Hibiscus rosa-sinensis*	Malvaceae	Shrub
253.	*Ixora coccinea*	Euphorbiaceae	Shrub
254.	*Jasminum sambac*	Oleaceae	Shrub
255.	*Lagerstroemia indica*	Lythraceae	Shrub
256.	*Lantana camara*	Verbenaceae	Shrub
257.	*Murraya panniculata*	Rutaceae	Shrub
258.	*Mussaenda frondosa*	Rubiaceae	Shrub
259.	*Mussaenda rosa*	Rubiaceae	Shrub
260.	*Nerium indicum*	Apocynaceae	Shrub
261.	*Nyctanthes arbor-trisitis*	Oleaceae	Shrub
262.	*Russelia juncea*	Scrophulariaceae	Shrub
263.	*Tabernaemontana divaricata*	Apocynaceae	Shrub
264.	*Tecoma stans*	Bignoniaceae	Shrub
265.	*Thevetia peruviana*	Apocynaceae	Shrub
266.	*Rosa chininsis var.minima*	Rosaceae	Shrub
267.	*Rosa damascena*	Rosaceae	Shrub
268.	*Rosa hybrida*	Rosaceae	Shrub
269.	*Rosa wichuriana*	Rosaceae	Shrub
270.	*Ageratum houstonianum*	Compositae	Annual
271.	*Althea rosea*	Malvaceae	Annual
272.	*Alyssum maritimum*	Cruiciferae	Annual
273.	*Antirrhinum majus*	Scrophulariaceae	Annual
274.	*Aster amellus*	Asteraceae	Annual
275.	*Calendula officinalis*	Compositae	Annual
276.	*Celosia cristata*	Amaranthaceae	Annual
277.	*Centaurea cyanus*	Asteraceae	Annual
278.	*Chrysanthemum coronarium*	Compositae	Annual
279.	*Cineraria deltoidea*	Asteraceae	Annual
280.	*Cosmos bipinnatus*	Compositae	Annual
281.	*Dahlia variabilis*	Compositae	Annual
282.	*Dahlia pinnata*	Asteraceae	Annual
283.	*Delphinium hybridum*	Ranunculaceae	Annual
284.	*Dendranthema grandiflora*	Asteraceae	Annual
285.	*Dianthus barbatus*	Caryophyllaceae	Annual
286.	*Dianthus caryophyllus*	Caryophyllaceae	Annual
287.	*Eschecholtzia californica*	Papaveraceae	Annual
288.	*Gaillardia pulchella*	Compositeae	Annual
289.	*Gazania linearis*	Asteraceae	Annual
290.	*Gerbera jamesonii*	Asteraceae	Annual
291.	*Gladiolus palustris*	Iridaceae	Annual
292.	*Gomphrena globosa*	Amaranthaceae	Annual
293.	*Helianthus annuus*	Asteraceae	Annual
294.	*Helichryum bracteatum*	Compositae	Annual
295.	*Iberis sempervirens*	Brassicaceae	Annual
296.	*Impatiens balsamina*	Balsaminaceae	Annual
297.	*Kochia scoparia*	Amaranthaceae	Annual
298.	*Lathyrus odoratus*	Leguminosae	Annual
299.	*Linum grandiflorum*	Linaceae	Annual

	Name	Family	Purpose
300.	*Mesembryanthemum criniflorum*	Aizoaceae	Annual
301.	*Myosotis sylvatica*	Boraginaceae	Annual
302.	*Nemasia strumosa*	Scrophuliaraceae	Annual
303.	*Nemesia caerulea*	Scrophulariaceae	Annual
304.	*Osteospermum fruticosum*	Asteraceae	Annual
305.	*Papaver rhoeas*	Papaveraceae	Annual
306.	*Pelargonium graveolens*	Geraniaceae	Annual
307.	*Petunia hybrida*	Solanaceae	Annual
308.	*Phlox paniculata*	Polemoniaceae	Annual
309.	*Polianthes tuberosa*	Asparagaceae	Annual
310.	*Portulaca grandiflora*	Portulaceae	Annual
311.	*Primula vulgaris*	Primulaceae	Annual
312.	*Salvia splendens*	Labiatae	Annual
313.	*Tagetes erecta*	Asteraceae	Annual
314.	*Tagetes patula*	Asteraceae	Annual
315.	*Tropaeolum majus*	Tropaeolaceae	Annual
316.	*Verbena hybrida*	Verbanaceae	Annual
317.	*Vinca rosea*	Apocynaceae	Annual
318.	*Viola tricolor var.hortensis*	Violaceae	Annual
319.	*Zinnia elegans*	Asteraceae	Annual
320.	*Avena sativa*	Poaceae	Crop
321.	*Brassica nigra*	Cruciferae	Crop
322.	*Cajanus cajan*	Fabaceae	Crop
323.	*Cicer arietinum*	Fabaceae	Crop
324.	*Glycine max*	Fabaceae	Crop
325.	*Guizotia abyssinica*	Compositeae	Crop
326.	*Lens culinaris*	Fabaceae	Crop
327.	*Linum usitatissimum*	Linaceae	Crop
328.	*Oryza sativa*	Poaceae	Crop
329.	*Pisum sativum*	Fabaceae	Crop
330.	*Sesamum indicum*	Pedaliaceae	Crop
331.	*Sorghum bicolor*	Poaceae	Crop
332.	*Trifolium alexandruinum*	Fabaceae	Crop
333.	*Triticum aestivum*	Poaceae	Crop
334.	*Vicia sativa*	Fabaceae	Crop
335.	*Vigna mungo*	Fabaceae	Crop
336.	*Vigna radiata*	Fabaceae	Crop
337.	*Vigna umbellata*	Fabaceae	Crop
338.	*Zea mays*	Poaceae	Crop
339.	*Abelmoschus moschatus* (L.) Medic.	Malvaceae	Medicinal
340.	*Abroma augusta (L.) LAF.*	Malvaceae	Medicinal
341.	*Abrus precatorius* L.	Fabaceae	Medicinal
342.	*Abrus precatorius* L.	Fabaceae	Medicinal
343.	*Abutilon hirsutum* G.Don.	Malvaceae	Medicinal
344.	*Abutilon indicum* L. Sweet.	Malvaceae	Medicinal
345.	*Abutilon instia* Willd.	Malvaceae	Medicinal
346.	*Abutilon tomentosum* W & A.	Malvaceae	Medicinal
347.	*Acacia auriculiformis* A. Cunn.	Mimosaceae	Medicinal
348.	*Acacia catechu*	Mimosaceae	Medicinal
349.	*Acacia concinna* D.C.	Mimosaceae	Medicinal

	Name	Family	Purpose
350.	*Acacia farnaciana* (L.) Willd.	Mimosaceae	Medicinal
351.	*Acacia holoceria*	Mimosaceae	Medicinal
352.	*Acacia leucophaea* Willd.	Mimosaceae	Medicinal
353.	*Acacia mangium*	Mimosaceae	Medicinal
354.	*Acacia nilotica*	Mimosaceae	Medicinal
355.	*Acacia pinnata* Willd.	Mimosaceae	Medicinal
356.	*Acacia tometosa* Willd.	Mimosaceae	Medicinal
357.	*Acalypha hispida* Burm.	Euphorbiaceae	Medicinal
358.	*Acalypha indica L*	Euphorbiaceae	Medicinal
359.	*Acalypha spicata*	Euphorbiaceae	Medicinal
360.	*Acalypha tricolor*	Euphorbiaceae	Medicinal
361.	*Achilea millefolium*	Amaranthaceae	Medicinal
362.	*Achyranthus aspera* L.	Amaranthaceae	Medicinal
363.	*Achyranthus bidenta*. Blume	Amaranthaceae	Medicinal
364.	*Acorus calamus* L.	Acoraceae	Medicinal
365.	*Adansoinia digitata* L.	Malvaceae	Medicinal
366.	*Adhatoda vasica* Nees.	Acanthaceae	Medicinal
367.	*Adiantum pedatum* L.	Pteridaceae	Medicinal
368.	*Adiantumcapillus-veneris* L.	Pteridaceae	Medicinal
369.	*Adina cordifolia*	Rubiaceae	Medicinal
370.	*Aegle marmelos* L.Corr.	Rutaceae	Medicinal
371.	*Aerva lanata* (L.) Juss.	Amaranthaceae	Medicinal
372.	*Aeschynomene aspera* L.	Fabaceae	Medicinal
373.	*Aeschynomene indica* L.	Fabaceae	Medicinal
374.	*Agathis spp.*	Araucariaceae	Medicinal
375.	*Agave americana* L.	Agavaceae	Medicinal
376.	*Agave cantata* Roxb.	Agavaceae	Medicinal
377.	*Agave sisalana* Perr.	Agavaceae	Medicinal
378.	*Ageratum conyzoides* L.	Asteraceae	Medicinal
379.	*Ageratum houstonium* Mill.	Asteraceae	Medicinal
380.	*Ailanthus excelsa* Roxb.	Simaroubaceae	Medicinal
381.	*Ailanthus triphysa* (Dennst.) Alston.	Simaroubaceae	Medicinal
382.	*AIlium porum*	Alliaceae	Medicinal
383.	*Albizia amara* (Roxb) Boir.	Mimosaceae	Medicinal
384.	*Albizia lebbek* L.	Mimosaceae	Medicinal
385.	*Albizia odoratissima* (L.F.) Bench.	Mimosaceae	Medicinal
386.	*Albizia procera* Roxb. Bench.	Mimosaceae	Medicinal
387.	*Alcasia macrorihiza* (L.) Schott.	Araceae	Medicinal
388.	*Alhagi camelorum* Fisch.	Fabaceae	Medicinal
389.	*Alisma plantago*	Alismataceae	Medicinal
390.	*Alium cepa* L.	Alliaceae	Medicinal
391.	*Allamanda cathartica* L.	Apocynaceae	Medicinal
392.	*Allium sativum* L.	Alliaceae	Medicinal
393.	*Alocasia cucullata* (lour) Schott.	Araceae	Medicinal
394.	*Alocasia indica* (Roxb) Schott.	Araceae	Medicinal
395.	*Aloe vera* (Syn.A.barbadensis)	Asphodelaceae	Medicinal
396.	*Alpenia officinarum* Hance.	Zingiberaceae	Medicinal
397.	*Alpinia calcarata*	Zingiberaceae	Medicinal
398.	*Alpinia galangal* (L.) Sw.	Zingiberaceae	Medicinal
399.	*Alpinia sanderae* Sand.	Zingiberaceae	Medicinal

	Name	Family	Purpose
400.	*Alpinia speciosa* (Wendl.) K. Schum.	Zingiberaceae	Medicinal
401.	*Alstonia scholaris* L.R.Br.	Apocynaceae	Medicinal
402.	*Alternanthera sessilis* L.	Amaranthaceae	Medicinal
403.	*Althaea officinalis* L.	Malvaceae	Medicinal
404.	*Althaea rosea* (L.) Cav.	Malvaceae	Medicinal
405.	*Alysicarpus longifolium* Wight. & Arn.	Fabaceae	Medicinal
406.	*Amaranthus blitum* L.var. oleracea	Amaranthaceae	Medicinal
407.	*Amaranthus cruentus* L.	Amaranthaceae	Medicinal
408.	*Amaranthus spinosus* L.	Amaranthaceae	Medicinal
409.	*Amaranthus tricolor* L.	Amaranthaceae	Medicinal
410.	*Ambrosia trifida*	Asteraceae	Medicinal
411.	*Ammannia baccifera* L.	Lythraceae	Medicinal
412.	*Ammi majus* L.	Apiaceae	Medicinal
413.	*Ammi visnaga*	Apiaceae	Medicinal
414.	*Amomum aromaticum* Roxb.	Zingiberaceae	Medicinal
415.	*Amorphophallus campanulatus* (Roxb)	Araceae	Medicinal
416.	*Amorphophallus sylvesticus* Kunth.	Araceae	Medicinal
417.	*Anacardium occidentale* L.	Anacdiaceae	Medicinal
418.	*Anagallis arvensis* L.	Primulaceae	Medicinal
419.	*Anamirta cocculus* (L.) Wight & Arn	Menispermaceae	Medicinal
420.	*Ananas comosus* (L) Merr.	Bromeliaceae	Medicinal
421.	*Andrographis paniculata*	Acanthaceae	Medicinal
422.	*Anethum graveolens* L.	Apiaceae	Medicinal
423.	*Anethum sowa* Kurz.	Apiaceae	Medicinal
424.	*Anisomeles indica* (L.) O.Kuntz.	Lamiaceae	Medicinal
425.	*Annona cherimolia* Mill.	Annonaceae	Medicinal
426.	*Annona muricata* L.	Annonaceae	Medicinal
427.	*Annona reticulata* L.	Annonaceae	Medicinal
428.	*Annona squamosa* L.	Annonaceae	Medicinal
429.	*Anogeissus accuminata* Roxb.	Combretaceae	Medicinal
430.	*Anogeissus latifolia* (D.C.) Wall	Combretaceae	Medicinal
431.	*Anogeissuspendula* Edgew.		Medicinal
432.	*Anthocephalus kadambu* Roxb. Miq.	Rubiaceae	Medicinal
433.	*Apium graveolens* L.	Apiaceae	Medicinal
434.	*Arachys hypojea*	Fabaceae	Medicinal
435.	*Araucaria bidwilli*	Araucariaceae	Medicinal
436.	*Arecacatechu* L.	Arecaceae	Medicinal
437.	*Argemonealba* Lestib.	Papaveraceae	Medicinal
438.	*Argemone mexicana* L.	Papaveraceae	Medicinal
439.	*Argyreia nurvosa* (Burn E.)	Convolvulaceae	Medicinal
440.	*Aristolochia brachiolata* Lamk.	Aristolochiaceae	Medicinal
441.	*Aristolochia indica* L.	Aristolochiaceae	Medicinal
442.	*Artabotrys uncinatus* (Lam.) Mern.	Annonaceae	Medicinal
443.	*Artemisia annua*	Asteraceae	Medicinal
444.	*Artemisia indica* Willd	Asteraceae	Medicinal
445.	*Artemisia maritima* L.	Asteraceae	Medicinal
446.	*Artemisia pallens* Wall.	Asteraceae	Medicinal
447.	*Artemisia vulgaris* L.	Asteraceae	Medicinal
448.	*A.nilagirica*(Clarke) Pamp.	Asteraceae	Medicinal
449.	*Artocarpus heterophyllus* lamk.	Moraceae	Medicinal

	Name	Family	Purpose
450.	*Artocarpus lakoocha* Roxb.	Moraceae	Medicinal
451.	*Arund donax* L	Poaceae	Medicinal
452.	*Asclepias curassavica* L.	Asclepiadaceae	Medicinal
453.	*Asparagus adscendens* Roxb.	Liliaceae	Medicinal
454.	*Asparagus filicinus* Ham.	Liliaceae	Medicinal
455.	*Asparagus gonocladus* Baker.	Liliaceae	Medicinal
456.	*Asparagus officinalis* L.	Liliaceae	Medicinal
457.	*Asparagus plumosus* Baker.	Liliaceae	Medicinal
458.	*Asparagus racemosus* Willd.	Liliaceae	Medicinal
459.	*Asparagus racemosus* willd.	Liliaceae	Medicinal
460.	*Asparagus sarmentosus* L.	Liliaceae	Medicinal
461.	*Asparagus spengeri* Regel.	Liliaceae	Medicinal
462.	Asphodelus *tenuifoli*. L.	Asphodelaceae	Medicinal
463.	*Asteracantha longifolia* L.	Acanthaceae	Medicinal
464.	*Astragales heratensis*	Fabaceae	Medicinal
465.	*Atropa belladona*	Solaneceae	Medicinal
466.	*Avena sativa* L.	Poaceae	Medicinal
467.	*Averrhoea carambola* L.	Averrhoeaceae	Medicinal
468.	Azanja lampas	Malvaceae	Medicinal
469.	*Azardirachta Indica*	Meliceae	Medicinal
470.	*Bacopa monneiri* (L) Penn.	Scrophulariaceae	Medicinal
471.	*Balanites aegyptiaca* (L.) Del.	Balaniteceae	Medicinal
472.	*Bambusa arundinacea* Retz.	Bambusaceae	Medicinal
473.	*Bambusa fulcata* Hort.	Bambusaceae	Medicinal
474.	*Bambusa polymorpha* Murno.	Bambusaceae	Medicinal
475.	*Bambusa tulda* Roxb.	Bambusaceae	Medicinal
476.	*Bambusa vulgaris* Schrader.	Bambusaceae	Medicinal
477.	*Barlaria alba* Lodd.	Acanthaceae	Medicinal
478.	*Barlaria cristana* L.	Acanthaceae	Medicinal
479.	*Basella alba* L.	Basellaceae	Medicinal
480.	*Bauhinia blackii*	Caesalpiniaceae	Medicinal
481.	*Bauhinia malabarica* Roxb.	Caesalpiniaceae	Medicinal
482.	*Bauhinia purpurea* L.	Caesalpiniaceae	Medicinal
483.	*Bauhinia tomentosa* L.	Caesalpiniaceae	Medicinal
484.	*Bauhinia vahlii* Wight & Arn.	Caesalpiniaceae	Medicinal
485.	*Bauhinia variegate* L.	Caesalpiniaceae	Medicinal
486.	*Begonia* L.	Begoniaceae	Medicinal
487.	*Benincasa hispida* (Thunb.) Cong.	Cucurbitaceae	Medicinal
488.	*Berberis aristata* D.C.	Berberidaceae	Medicinal
489.	*Beta vulgaris*	Chenopodiaceae	Medicinal
490.	*Biden pilosa* L.	Asteraceae	Medicinal
491.	*Biophytum sensitivum* D.C.	Oxalidaceae	Medicinal
492.	*Bixa orellana* L.	Bixaceae	Medicinal
493.	*Blepharis edulis* Pers.	Acanthaceae	Medicinal
494.	*Bluemea balsaifera* (L) D.C.	Asteraceae	Medicinal
495.	*Bluemea lacera* D.C.	Asteraceae	Medicinal
496.	*Boerrhavia diffusa*	Nyctaginaceae	Medicinal
497.	*Boerrhavia repends*	Nyctaginaceae	Medicinal
498.	*Bombax ceiba* L.	Bombacaceae	Medicinal
499.	*Borassus flabellifer L.*	Arecaceae	Medicinal

	Name	Family	Purpose
500.	*Boswellia serrata* Roxb.	Berseraceae	Medicinal
501.	*Brassica juncea* L.	Brassicaceae	Medicinal
502.	*Bridelia squamosa* Gehrm.	Euphorbiaceae	Medicinal
503.	*Brownea coccinea*	Fabaceae	Medicinal
504.	*Bryonia laciniosa* L.	Cucurbitaceae	Medicinal
505.	*Bryophyllum pinnata*	Crassulaceae	Medicinal
506.	*Buchanania lanzan* Spreng.	Anacardiaceae	Medicinal
507.	*Bursera panicillata* (Sesse & Moc DC)	Burseraceae	Medicinal
508.	*Butea monosperma* Lamk. Tanbert.	Fabaceae	Medicinal
509.	*Butea superba* Roxb.	Fabaceae	Medicinal
510.	*Cactus grandiflorus*	Cactaceae	Medicinal
511.	*Caesalpinia bonducella* fleming	Caesalpiniaceae	Medicinal
512.	*Caesalpinia coriaria* willd.	Caesalpiniaceae	Medicinal
513.	*Caesalpinia crista* Thumb.	Caesalpiniaceae	Medicinal
514.	*Caesalpinia digyna* Rottle.	Caesalpiniaceae	Medicinal
515.	*Caesalpinia pulcherrima* Swartz.	Caesalpiniaceae	Medicinal
516.	*Caesalpinia sappan* L.	Caesalpiniaceae	Medicinal
517.	*Caesulia axillarisi* Roxb.	Asteraceae	Medicinal
518.	*Caladium seguinum*	Araceae	Medicinal
519.	*Calandrinia ciliata*	Portulacaceae	Medicinal
520.	*Calendula officinalis* L.	Asteraceae	Medicinal
521.	*Callicarpa macrophylla*	Verbenaceae	Medicinal
522.	*Callistemon lanceolatus*	Myrtaceae	Medicinal
523.	*Calotropis gigantia* (L) R. Br.	Asclepiadaceae	Medicinal
524.	*Calotropis procera* (Ait.) R.Br.	Asclepiadaceae	Medicinal
525.	*Camellia sinensis* (L.)	Theaceae	Medicinal
526.	*Cananabis sativa*	Cananabinaceae	Medicinal
527.	*Canarium commune* L.	Berseraceae	Medicinal
528.	*Canarium strictum* Roxb.	Berseraceae	Medicinal
529.	*Canavalia ansiformis* D.C.	Fabaceae	Medicinal
530.	*Canavlia gludiata* (Jacq.) D.C.	Fabaceae	Medicinal
531.	*Cannabis indica*	Cannabeceae	Medicinal
532.	*Canscora decussate* Schutt.	Gentieanaceae	Medicinal
533.	*Capparis deciduas* Forsk	Capparidaceae	Medicinal
534.	*Capparis grandis*	Capparidaceae	Medicinal
535.	*Capsicum annum* Linn.	Solanaceae	Medicinal
536.	*Cardiospermum halicacabum* L.	Sapindaceae	Medicinal
537.	*Careya arborea* Roxb.	Barringtoniaceae	Medicinal
538.	*Carica Papaya* L.	Caricaceae	Medicinal
539.	*Carissa carandas* L.	Apocynaceae	Medicinal
540.	*Carissa grandiflora*	Apocynaceae	Medicinal
541.	*Carissa spinarum* L.	Apocynaceae	Medicinal
542.	*Carrisa spinosa*	Apocynaceae	Medicinal
543.	*Carthamus tinctorius* L.	Asteraceae	Medicinal
544.	*Carum carvi* L.	Apiaceae	Medicinal
545.	*Caryota urens* L.	Arecaceae	Medicinal
546.	*Cassia absus* L.	Caesalpiniaceae	Medicinal
547.	*Cassia alata* L.	Caesalpiniaceae	Medicinal
548.	*Cassia angustifolia* Vahl.	Caesalpiniaceae	Medicinal
549.	*Cassia artimisoides* Gaudich.Ex. Dc.	Caesalpiniaceae	Medicinal

	Name	Family	Purpose
550.	*Cassia auriculata* L.	Caesalpiniaceae	Medicinal
551.	*Cassia fistula* L.	Caesalpiniaceae	Medicinal
552.	*Cassia glauca* Lamk.	Caesalpiniaceae	Medicinal
553.	*Cassia grandis* L.F.	Caesalpiniaceae	Medicinal
554.	*Cassia javanica* L.	Caesalpiniaceae	Medicinal
555.	*Cassia mimosifolia*	Caesalpiniaceae	Medicinal
556.	*Cassia seracea*	Caesalpiniaceae	Medicinal
557.	*Cassia siamea*	Caesalpiniaceae	Medicinal
558.	*Cassia sophora* L	Caesalpiniaceae	Medicinal
559.	*Cassia tora* L.	Caesalpiniaceae	Medicinal
560.	*Cassytha filifomis* L	Cassythaceae	Medicinal
561.	*Casuarina equisetifolia* J.R. & G.	Casuarinaceae	Medicinal
562.	*Catharanthus roseus* L.G. Don.	Apocynaceae	Medicinal
563.	*Ceiba pentandra*	Bombacaceae	Medicinal
564.	*Celastruspaniculatus* Willd.	Celastraceae	Medicinal
565.	*Celosiaargentia* L	Amaranthaceae	Medicinal
566.	*Celosia cristata* L.	Amaranthaceae	Medicinal
567.	*Centaurea cynanus* L.	Asteraceae	Medicinal
568.	*Centella asiatica*	Hydrocotylaceae	Medicinal
569.	*Centipeda minima* (L.) A.Br. Aschers	Asteraceae	Medicinal
570.	*Centratherum anthelminticum*	Asteraceae	Medicinal
571.	*Cephalandra indica*	Cucurbitaceae	Medicinal
572.	*Cestrum aurantiacum* Linidl.	Solanaceae	Medicinal
573.	*Cestrum diurnum*	Solanaceae	Medicinal
574.	*Cestrum nocturnum* L.	Solanaceae	Medicinal
575.	*Cestrum parqui* L.	Solanaceae	Mêdicinal
576.	*Chenopodium album* L.	Chenopodiaceae	Medicinal
577.	*Chenopodium murate* L.	Chenopodiaceae	Medicinal
578.	*Chlorophytum arundinaceum*	Liliaceae	Medicinal
579.	*Chlorophytum borivllianum*	Liliaceae	Medicinal
580.	*Chlorophytum laxum*	Liliaceae	Medicinal
581.	*Chlorophytum tuberosum*	Liliaceae	Medicinal
582.	*Chlorophytum variegates*	Liliaceae	Medicinal
583.	*Chloroxylon switenia* D.C.	Flindersiaceae	Medicinal
584.	*Chrysanthemum coronarium* L.	Asteraceae	Medicinal
585.	*Chrysanthemum indicum* L.	Asteraceae	Medicinal
586.	*Chukrosia tabularis A. Juss.*	Malvaceae	Medicinal
587.	*Cicca acida* (L.) Merr.	Euphorbiaceae	Medicinal
588.	*Cichorium endive* L.	Asteraceae	Medicinal
589.	*Cichorium intybus*	Asteraceae	Medicinal
590.	*Cinnamomum camphora* (L.) Nees.	Lauraceae	Medicinal
591.	*Cinnamomum zeylanica*	Lauraceae	Medicinal
592.	*Cinnamonum tamala*	Lauraceae	Medicinal
593.	*Cissampelos pareira* L.	Menispermaceae	Medicinal
594.	*Cissus quadrangularis* L.	Vitaceae	Medicinal
595.	*Cissus repens* Lamk.	Vitaceae	Medicinal
596.	*Citrullus colocynthis* L.	Cucurbitaceae	Medicinal
597.	*Citrullus lanatus* var. fistulosus	Cucurbitaceae	Medicinal
598.	*Citrus aurantifolia*	Rutaceae	Medicinal
599.	*Citrus limon* (L.) Barm	Rutaceae	Medicinal

	Name	Family	Purpose
600.	*Citrus medica* L.	Rutaceae	Medicinal
601.	Citrus reticulata	Rutaceae	Medicinal
602.	*Citrus sinensis*	Rutaceae	Medicinal
603.	*Clarkia elagans* Dougl.	Onagraceae	Medicinal
604.	*Cleistanthus collinus* Roxb. Benth.	Euphorbiaceae	Medicinal
605.	*Clematis goourisana* Roxb.	Ranunculaceae	Medicinal
606.	*Cleome brachycarpa* Vahl.	Cleomaceae	Medicinal
607.	*Cleome simplifolia* Hook. F. &T.	Cleomaceae	Medicinal
608.	*Cleome spinosa* L.	Cleomaceae	Medicinal
609.	*Cleome viscosa* L	Cleomaceae	Medicinal
610.	*Clerodendrum indicum* (L.) Kuntaz.	Verbenaceae	Medicinal
611.	*Clerodendrum inerme*	Verbenaceae	Medicinal
612.	*Clerodendrum infortunatum* Linn.	Verbenaceae	Medicinal
613.	*Clerodendrum japanicum* (Thunb.)	Verbenaceae	Medicinal
614.	*Clerodendrum phlomidis* L.F.	Verbenaceae	Medicinal
615.	*Clerodendrum serratum* (L.) Spreng.	Verbenaceae	Medicinal
616.	*Clerodendrum thomsonmiae*	Verbenaceae	Medicinal
617.	*Clitoria mariana* L.	Fabaceae	Medicinal
618.	*Clitoria ternatea* L.	Fabaceae	Medicinal
619.	*Coccinia grandis*	Euphorbiaceae	Medicinal
620.	*Cocculus hirsutus*	Menispermaceae	Medicinal
621.	*Cochlospermum religiosperm* (l.) Alston.	Cochlosperamaceae	Medicinal
622.	*Cocus nucefera* L.	Arecaceae	Medicinal
623.	*Coffea arabica* L.	Rubiaceae	Medicinal
624.	*Coix gigantea* Koen ex Roxb.	Poaceae	Medicinal
625.	*Coix lacrima* Jobi.	Poaceae	Medicinal
626.	*Coleus aromaticus*	Lamiaceae	Medicinal
627.	*Coleus barabatus* Benth.	Lamiaceae	Medicinal
628.	*Coleus blumei* Benth.	Lamiaceae	Medicinal
629.	*Colocasia esculanta* (L.)	Araceae	Medicinal
630.	*Commelina bengalensis* Wall.	Commelinaceae	Medicinal
631.	*Commelina oblique* Harn	Commelinaceae	Medicinal
632.	*Commiphora bhandari* Com. Nov.	Burseraceae	Medicinal
633.	*Commiphora wightii* (Arnott.)	Burseraceae	Medicinal
634.	*Convolvlus argentens* Wall	Convolvlaceae	Medicinal
635.	*Convolvlus arvensis* L.	Convolvlaceae	Medicinal
636.	*Convolvlus microphyllus* Sieb.	Convolvlaceae	Medicinal
637.	*Convolvlus pluricaulis*	Convolvlaceae	Medicinal
638.	*Convolvlus scamonia* L	Convolvlaceae	Medicinal
639.	*Coptis teeta*	Ranunculaceae	Medicinal
640.	*Corchorus acutangulus* Lamk.	Tiliaceae	Medicinal
641.	*Corchorus antichorus* Roeusch.	Tiliaceae	Medicinal
642.	*Corchorus astuan*	Tiliaceae	Medicinal
643.	*Corchorus capsularis* L.	Tiliaceae	Medicinal
644.	*Corchorus fascicularis*	Tiliaceae	Medicinal
645.	*Corchorus olitorius* L.	Tiliaceae	Medicinal
646.	*Corchorus tridens*	Tiliaceae	Medicinal
647.	*Corchorus trilocularis*	Tiliaceae	Medicinal
648.	*Cordia dischotoma* var. Wallichi	Ehretiaceae	Medicinal
649.	*Cordia macleoidii*	Ehretiaceae	Medicinal

	Name	Family	Purpose
650.	*Cordia sebestena* L.	Ehretiaceae	Medicinal
651.	*Coriandrum didymus*	Apiaceae	Medicinal
652.	*Coriandrum sativum* L.	Apiaceae	Medicinal
653.	*Costus igneous*	Costaceae	Medicinal
654.	*Costus speciosus* Smith.	Costaceae	Medicinal
655.	*Cotula anthimoides* L.	Asteraceae	Medicinal
656.	*Cotulahemisphaerica*	Asteraceae	Medicinal
657.	*Crataevanurvala* Buch Ham.	Capparidaceae	Medicinal
658.	*Crinumasiaticum* L.	Amaryllidaceae	Medicinal
659.	*Crinumdefixum* Ker. Gawl.	Amaryllidaceae	Medicinal
660.	*Crinumlatifolium* L	Amaryllidaceae	Medicinal
661.	*Crinum powelli* Hort. Var album Hort.	Amaryllidaceae	Medicinal
662.	*Crinumzeylanicum* L.	Amaryllidaceae	Medicinal
663.	*Crocus sativus* L.	Iridaceae	Medicinal
664.	*Crotalaria juncea* L.	Fabaceae	Medicinal
665.	*Crotalaria medicagenia* Lamk.	Fabaceae	Medicinal
666.	*Crotolaria mucroonata* Desf.	Fabaceae	Medicinal
667.	*Crotolaria retusa* L.	Fabaceae	Medicinal
668.	*Crotolaria sericea* L.	Fabaceae	Medicinal
669.	*Crotonaromaticus* L	Euphorbiaceae	Medicinal
670.	*Crotonbonplan* dianum	Euphorbiaceae	Medicinal
671.	*Crotonoblongifolius* Roxb.	Euphorbiaceae	Medicinal
672.	*Crotontiglium* L.	Euphorbiaceae	Medicinal
673.	*Cryptolepis buchanani* R & S.	Periplocaceae	Medicinal
674.	*Cucubitamaxima* Duch.	Cucurbitaceae	Medicinal
675.	*Cucumisanguria* L.	Cucurbitaceae	Medicinal
676.	*Cucumismelo* L. var. Ogrestis	Cucurbitaceae	Medicinal
677.	*Cucumismelo* var. *mommeordica*	Cucurbitaceae	Medicinal
678.	*Cucumismelo* var. *utilissimus*	Cucurbitaceae	Medicinal
679.	*Cucumissativa* L.	Cucurbitaceae	Medicinal
680.	*Cuminumcyminum* L.	Apiaceae	Medicinal
681.	*Cupressuslusitanica* Mill.	Cupressaceae	Medicinal
682.	*Cupressustorulosa* D. Don.	Cupressaceae	Medicinal
683.	*Curculigolatifolia* Dry	Hypoxidaceae	Medicinal
684.	*Curculigoorchioides* Gaertn.	Hypoxidaceae	Medicinal
685.	*Curcumaangustifolia* Roxb	Zingiberaceae	Medicinal
686.	*Curcumaaromatica* Salisb.	Zingiberaceae	Medicinal
687.	*Curcumacaesia* Roxb.	Zingiberaceae	Medicinal
688.	*Curcumadomestica* Vahl.	Zingiberaceae	Medicinal
689.	*Curcumazedoaria* (Christm.) Rose.	Zingiberaceae	Medicinal
690.	*Cuscutachinesis*	Cuscutaceae	Medicinal
691.	*Cuscutaepathymum*	Cuscutaceae	Medicinal
692.	*Cuscutaepilinum*	Cuscutaceae	Medicinal
693.	*Cuscutahyaline* Roth.	Cuscutaceae	Medicinal
694.	*Cuscutareflexaa* Roxb.	Cuscutaceae	Medicinal
695.	*Cyamopsistetragonoloba* (L.) Taub.	Fabaceae	Medicinal
696.	*Cyanaracolymus*	Asteraceae	Medicinal
697.	*Cyanotisaxillaris* Roem.	Commelinaceae	Medicinal
698.	*Cycascircinalis* L.	Cycadaceae	Medicinal
699.	*Cycasrevolute* Thunb.	Cycadaceae	Medicinal

	Name	Family	Purpose
700.	*Cycaszamia*	Cycadaceae	Medicinal
701.	*Cymbopogoncaesius* (Nees) Stapf.	Poaceae	Medicinal
702.	*Cymbopogoncitratus* D.C. Stapf.	Poaceae	Medicinal
703.	*Cymbopogonflexuosus*	Poaceae	Medicinal
704.	*Cymbopogonjwarancusa* Jones Scults.	Poaceae	Medicinal
705.	*Cymbopogonmartini* Roxb. Wats.	Poaceae	Medicinal
706.	*Cymbopogonnardus* (L) Renole.	Poaceae	Medicinal
707.	*Cymbopogonpendulous* Wats.	Poaceae	Medicinal
708.	*Cymbopogonschoenaznthus* L. Spreng.	Poaceae	Medicinal
709.	Cymbopogon winterianus Jowitt.	Poaceae	Medicinal
710.	*Cynara scolymus* L.	Asteraceae	Medicinal
711.	*Cynodon dactylon* Pers.	Poaceae	Medicinal
712.	*Cyperus articulates* L.	Cyperaceae	Medicinal
713.	*Cyperus bulbosus*	Cyperaceae	Medicinal
714.	*Cyperus esculentus* L.	Cyperaceae	Medicinal
715.	*Cyperus rotundus* L.	Cyperaceae	Medicinal
716.	*Cyperus scariosus* R.Br.	Cyperaceae	Medicinal
717.	*Cyphomandra betacea*	Solanaceae	Medicinal
718.	*Daedalacanthus roseus* T. Anders.	Acantheceae	Medicinal
719.	*Daemia extensa* Br.	Asclepidaceae	Medicinal
720.	*Dalbergia latifolia*	Fabaceae	Medicinal
721.	*Dalbergia sisso* Roxb.	Fabaceae	Medicinal
722.	*Datura innoxia* Mill.	Solanaceae	Medicinal
723.	*Datura metel* L	Solanaceae	Medicinal
724.	*Datura strantonium* L	Solanaceae	Medicinal
725.	*Daucus carota* L	Apiaceae	Medicinal
726.	*Delonix regia* (Boj.) Raf.	Caesalpiniaceae	Medicinal
727.	*Delphinium brunonianum* Rogu.	Ranunculaceae	Medicinal
728.	*Dendrocalamus strictus* Nees.	Poaceae	Medicinal
729.	*Dendrophthoe falcata*	Loranthaceae	Medicinal
730.	*Desmodium diffusum* D.C.	Fabaceae	Medicinal
731.	*Desmodium gangaticunt* D.C.	Fabaceae	Medicinal
732.	*Desmodium polycarpus* D.C	Fabaceae	Medicinal
733.	*Desmodium tomentosa* Heyne.	Fabaceae	Medicinal
734.	*Desmodium triflorum* (L.) D.C.	Fabaceae	Medicinal
735.	*Desmostachia bipinnata* (L) Stapf.	Poaceae	Medicinal
736.	*Dianthus caryophylhts* L	Caryophyllaceae	Medicinal
737.	*Dichrostachys cinerea*	Mimosaceae	Medicinal
738.	*Digera arvensis* Forsk.	Amaranthaceae	Medicinal
739.	*Digera muricata* (L) Mart.	Amaranthaceae	Medicinal
740.	*Dillenia indica* L	Dilleniaceae	Medicinal
741.	*Dioscorea alata* L.	Dioscoreaceae	Medicinal
742.	*Dioscorea bulbifera* L.	Dioscoreaceae	Medicinal
743.	*Dioscorea daemon*	Dioscoreaceae	Medicinal
744.	*Dioscorea deltoidea* Wall. ex Kunth.	Dioscoreaceae	Medicinal
745.	*Dioscorea esculenta* (Lour).Burkill.	Dioscoreaceae	Medicinal
746.	*Dioscorea floribunda*	Dioscoreaceae	Medicinal
747.	*Dioscorea hamiltonii* Hook f.	Dioscoreaceae	Medicinal
748.	*Dioscorea pentaphylla* L	Dioscoreaceae	Medicinal
749.	*Diospyros kaki* L	Ebenaceae	Medicinal

	Name	Family	Purpose
750.	*Diospyros lotus* L.	Ebenaceae	Medicinal
751.	*Diospyros melanoxylon* Roxb.	Ebenaceae	Medicinal
752.	*Diospyros montena*	Ebenaceae	Medicinal
753.	*Diospyros tomentosa* Roxb.	Ebenaceae	Medicinal
754.	*Dodonaea viscosa* (L.) Jacq.	Sapindaceae	Medicinal
755.	*Dolichos falcatus*	Fabaceae	Medicinal
756.	*Dolichos uniflorus* Lam	Fabaceae	Medicinal
757.	*Drosera indica* L	Droseraceae	Medicinal
758.	*Dudymocarpus pedicillata* R. Br.	Gesneriaceae	Medicinal
759.	*Echinopes echinatus* D.C.	Asteraceae	Medicinal
760.	*Eclipta alba* Hassk.	Asteraceae	Medicinal
761.	*Elaeies guinensis*	Arecaceae	Medicinal
762.	*Elaeodendron glaucum* Pers.	Celastraceae	Medicinal
763.	*Eleocarpus sphaericus* Gaertn.	Eleocarpaceae	Medicinal
764.	*Elephantopus scaber*	Asteraceae	Medicinal
765.	*Elettaria cardamomum*	Zingiberaceae	Medicinal
766.	*Embelia ribes* Burm F.	Myrsinaceae	Medicinal
767.	*Emblica officinalis*	Euphorbiaceae	Medicinal
768.	*Enicostimma hyssopifolium* (Wild.)	Gentianaceae	Medicinal
769.	*Equisetum arvens* L. Laden	Equisetaceae	Medicinal
770.	*Erianthus arundinaceus*	Poaceae	Medicinal
771.	*Erigeron canadensis* L.	Asteraceae	Medicinal
772.	*Eriolaena quinquelocularis* Wt.	Sterculiaceae	Medicinal
773.	*Eruca vesicaria* (L.) Car.	Brassicaceae	Medicinal
774.	*Ervthrina variegate* L.	Fabaceae	Medicinal
775.	*Erythrina blakei* Hort.ex.Parker.	Fabaceae	Medicinal
776.	*Erythrina indica* Lam.	Fabaceae	Medicinal
777.	*Eucalyptus citriodora* Hook.	Myrtaceae	Medicinal
778.	*Eucalyptus globulus* Labill.	Myrtaceae	Medicinal
779.	*Eucalyptus melanopfiloia* F.Muell.	Myrtaceae	Medicinal
780.	*Eucalyptus rostrata* Schlecht.	Myrtaceae	Medicinal
781.	*Eugenia uniflora*	Myrtaceae	Medicinal
782.	*Eupatorium triplinerve* Vahl.	Asteraceae	Medicinal
783.	*Euphorbia antiquorum* L	Euphorbiaceae	Medicinal
784.	*Euphorbia dracunculoides*	Euphorbiaceae	Medicinal
785.	*Euphorbia geniculata* Orteg	Euphorbiaceae	Medicinal
786.	*Euphorbia hirta*.L.	Euphorbiaceae	Medicinal
787.	*Euphorbia hypercifolia*	Euphorbiaceae	Medicinal
788.	*Euphorbia microphylla*	Euphorbiaceae	Medicinal
789.	*Euphorbia nervifolia* Auct.Pl.non L.	Euphorbiaceae	Medicinal
790.	*Euphorbia pulcherima* Willd.ex. Koltz.	Euphorbiaceae	Medicinal
791.	*Euphorbia thymifolia*	Euphorbiaceae	Medicinal
792.	*Euphorbia tirucalli* L.	Euphorbiaceae	Medicinal
793.	*Euryale ferox*	Eurylaceae	Medicinal
794.	*Evolvulus alsinoides* Wall.	Convolvulaceae	Medicinal
795.	*Excoecaria bicolor*	Euphorbiaceae	Medicinal
796.	*Ferula galbeniflua*	Apiaceae	Medicinal
797.	*Ficus bengalensis* L.	Moraceae	Medicinal
798.	*Ficus benjamina*	Moraceae	Medicinal
799.	*Ficus carica* L	Moraceae	Medicinal

	Name	Family	Purpose
800.	*Ficus elastica* Roxb.	Moraceae	Medicinal
801.	*Ficus glomerata* Roxb.	Moraceae	Medicinal
802.	*Ficus hispidu* L.f.	Moraceae	Medicinal
803.	*Ficus infectoria* Roxb.	Moraceae	Medicinal
804.	*Ficus krishnae* C. DC.	Moraceae	Medicinal
805.	*Ficus palmata* Forsk.	Moraceae	Medicinal
806.	*Ficus retusa* L	Moraceae	Medicinal
807.	*Ficus rumphii* Blume.	Moraceae	Medicinal
808.	*Flacourtia cataphrcta* Roxh.	Flacourtiaceae	Medicinal
809.	*Flacourtia indica* (Burm) F.Merr.	Flacourtiaceae	Medicinal
810.	*Fragaria vesca* L.	Rosaceae	Medicinal
811.	*Fumaria indica*	Fumariaceae	Medicinal
812.	*Fumaria of cinalis* L.	Fumariaceae	Medicinal
813.	*Funiculum vulgare* Mill.	Apiaceae	Medicinal
814.	*Gardenia campanulata* Roxb.	Rubiaceae	Medicinal
815.	*Gardenia gummifera* L.F.	Rubiaceae	Medicinal
816.	*Gardenia jasminoides* Ellis.	Rubiaceae	Medicinal
817.	*Gardenia latifolia* Ait.	Rubiaceae	Medicinal
818.	*Gardenia resinifera* Roth.	Rubiaceae	Medicinal
819.	*Geranium nepalense* Sweet.	Geraniaceae	Medicinal
820.	*Gliricidia maculata* (H.B.K.) Steud.	Fabaceae	Medicinal
821.	*Gloriosa superba* L.	Liliaceae	Medicinal
822.	*Glycosmis pentaphylla* Retz.D.C.	Rutaceae	Medicinal
823.	*Glycyrrhiza glabra* L.	Fabaceae	Medicinal
824.	*Gmelina arborea* L.	Verbenaceae	Medicinal
825.	*Gomphrena celoseodes*	Amaranthaceae	Medicinal
826.	*Gomphrena pungens*	Amaranthaceae	Medicinal
827.	*Gossypium arboreum*	Malvaceae	Medicinal
828.	*Gossypium hirsutum*	Malvaceae	Medicinal
829.	*Grevillea robusta* A. Cunn.	Proteaceae	Medicinal
830.	*Grewia oppositifolia* Buch.	Malvaceae	Medicinal
831.	*Grewia populifolia* Vahl	Malvaceae	Medicinal
832.	*Grewia subinaequalis* D.C.	Malvaceae	Medicinal
833.	*Grewia tenure*	Malvaceae	Medicinal
834.	*Gymnema sylvestre* (Raz.) R.Br.	Asclepiadaceae	Medicinal
835.	*Gynandropsis gynandra*	Cleomaceae	Medicinal
836.	*Hedychium coronarium* Koenig. ex.Retz.	Zingiberaceae	Medicinal
837.	*Hedychium flavum* Oxb.	Zingiberaceae	Medicinal
838.	*Hedychium spicattan* Buch.Ham	Zingiberaceae	Medicinal
839.	*Helianthus annuus* L.	Asteraceae	Medicinal
840.	*Helianthus tuberosus* L	Asteraceae	Medicinal
841.	*Helicteris isora*	Sterculiaceae	Medicinal
842.	*Heliotroipium ellipticum* Ledebour.	Boraginaceae	Medicinal
843.	*Heliotropium indicum*	Boraginaceae	Medicinal
844.	*Heliotropium strigosum* Willd.	Boraginaceae	Medicinal
845.	*Hernidesmus indicus* R.Br.	Periplocaceae	Medicinal
846.	*Heteropogon contortus* (L.) P.Beauv.	Poaceae	Medicinal
847.	*Hevea brasiliensis* (Willd.ex.Adr.de.Juss)	Euphorbiaceae	Medicinal
848.	*Hibiscus liliaceus* L.	Malvaceae	Medicinal
849.	*Hibiscus mulabilis* L.	Malvaceae	Medicinal

	Name	Family	Purpose
850.	*Hibiscus rosa* sinensis L.	Malvaceae	Medicinal
851.	*Hibiscus rosa* subdarifa L.	Malvaceae	Medicinal
852.	*Hibiscus surraliensis* L.	Malvaceae	Medicinal
853.	*Hibiscus tiliaceus* L.	Malvaceae	Medicinal
854.	*Hibiscus vitifolius* L	Malvaceae	Medicinal
855.	*Hiptage berighalensis* (L.) Kurz.	Malpighiaceae	Medicinal
856.	*Holarrhena antidysentrica*	Apocyneceae	Medicinal
857.	*Holoptelea integrifolia* (Roxb).	Ulmaceae	Medicinal
858.	*Holostemma rheedei* Wall.	Asclepiadaceae	Medicinal
859.	*Hopea odorata* Roxb.	Dipterocarpaceae	Medicinal
860.	*Hydnocarpus kurzii* Warb.	Flacourtiaceae	Medicinal
861.	*Hygrophyla spinosa* T.Abders Buchtiam.	Acathaceae	Medicinal
862.	*Hyocyamus muticus*	Solanaceae	Medicinal
863.	*Hyocyamus niger* L.	Solanaceae	Medicinal
864.	*Hypophorbe lagenicaulis*	Arecaceae	Medicinal
865.	*Hyptis suaveolens* L.	Lamiaceae	Medicinal
866.	*Impatiens balsamica* L.	Balasaminaceae	Medicinal
867.	*Imperata arundinacea*	Poaceae	Medicinal
868.	*Indigofera linifolia* Retz.	Fabaceae	Medicinal
869.	*Indigofera trita* L.F.	Fabaceae	Medicinal
870.	*Inula racemosa*	Asteraceae	Medicinal
871.	*Ipoemoea nil* (L). Roth.	Cunvolvulaceae	Medicinal
872.	*Ipomoea alba* L.	Cunvolvulaceae	Medicinal
873.	*Ipomoea aquatica* Forsk.	Cunvolvulaceae	Medicinal
874.	*Ipomoea biloba* Forsk.	Cunvolvulaceae	Medicinal
875.	*Ipomoea Dasysperma* Jacq.	Cunvolvulaceae	Medicinal
876.	*Ipomoea eriocarpa* R.Br.	Cunvolvulaceae	Medicinal
877.	*Ipomoea hederacea* Jacq.	Cunvolvulaceae	Medicinal
878.	*Ipomoea hederifolia* L.	Cunvolvulaceae	Medicinal
879.	*Ipomoea hispida*	Cunvolvulaceae	Medicinal
880.	*Ipomoea mauritiana* Jacq.	Cunvolvulaceae	Medicinal
881.	*Ipomoea obsewea* Ker	Cunvolvulaceae	Medicinal
882.	*Ipomoea palmata* Forsk.	Cunvolvulaceae	Medicinal
883.	*Ipomoea paniculata*	Cunvolvulaceae	Medicinal
884.	*Ipomoea pestigridis* L.	Cunvolvulaceae	Medicinal
885.	*Ipomoea purpurea* (L) Roth.	Cunvolvulaceae	Medicinal
886.	*Ipomoea tricolour* Cav.	Cunvolvulaceae	Medicinal
887.	*Ipomoea turbinata* Lag.	Cunvolvulaceae	Medicinal
888.	*Ipomoea uniflora* Rooem. & Schult.	Cunvolvulaceae	Medicinal
889.	*Ipornoea batatas* Forsk.	Cunvolvulaceae	Medicinal
890.	*Ixora coccinea* L.	Rubiaceae	Medicinal
891.	*Ixora parviflora* Vahl.	Rubiaceae	Medicinal
892.	*Jacaranda mimosifolia* D.Don.	Bignoniaceae	Medicinal
893.	*Jasminum angustifolia* Vahl.	Oleaceae	Medicinal
894.	*Jasminum arborescens* Roxb.	Oleaceae	Medicinal
895.	*Jasminum auriculatum* Vahl.	Oleaceae	Medicinal
896.	*Jasminum calophyllum* Wallet Dc.	Oleaceae	Medicinal
897.	*Jasminum flexile* Vahl	Oleaceae	Medicinal
898.	*Jasminum frunflorum* L.	Oleaceae	Medicinal
899.	*Jasminum humile* L.	Oleaceae	Medicinal

	Name	Family	Purpose
900.	*Jasminum multiflorum*	Oleaceae	Medicinal
901.	*Jasminum officinale* L.	Oleaceae	Medicinal
902.	*Jasminum sanbac* (L.) Ait	Oleaceae	Medicinal
903.	*Jatropha curcas* L.	Euphorbiaceae	Medicinal
904.	*Jatropha glandulifera* Roxb.	Euphorbiaceae	Medicinal
905.	*Jatropha gossypifolia* L.	Euphorbiaceae	Medicinal
906.	*Jatropha multifida* L	Euphorbiaceae	Medicinal
907.	*Jatropha padaraefolia* Andr.	Euphorbiaceae	Medicinal
908.	*Jatropha podagrica* Hook.	Euphorbiaceae	Medicinal
909.	*Juniperus commuris* L. Hauber.	Cuppressaceae	Medicinal
910.	*Jussiaea repens* L.	Onagraceae	Medicinal
911.	*Jussiaea tenella* Burmf.	Onagraceae	Medicinal
912.	*Jussiaea uffruticosa* L.	Onagraceae	Medicinal
913.	*Justicia diffusa* Willd.	Acanthaceae	Medicinal
914.	*Justicia gendarussa* L.F. Neel.	Acanthaceae	Medicinal
915.	*Justicia procumbens* L.	Acanthaceae	Medicinal
916.	*Kalanchoe integra* (Mcedic) Kuntz.	Crassulaceae	Medicinal
917.	*Kalanchoe laciniata* D.C.	Crassulaceae	Medicinal
918.	*Kalanchoe pinnata*	Crassulaceae	Medicinal
919.	*Kalanchoe tigrinum*	Crassulaceae	Medicinal
920.	*Kigelia pinnata* D.C.	Bignoniaceae	Medicinal
921.	*Kochia indica* Wight.	Chenopodiaceae	Medicinal
922.	*Kochia scoparia* Schrad.	Chenopodiaceae	Medicinal
923.	*Kydia calycina*	Malvaceae	Medicinal
924.	*Kyllinga triceps* Roxb.	Cyperaceae	Medicinal
925.	*Lablab purpureus*	Fabaceae	Medicinal
926.	*Lactuca indica* L.	Asteraceae	Medicinal
927.	*Lactuca remotifolia* D.C.	Asteraceae	Medicinal
928.	*Lactuca runcinata* D.C.	Asteraceae	Medicinal
929.	*Lactuca saliva* L. Salad	Asteraceae	Medicinal
930.	*Lagenaria vulgaris* Ser.	Cucurbitaceae	Medicinal
931.	*Lagerstromia indica*	Lythraceae	Medicinal
932.	*Lagerstromia parviflora*	Lythraceae	Medicinal
933.	*Lagerstromia speciosa* (L.) Pers.	Lythraceae	Medicinal
934.	*Lathyrus aphuca* L.	Fabaceae	Medicinal
935.	*Lathyrus odoratus*	Fabaceae	Medicinal
936.	*Lathyrus sativus* L.	Fabaceae	Medicinal
937.	*Launaea aspleniifolia* Hook f.	Asteraceae	Medicinal
938.	*Launaea niulicaulus* Hook f.	Asteraceae	Medicinal
939.	*Lavendula angustifolia* Mill.	Lamiaceae	Medicinal
940.	*Lavendula bipinnata* Kuntaze.	Lamiaceae	Medicinal
941.	*Lawsonia inermis* L.	Lythraceae	Medicinal
942.	*lberis amara* L	Brassicaceae	Medicinal
943.	*lchnocarpus fruitescens* R.Br.	Apocynaceae	Medicinal
944.	*Leonotis nepetiolia*	Lamiaceae	Medicinal
945.	*Lepidagathis cristra* Willd.	Acanthaceae	Medicinal
946.	*Lepidium sativiun* L.	Brassicaceae	Medicinal
947.	*Leucas aspera*	Lamiaceae	Medicinal
948.	*Leucas aspera* (Willd.) Sperng.	Lamiaceae	Medicinal
949.	*Leucas cephalotes* (Roth.) Sperng.	Lamiaceae	Medicinal

	Name	Family	Purpose
950.	*Leucas lavendulifolia* Rees.	Lamiaceae	Medicinal
951.	*Licopodium clavatum* L.	Lycopodiaceae	Medicinal
952.	*Lilium candidum* L.	Liliaceae	Medicinal
953.	*Lilium Nanteuni* Wall.	Liliaceae	Medicinal
954.	*Lilium tigrinum*	Liliaceae	Medicinal
955.	*Limnophylla aromatica* (Lam.) Merrill.	Scrophulaviaceae	Medicinal
956.	*Limnophylla indica* (L.) Druce.	Scrophulaviaceae	Medicinal
957.	*Limnophylla rugosa* (Roth) Merr.	Scrophulaviaceae	Medicinal
958.	*Limonia acidissima* L.	Rutaceae	Medicinal
959.	*Lindenburgia indica* Kuntz.	Scrophulaviaceae	Medicinal
960.	*Lindernia crustacea* (L.) Muell.	Scrophulaviaceae	Medicinal
961.	*Lindernia oppositifolia* Retz.	Scrophulaviaceae	Medicinal
962.	*Lippia nodiflfora* Rich.	Verbanceae	Medicinal
963.	*Litchi chinensis* (Gaertn.) Sonn.	Sapindaceae	Medicinal
964.	*Litsea polyantha* Jurs.	Lauraceae	Medicinal
965.	*lndigofera glandulosa*	Fabaceae	Medicinal
966.	*lndigofera hirsuta* L	Fabaceae	Medicinal
967.	*lndigofera suffruticosa* Mill	Fabaceae	Medicinal
968.	*lndigofera tinctoria* L	Fabaceae	Medicinal
969.	*Lolium temulentum* L.	Poaceae	Medicinal
970.	*Loranthus globosus* Roxb.	Loranthaceae	Medicinal
971.	*Loranthus longiflorus* Rerr.	Loranthaceae	Medicinal
972.	*Luffa accutangula* Roxb.	Cucurbitaceae	Medicinal
973.	*Luffa aegyptiaca* Mill.	Cucurbitaceae	Medicinal
974.	*Luffa echinata* Roxb.	Cucurbitaceae	Medicinal
975.	*Lycopodium cernuum* Linn.	Lycopodiaceae	Medicinal
976.	*Madhuca india* J.F. Emel.	Sapotaceae	Medicinal
977.	*Magnolia grandiflorai* L.	Magnoliaceae	Medicinal
978.	*Malachra capitata* L.	Malvaceae	Medicinal
979.	*Mallotus esculenta* Granth	Euphorbiaceae	Medicinal
980.	*Mallotus philippensis* Lamk Muel.	Euphorbiaceae	Medicinal
981.	*Malus pumilla* Mill.	Rosaceae	Medicinal
982.	*Malva parvifolia*	Malvaceae	Medicinal
983.	*Malva rotundifolia*	Malvaceae	Medicinal
984.	*Malvastrum coromandelianum* (L.)	Malvaceae	Medicinal
985.	*Manihot esculenta*	Euphorbiaceae	Medicinal
986.	*Manilkara kauki* L.Dub.	Sapotaceae	Medicinal
987.	*Mansoa alliaceae*	Bignoniaceae	Medicinal
988.	*Maranta arundenacea* L.	Marantaceae	Medicinal
989.	*Marsdenia tenacissima* Roxb.Moon.	Asclepiadaceae	Medicinal
990.	*Marsilea minuta* L.	Marsileaceae	Medicinal
991.	*Martynia annua* L.	Martyniaceae	Medicinal
992.	*Marus nigra* L.	Moraceae	Medicinal
993.	*Matricaria chamomilla* L.	Asteraceae	Medicinal
994.	*Medicago hispidea* Gaetn.	Fabaceae	Medicinal
995.	*Medicago saliva*	Fabaceae	Medicinal
996.	*Melia azadirach* L.	Meliaceae	Medicinal
997.	*Melilotus alba* Medik.	Fabaceae	Medicinal
998.	*Melitotus parviflora* Desf.	Fabaceae	Medicinal
999.	*Mentha arvensis* L.	Lamiaceae	Medicinal

	Name	Family	Purpose
1000.	*Mentha piperita* L.	Lamiaceae	Medicinal
1001.	*Mentha spicata* L.	Lamiaceae	Medicinal
1002.	*Mentha viridis*	Lamiaceae	Medicinal
1003.	*Merrimia dissecta* (Jacq.) Hallrerf.	Convolvulaceae	Medicinal
1004.	*Merrimia emarginata* Hallier	Convolvulaceae	Medicinal
1005.	*Merrimia tubrosa* (L.) Rendle.	Convolvulaceae	Medicinal
1006.	*Mesua ferrea* L.	Guttiferae	Medicinal
1007.	*Metroxylon sagu* Roxb.	Arecaceae	Medicinal
1008.	*Michelia champaca*	Magnoliaceae	Medicinal
1009.	*Miliusa tomentosa* Roxb ex.J.Sinclair	Annonaceae	Medicinal
1010.	*Millingtonia hortensis* L.F.	Vignoniaceae	Medicinal
1011.	*Millusa ovalifolia* Kurz.	Annonaceae	Medicinal
1012.	*Mimosa hamata* Willd.	Mimosaceae	Medicinal
1013.	*Mimosa pudica* L.	Mimosaceae	Medicinal
1014.	*Mimusops elengi* L.	Sapotaceae	Medicinal
1015.	*Mirabilis jalapa* L.	Nyctaginaceae	Medicinal
1016.	*Mitragyna parvifolia* Roxb. Kurth.	Rubiaceae	Medicinal
1017.	*Mollugo cerviana* (L.) Ser.	Molluginaceae	Medicinal
1018.	*Mollugo hirta* Thumb.	Molluginaceae	Medicinal
1019.	*Mollugo lotoides* L.	Molluginaceae	Medicinal
1020.	*Mollugo nudicaulis* Lam.	Molluginaceae	Medicinal
1021.	*Mollugo oppositifolia* L.	Molluginaceae	Medicinal
1022.	*Mollugo pentaphylla* L.	Molluginaceae	Medicinal
1023.	*Mollugo stricta*	Molluginaceae	Medicinal
1024.	*Momordica balsamina* L	Cucurbitaceae	Medicinal
1025.	*Momordica charantia* L.	Cucurbitaceae	Medicinal
1026.	*Momordica cochinchinensis*	Cucurbitaceae	Medicinal
1027.	*Momordica cymbelaria* Fenze.	Cucurbitaceae	Medicinal
1028.	*Momordica dioca* Roxb.	Cucurbitaceae	Medicinal
1029.	*Momordica longifolia*	Cucurbitaceae	Medicinal
1030.	*Momordica tubrosa* Cogn.	Cucurbitaceae	Medicinal
1031.	*Morinda citrifolia* L.	Rubiaceae	Medicinal
1032.	*Moringa oleifera*	Moringaceae	Medicinal
1033.	*Morus alba* L.	Moraceae	Medicinal
1034.	*Morus serrata* Roxb.	Moraceae	Medicinal
1035.	*Mucuna bracteata* D.C.	Fabaceae	Medicinal
1036.	*Mucuna capitata* Wight. & Arn.	Fabaceae	Medicinal
1037.	*Mucuna pruriens* (L) D.C.	Fabaceae	Medicinal
1038.	*Murraya koenigii* (L.) Spreng.	Rutaceae	Medicinal
1039.	*Murraya paniculata* (L.) Jack.	Rutaceae	Medicinal
1040.	*Musa paradisiaca* L.	Musaceae	Medicinal
1041.	*Mussaenda roxburghii* Hook f.	Rubiaceae	Medicinal
1042.	*Nardostachys grandiflora* D.C.	Veleriaceae	Medicinal
1043.	*Nasturtium officinalis* R.Br.	Brassicaceae	Medicinal
1044.	*Nelumbo nucifera*	Nelumbonaceae	Medicinal
1045.	*Neptunia prostrata* (Lamk).	Mimosaceae	Medicinal
1046.	*Nerium indicum* Mill.	Apocynaceae	Medicinal
1047.	*Nerium oleander* L.	Apocynaceae	Medicinal
1048.	*Nicotiana plumbagoginifolia*	Solanaceae	Medicinal
1049.	*Nicotiana tabacum*	Solanaceae	Medicinal

	Name	Family	Purpose
1050.	*Nigella domascena* L.	Ranunculaceae	Medicinal
1051.	*Nigella sativa* L.	Ranunculaceae	Medicinal
1052.	*Nyctanthes arbortristis* L.	Nyctanthaceae	Medicinal
1053.	*Nymphaea alba* L.	Nymphaeaceae	Medicinal
1054.	*Nymphaea esculenta* Roxb.	Nymphaeaceae	Medicinal
1055.	*Nymphaea nouchali* Burm F.	Nymphaeaceae	Medicinal
1056.	*Nymphaea stellalata* Willd.	Nymphaeaceae	Medicinal
1057.	*Nymphaea tetragona* Georgi.	Nymphaeaceae	Medicinal
1058.	*Ocimum basilicum* Sims.	Lamiaceae	Medicinal
1059.	*Ocimum canum* Sims.	Lamiaceae	Medicinal
1060.	*Ocimum graticimum* L.	Lamiaceae	Medicinal
1061.	*Ocimum kilmandscharicum* Guerke.	Lamiaceae	Medicinal
1062.	*Ocimum sanctum* L.	Lamiaceae	Medicinal
1063.	*Oenothera lamarkiana*	Onagraceae	Medicinal
1064.	*Oldenlandia biflora* L	Rubiaceae	Medicinal
1065.	*Oldenlandia corymbosa* L.	Rubiaceae	Medicinal
1066.	*Oldenlandia diffusa* Willd. Roxb.	Rubiaceae	Medicinal
1067.	*Oldenlandia herbacea* Roxb.	Rubiaceae	Medicinal
1068.	*Oldenlandia hispida* Benth	Rubiaceae	Medicinal
1069.	*Oldenlandia umbellata*	Rubiaceae	Medicinal
1070.	*Onosma bractatum* Wall.	Boraginaceae	Medicinal
1071.	*Onosma echioides* L.	Boraginaceae	Medicinal
1072.	*Operculina turpenthum* (L.) S. Manso.	Convolvulaceae	Medicinal
1073.	*Opuntia cochinellifera* Mill.	Cactiaceae	Medicinal
1074.	*Opuntia dellenii.* Haw.	Cactiaceae	Medicinal
1075.	*Opuntia vulgaris* Mill.	Cactiaceae	Medicinal
1076.	*Origanum majorana*	Lamiaceae	Medicinal
1077.	*Orobanche aegyptiaca* Pers.	Orobanchaceae	Medicinal
1078.	*Oroxylum indicum*	Bignoniaceae	Medicinal
1079.	*Ougeinia oojeinensis* (Roxb.) Hochr.	Fabaceae	Medicinal
1080.	*Oxalis carniculata* L.	Oxalidaceae	Medicinal
1081.	*Oxalis latifolia* H.B. & K.	Oxalidaceae	Medicinal
1082.	*Oxystelma sccamone* (L.)	Asclepiadaceae	Medicinal
1083.	*Paederia foetida* L.	Rubiaceae	Medicinal
1084.	*Pandanus andamanensium*	Pandanaceae	Medicinal
1085.	*Pandanus furcatus* Roxb.	Pandanaceae	Medicinal
1086.	*Pandanus odoratissimus* F.	Pandanaceae	Medicinal
1087.	*Pandanus utilis* Bory.	Pandanaceae	Medicinal
1088.	*Panicum anitidotale* Retz.	Poaceae	Medicinal
1089.	*Papaver somniferum* L.	Papaveraceae	Medicinal
1090.	*Parkia biglandulosa*	Mimosaceae	Medicinal
1091.	*Parkinsonia aculeata* L.	Caesalpiniaceae	Medicinal
1092.	*Parthenium hysterophorus L.*	Asteraceae	Medicinal
1093.	*Paspalum scrohiculatum* L.	Poaceae	Medicinal
1094.	*Passiflora edulis* Sims.	Passifloraceae	Medicinal
1095.	*Passiflora foetida* L.	Passifloraceae	Medicinal
1096.	*Passiflora incarnata* L	Passifloraceae	Medicinal
1097.	*Pavonia odorata* Willd.	Malvaceae	Medicinal
1098.	*Pedalium murex* L	Pedaliaceae	Medicinal
1099.	*Peganum hermala*	Zygophyllaceae	Medicinal

	Name	Family	Purpose
1100.	*Pelargonium graveolens* (L.) Herit.	Geraniaceae	Medicinal
1101.	*Pellaea calomelanos* Link.	Sinopteridaceae	Medicinal
1102.	*Peltophorum pterocarpum*	Caesalpiniaceae	Medicinal
1103.	*Pennisetum typhoides* (Burm. F.) Stapf.	Poaceae	Medicinal
1104.	*Pergularia daemia* Forsk.	Asclepiadaceae	Medicinal
1105.	*Perilla fruitiscens*	Lamiaceae	Medicinal
1106.	*Peristrophe bicalyculata* Nees.	Acanthaceae	Medicinal
1107.	*Peuraria tubrosa* Roxb. ex. Willd.	Fabaceae	Medicinal
1108.	*Phaseolus trilobatus* (L.)Schreb.	Fabaceae	Medicinal
1109.	*Phaseolus vulgaris* L.	Fabaceae	Medicinal
1110.	*Phoenix acaulis* Roxb.	Arecaceae	Medicinal
1111.	*Phoenix sylvestris* (L) Roxb.	Arecaceae	Medicinal
1112.	*Phragmitis communis* Trin.	Poaceae	Medicinal
1113.	*Phragmitis karka* (Retz.)Trin ex. Steud.	Poaceae	Medicinal
1114.	*Phyllanathus simplex* Retz.	Euphorbiaceae	Medicinal
1115.	*Phyllanthgus madraspaterri*	Euphorbiaceae	Medicinal
1116.	*Phyllanthus urinaria* L.	Euphorbiaceae	Medicinal
1117.	*Physalis alkekenji* L.	Solanaceae	Medicinal
1118.	*Physalis minima* L.	Solanaceae	Medicinal
1119.	*Physalis peruviana* L.	Solanaceae	Medicinal
1120.	*Pimenta dioca* (L) Merrill.	Myrtaceae	Medicinal
1121.	*Pimpinella anisurn* L.	Apiaceae	Medicinal
1122.	*Pinus roxburghii*	Pinaceae	Medicinal
1123.	*Piperbetle* L.	Piperaceae	Medicinal
1124.	*Piperlongum* L.	Piperaceae	Medicinal
1125.	*Piper nigrum*	Piperaceae	Medicinal
1126.	*Pistia stratiotes* L.	Araceae	Medicinal
1127.	*Pitchecolobium dulce* Benth.	Mimosaceae	Medicinal
1128.	*Plantago ovta* Forsk.	Plantaginaceae	Medicinal
1129.	*Pluchia lanceolata* Oliver & Hiern.	Asteraceae	Medicinal
1130.	*Plumbago indica* L.	Plumbaginaceae	Medicinal
1131.	*Plumbago zeylanica* L.	Plumbaginaceae	Medicinal
1132.	*Plumeria alba* L.	Apocynaceae	Medicinal
1133.	*Plumeria rubra* forma acuminate	Apocynaceae	Medicinal
1134.	*Plumeria rubra* L.	Apocynaceae	Medicinal
1135.	*Pogostemon cabin* (Blanco) Benth.	Lamiaceae	Medicinal
1136.	*Pogostemon patchouli* Hook. F.	Lamiaceae	Medicinal
1137.	*Polyalthea longifolia* Benth.	Annonaceae	Medicinal
1138.	*Polyalthea Suberosa* Thw.	Annonaceae	Medicinal
1139.	*Polygonum amphibium* L.	Polygonaceae	Medicinal
1140.	*Polygonum amplex* caule D. Don	Polygonaceae	Medicinal
1141.	*Polygonum hydropiper* L.	Polygonaceae	Medicinal
1142.	*Polygonum multiflorum* Thumb.	Polygonaceae	Medicinal
1143.	*Polygonum punctatum* Ham.	Polygonaceae	Medicinal
1144.	*Pongamia pinnata* L. Pierre.	Fabaceae	Medicinal
1145.	*Portulaca grandiflora Hook.*	Portulacaceae	Medicinal
1146.	*Portulaca oleracea*	Portulacaceae	Medicinal
1147.	*Pothos cathcarti* Schott.	Araceae	Medicinal
1148.	*Premna mucronata* Roxb.	Verbenaceae	Medicinal
1149.	*Premna obtusifolia.*	Verbenaceae	Medicinal

	Name	Family	Purpose
1150.	*Prunus domestica* L.	Rosaceae	Medicinal
1151.	*Psidium guajava L.*	Myrtaceae	Medicinal
1152.	*Psoralea corylifolia* L.	Fabaceae	Medicinal
1153.	*Pterocarpus marsupium* L.	Fabaceae	Medicinal
1154.	*Pterocarpus santalinus* L.	Fabaceae	Medicinal
1155.	*Pterospermum acerifolium* Wild.	Sterculiaceae	Medicinal
1156.	*Pterospermum canescens* Roxb.	Sterculiaceae	Medicinal
1157.	*Punica granatum* L.	Punicaceae	Medicinal
1158.	*Purtanjiva roxburghii* Wall.	Euphorbiaceae	Medicinal
1159.	*Pygmaeopremnha herbacea*	Verbenaceae	Medicinal
1160.	*Pyrostegia venusa*	Bignoniaceae	Medicinal
1161.	*Quisqualis indica L.*	Combetaceae	Medicinal
1162.	*Radermachera xylocarpa*	Bignoniaceae	Medicinal
1163.	*Randia spinosa* Poir.	Rubiaceae	Medicinal
1164.	*Ranunculus sceleratus* L.	Ranunculaceae	Medicinal
1165.	*Raphanus sativus* L.	Brassicaceae	Medicinal
1166.	*Rauvolfia serpentine* L. Benth	Apocynaceae	Medicinal
1167.	*Rauvolfia teraphylla* L.	Apocynaceae	Medicinal
1168.	*Rhus parviflora*		Medicinal
1169.	*Rhynchosia capitata* D.C.	Fabceae	Medicinal
1170.	*Rhynchosia minima* (L.) D.C.	Fabceae	Medicinal
1171.	*Ricinus communis* L.	Euphorbiaceae	Medicinal
1172.	*Rivinia humilis*	Protiveraceae	Medicinal
1173.	*Rorirppa indica* (L.) Hern.	Brassicaceae	Medicinal
1174.	*Rosa alba* L	Rosaceae	Medicinal
1175.	*Rosa banksiae* R.Br.	Rosaceae	Medicinal
1176.	*Rosa bourboniana* Desportes.	Rosaceae	Medicinal
1177.	*Rosa braunonii* Lindl	Rosaceae	Medicinal
1178.	*Rosa centifolia* L	Rosaceae	Medicinal
1179.	*Rosa chinensis* Jacq.	Rosaceae	Medicinal
1180.	*Rosa damascena* Mill.	Rosaceae	Medicinal
1181.	*Rosa gallica* L.	Rosaceae	Medicinal
1182.	*Rosa india* Hook F.	Rosaceae	Medicinal
1183.	*Rosa multiflora* Thunb.	Rosaceae	Medicinal
1184.	*Rosmarinus* officinalis L.	Lamiaceae	Medicinal
1185.	*Rubia cordifolia* L. Sansu. Hook.R.	Rubiaceae	Medicinal
1186.	*RueIlia tubrosa* L.	Acanthaceae	Medicinal
1187.	*Rumex acetocella* L.	Polygonaceae	Medicinal
1188.	*Rumex acetosa* L.	Polygonaceae	Medicinal
1189.	*Rumex dentatus* L	Polygonaceae	Medicinal
1190.	*Rungia pectinata* (L.) Nees.	Acanthaceae	Medicinal
1191.	*Rungia repens* (L.) Nees.	Acanthaceae	Medicinal
1192.	*Ruta graveolens* L.	Rutaceae	Medicinal
1193.	*Saccharum munja* Roxb.	Poaceae	Medicinal
1194.	*Saccharum officinarum* L.	Poaceae	Medicinal
1195.	*Saccharum spontaneum* L.	Poaceae	Medicinal
1196.	*Sagittaria sagitifolia* L.	Alismataceae	Medicinal
1197.	*Salsola baryosma* (R. & S.) Dandy.	Chenopodiaceae	Medicinal
1198.	*Salvadora oleoides* Decne.	Salvadoraceae	Medicinal
1199.	*Salvadora persica* L.	Salvadoraceae	Medicinal

	Name	Family	Purpose
1200.	*Salvadora plebeia* Br.	Salvadoraceae	Medicinal
1201.	*Salvia aegyptiaca* L.	Lamiaceae	Medicinal
1202.	*Salvia hispanica*	Lamiaceae	Medicinal
1203.	*Salvia officinalis* L	Lamiaceae	Medicinal
1204.	*Salvia sclarez* L	Lamiaceae	Medicinal
1205.	*Salvia splendens* Ker Gawl.	Lamiaceae	Medicinal
1206.	*Sansevieria roxburghiana* Schutt.	Agavaceae	Medicinal
1207.	*Sansevieria zeylanica* Willd.	Agavaceae	Medicinal
1208.	*Santalum album* L.	Santalaceae	Medicinal
1209.	*Sapindus emerginatus* Vahl.	Sapindaceae	Medicinal
1210.	*Sapindus mukorossi* Gaertn.	Sapindaceae	Medicinal
1211.	*Saraca asoca* (Roxb.) de.Wild.	Caesalpiniaceae	Medicinal
1212.	*Sarcostemma acidum*	Apocynaceae	Medicinal
1213.	*Schelichera oleosa* (Lour) Oken.	Sapindaceae	Medicinal
1214.	*Scindapsus officinalis*Schott. (Syn Pothos officinalis)	Araceae	Medicinal
1215.	*Scirpus grossus* L.F.	Cyperaceae	Medicinal
1216.	*Scirpus lacustris* L.	Cyperaceae	Medicinal
1217.	*Sclera pergracillis* Nees. Kunth.	Cyperaceae	Medicinal
1218.	*Scoparia dulcis* L.	Flacourtiaceae	Medicinal
1219.	*Semecarpus anacardium* L.F.	Anacardiaceae	Medicinal
1220.	*Senecio vulgaris* L.	Asteraceae	Medicinal
1221.	*Sesamum indicum* L.	Pedaliaceae	Medicinal
1222.	*Sesbania aculeata* Pers.	Fabaceae	Medicinal
1223.	*Sesbania bispinosa* Jacq.W.F.	Fabaceae	Medicinal
1224.	*Sesbania grandiflora* (L.) Pair.	Fabaceae	Medicinal
1225.	*Sesbania sesban* (L.) Merr.	Fabaceae	Medicinal
1226.	*Setaria italica* (L.) Beauv.	Poaceae	Medicinal
1227.	*Shorea robusta* Gaertn.	Dipterocarpaceae	Medicinal
1228.	*Sida alba* L. SynS. Spinosa L.	Malvaceae	Medicinal
1229.	*Sida carpinifolia* Mast	Malvaceae	Medicinal
1230.	*Sida cordata* (Burms. F.) Borssum	Malvaceae	Medicinal
1231.	*Sida cordifolia* L.	Malvaceae	Medicinal
1232.	*Sida rhombifolia* L.	Malvaceae	Medicinal
1233.	*Silybum marianum*	Asteraceae	Medicinal
1234.	*Sisymbrium irio* L.	Brassicaceae	Medicinal
1235.	*Smilax china* Linn.	Smilacaceae	Medicinal
1236.	*Smilax zeylanica* L.	Smilacaceae	Medicinal
1237.	*Smithia sensitiva* Ait	Fabaceae	Medicinal
1238.	*Solanum grandiflorum*	Solanaceae	Medicinal
1239.	*Solanum indicum*	Solanaceae	Medicinal
1240.	*Solanum lacianiatum*	Solanaceae	Medicinal
1241.	*Solanum pseudo* capsicum	Solanaceae	Medicinal
1242.	*Solanum sissymbrifolium*	Solanaceae	Medicinal
1243.	*Solanum stramonifolium* Jacq.	Solanaceae	Medicinal
1244.	*Solanum surattense* Burmf.	Solanaceae	Medicinal
1245.	*Solanum torvum* Swartz.	Solanaceae	Medicinal
1246.	*Solanum viarum* Dunal.	Solanaceae	Medicinal
1247.	*Solidago canadensis* L.	Asteraceae	Medicinal
1248.	*Sonchus arvensis*	Asteraceae	Medicinal

	Name	Family	Purpose
1249.	*Sonchus oleraceus* L	Asteraceae	Medicinal
1250.	*Soymida febrifuga*& Juss.	Meliaceae	Medicinal
1251.	*Spathodea campanulata* Beauv.	Bignoniaceae	Medicinal
1252.	*Spatholobus roxburghii*	Fabaceae	Medicinal
1253.	*Spermadictylon sauveolens* Roxb.	Rubiaceae	Medicinal
1254.	*Sphaeranthus. indicus*	Asteraceae	Medicinal
1255.	*Spilanthus oleracea* Murr.	Asteraceae	Medicinal
1256.	*Spilanthus paniculata* Wall ex. Dc.	Asteraceae	Medicinal
1257.	*Spinacia oleracea*	Chenopodiaceae	Medicinal
1258.	Sterculia villosa Roxb.	Sterculiaceae	Medicinal
1259.	*Sterculia wrens* Roxb.	Sterculiaceae	Medicinal
1260.	*Stereospermum coleus*	Bignoniaceae	Medicinal
1261.	*Stereosperum suaveolens* D.C.	Bignoniaceae	Medicinal
1262.	*Strychnos nux-vomica* L.	Strychnaceae	Medicinal
1263.	*Strychnos potatorum* L.F.	Strychnaceae	Medicinal
1264.	*Swertia chirata* (Roxb. ex.flem.)	Gentianaceae	Medicinal
1265.	*Swietenia mahagoni* (L) Jack.	Meliaceae	Medicinal
1266.	*Symplocas racemosa* Roxb	Symplocaceae	Medicinal
1267.	*Syzygium aromaticum*	Myrtaceae	Medicinal
1268.	*Syzygium aromaticus*	Myrtaceae	Medicinal
1269.	*Syzygium cumini* (L.) Skeels.	Myrtaceae	Medicinal
1270.	*Syzygium fruiticosum* Rook D.C.	Myrtaceae	Medicinal
1271.	*Syzygium samaragense*	Myrtaceae	Medicinal
1272.	*Tabebuia pentaphylla* (L) Hemsl.	Bignoniaceae	Medicinal
1273.	*Tabebuia rosea* Bertol D.C.	Bignoniaceae	Medicinal
1274.	*Tabebuia spectahilis* N.Chils.	Bignoniaceae	Medicinal
1275.	*Tabernaemontanadivaricata* (L.) R.Br.	Apocynaceae	Medicinal
1276.	*Tacoma stuns* (L) H.B. & K.	Bignoniaceae	Medicinal
1277.	*Tagetes erecta* L.	Asteraceae	Medicinal
1278.	*Tagetes lucidu* Cav	Asteraceae	Medicinal
1279.	*Tagetes minuta* L.	Asteraceae	Medicinal
1280.	*Tagetes patula* L.	Asteraceae	Medicinal
1281.	*Tagetes tenuifolia* Cav	Asteraceae	Medicinal
1282.	*Tamarindus indica* L.	Mimosaceae	Medicinal
1283.	*Tamarix troupii* Hole.	Tamaricaceae	Medicinal
1284.	*Taxodium distichum* L.Rich.	Taxodiaceae	Medicinal
1285.	*Tecomella undulata* G.Don.	Bignoniaceae	Medicinal
1286.	*Tectona grandis* L.F.	Verbenaceae	Medicinal
1287.	*Tephorosia candida* Roxb D.C.	Fabaceae	Medicinal
1288.	*Tephrosia purpurea* (L) Pers.	Fabaceae	Medicinal
1289.	*Tephrosia tinctoria* Pers.	Fabaceae	Medicinal
1290.	*Tephrosia vilosa* Pers.	Fabaceae	Medicinal
1291.	*Terminalia arjuna* (Roxb) Wight.	Combretaceae	Medicinal
1292.	*Terminalia bellirica* (Gaertn) Roxb.	Combretaceae	Medicinal
1293.	*Terminalia catappa* L.	Combretaceae	Medicinal
1294.	*Terminalia chebula* Retz.	Combretaceae	Medicinal
1295.	*Terminalia tomentosa*	Combretaceae	Medicinal
1296.	*Thalictrum foliolosum* D.C.	Ranunculaceae	Medicinal
1297.	*Theobroma cacao L.*	Sterculiaceae	Medicinal
1298.	*Thespesia popuMea* (L.)	Malvaceae	Medicinal

	Name	Family	Purpose
1299.	*Thevetia peruviana* (Pers.) K. Schum	Apocynaceae	Medicinal
1300.	*Thuja occidentalis* L.	Cupressaceae	Medicinal
1301.	*Thuja orientalis* L.	Cupressaceae	Medicinal
1302.	*Thunbergia alata* Boj. Exsim.	Thunbergiaceae	Medicinal
1303.	*Thunbergia faragrans* Roxb.	Thunbergiaceae	Medicinal
1304.	*Thunbergia grandiflora* Roxb.	Thunbergiaceae	Medicinal
1305.	*Thymus serpyllum* L.	Lamiaceae	Medicinal
1306.	*Tinospora cordifolia* (Wind) Meers.	Menispermaceae	Medicinal
1307.	*Tinospora crispa* (L) Miers.	Menispermaceae	Medicinal
1308.	*Tinospora sinensis* (Lour.) Merr.	Menispermaceae	Medicinal
1309.	*Toona ciliata*	Meliaceae	Medicinal
1310.	*Trachyspermum ammi* (L) Spraoue.	Apiaceae	Medicinal
1311.	*Trachyspermum roxburghianum*	Apiaceae	Medicinal
1312.	*Tradescantia discolour* L Herit.	Commelinaceae	Medicinal
1313.	*Trapa natans* L.	Trapaceae	Medicinal
1314.	*Triadica sebifera* (L.)Small.	Euphorbiaceae	Medicinal
1315.	*Trianthema portulacastrum*	Aizoaceae	Medicinal
1316.	*Tribulus terestris* L.	Zygophyllaceae	Medicinal
1317.	*Trichodesma amplexicanle* Roth.	Boraginaceae	Medicinal
1318.	*Trichodesma indicum* Br.	Boraginaceae	Medicinal
1319.	*Trichodesma zeylanicum* Br.	Boraginaceae	Medicinal
1320.	*Tricholepis glabrrima* D.C.	Asteraceae	Medicinal
1321.	*Trichosanthes anguina* L	Cucurbitaceae	Medicinal
1322.	*Trichosanthes cucumerina* L.	Cucurbitaceae	Medicinal
1323.	*Trichosanthus dioica Roxb.*	Cucurbitaceae	Medicinal
1324.	*Trichosanthus laciniosa* Wall.	Cucurbitaceae	Medicinal
1325.	*Tridax procumbens*	Asteraceae	Medicinal
1326.	*Trigonella foenum* graecum L.	Fabaceae	Medicinal
1327.	*Triumfetta pentandra* A. Rich.	Tiliaceae	Medicinal
1328.	*Triumphetta rhomboidea* Jacq.	Tiliaceae	Medicinal
1329.	*Tylophora fasciculata* Buch. Ham.	Asclepiadaceae	Medicinal
1330.	*Tylophora indica* (Burm.f.)	Asclepiadaceae	Medicinal
1331.	*Typha angusta* Bory. & Chaub.	Typhaceae	Medicinal
1332.	*Uraria lobata* L	Fabaceae	Medicinal
1333.	*Uraria picta* Desv.	Fabaceae	Medicinal
1334.	*Urena lobata* L var. Cinuata King.	Malvaceae	Medicinal
1335.	*Urena repanda* Roxb ex.sm.	Malvaceae	Medicinal
1336.	*Urenia lobata* L.	Malvaceae	Medicinal
1337.	*Urginea indica* (Roxb) Kunth	Liliaceae	Medicinal
1338.	*Urtica dioca* L.	Urticaceae	Medicinal
1339.	*Valeriana jatamanshi* Janes.	Valeriaceae	Medicinal
1340.	*Vallisnaria spiralis* L.	Hydrocharitaceae	Medicinal
1341.	*Vanda parviflora*	Orchidaceae	Medicinal
1342.	*Vanda tessellata*	Orchidaceae	Medicinal
1343.	*Ventilago denticulata* Willd.	Rhamnaceae	Medicinal
1344.	*Ventilago madraspatana* Gaertn.	Rhamnaceae	Medicinal
1345.	*Verbascum thaspus* L	Scrophulaiaceae	Medicinal
1346.	*Verbena officinalis* L	Verbenaceae	Medicinal
1347.	*Vernonia anthelmintica* Willd.	Asteraceae	Medicinal
1348.	*Vernonia cinerea* (L) Less.	Asteraceae	Medicinal

	Name	Family	Purpose
1349.	*Vetiveria zizanioides* (L.) Mash.	Poaceae	Medicinal
1350.	*Vicia hirsuta* (L.) S.F.	Fabaceae	Medicinal
1351.	*Vicia sativa* L.	Fabaceae	Medicinal
1352.	*Victoria regia* lindl.	Euryalaceae	Medicinal
1353.	*Vigna aconitifolia* (Jacq.) Marechal.	Fabaceae	Medicinal
1354.	*Vigna trilobata* (L) Verdcourt.	Fabaceae	Medicinal
1355.	*Vigna umbellata* (Thumb)	Fabaceae	Medicinal
1356.	*Vinca minor* L.	Apocynaceae	Medicinal
1357.	*Viola odorata*	Violaceae	Medicinal
1358.	*Viola serpens* Wall ex.Ging.	Violaceae	Medicinal
1359.	*Vitex glabra* R.Br.	Verbenaceae	Medicinal
1360.	*Vitex negundo* L.	Verbenaceae	Medicinal
1361.	*Vitex peduncularis* Wall.	Verbenaceae	Medicinal
1362.	*Vitis cetosa*	Vitaceae	Medicinal
1363.	*Vitis indica* L	Vitaceae	Medicinal
1364.	*Vitis latifolia* Roxb	Vitaceae	Medicinal
1365.	*Vitis pedata* Vahi.	Vitaceae	Medicinal
1366.	*Vitis trifolia* L.	Vitaceae	Medicinal
1367.	*Volutarella divericata* Benth.	Asteraceae	Medicinal
1368.	*Withania somnifera* Dunal.	Solanaceae	Medicinal
1369.	*Woodfordia fruitcosa* L. Kurtz.	Lythraceae	Medicinal
1370.	*Wrightia tinctoria* R.Br.	Apocynaceae	Medicinal
1371.	*Wrightia tomentosa* (Roxb.)	Apocynaceae	Medicinal
1372.	*Xanthium strumarium* L.	Asteraceae	Medicinal
1373.	*Yucca aloifolia* L.	Agavaceae	Medicinal
1374.	*Yucca filanentosa* L.	Agavaceae	Medicinal
1375.	*Yucca gloriosa*	Agavaceae	Medicinal
1376.	*Zingiber cassumunar* Roxb.	Zingiberaceae	Medicinal
1377.	*Zingiber officinalis* Rose.	Zingiberaceae	Medicinal
1378.	*Ziziphus mauritiana*	Rhamnaceae	Medicinal
1379.	*Ziziyphus nummularia*	Rhamnaceae	Medicinal
1380.	*Ziziyphus xylopyra* Wills.	Rhamnaceae	Medicinal
1381.	*Zizyphus oenopila* Mill.	Rhamnaceae	Medicinal
1382.	*Abelmoschus fioulneus*	Malvaceae	Weed
1383.	*Abelmoschus manihot*	Malvaceae	Weed
1384.	*Abelmoschus moschatus*	Malvaceae	Weed
1385.	*Abutilon graveolens* Wight & Arnott	Malvaceae	Weed
1386.	*Achyranthes aspera L.*	Amaranthaceae	Weed
1387.	*Acrachneracemosa*	Poaceae	Weed
1388.	*Aeschynomene indica* (Linn.)	Fabaceae	Weed
1389.	*Ageratum conyzoides*	Asteraceae	Weed
1390.	*Agropyron repens*	Asteraceae	Weed
1391.	*Alternanthera philoxeroides*	Amaranthaceae	Weed
1392.	*Alternanthera sessilis*	Amaranthaceae	Weed
1393.	*Alternanthera spp*	Amaranthaceae	Weed
1394.	*Alysicarpus rugousus* (Wild.) DC	Fabaceae	Weed
1395.	*Alysicarpus tetragonolobus* Edgew	Fabaceae	Weed
1396.	*Amaranthus spinosus*	Amaranthaceae	Weed
1397.	*Amaranthus viridis*	Amaranthaceae	Weed
1398.	*Ammannia baccifera*	Lythraceae	Weed

	Name	Family	Purpose
1399.	*Anagallis arvensis*	Primulaceae	Weed
1401.	*Argemone mexicana*	Papaveraceae	Weed
1402.	*Asphodelus tenuifolius*	Asphodelaceae	Weed
1403.	*Avena fatua*	Poaceae	Weed
1404.	*Avena ludoviciana*	Poaceae	Weed
1405.	*Azolla filiculoides*	Salviniaceae	Weed
1406.	*Brachiaria mutica*	Poaceae	Weed
1407.	*Brachiaria reptans*	Poaceae	Weed
1408.	*Caesulia axillaris*	Asteraceae	Weed
1409.	*Calotropis procera*	Apocynaceae	Weed
1410.	*Carthamus oxyacantha*	Asteraceae	Weed
1411.	*Cassia tora*	Fabaceae	Weed
1412.	*Ceratophyllum demersum*	Ceratophyllaceae	Weed
1413.	*Chenopodium album*	Amaranthaceae	Weed
1414.	*Chenopodium murale*	Amaranthaceae	Weed
1415.	*Cichorium intybus*	Asteraceae	Weed
1416.	*Cirsium arvense*	Asteraceae	Weed
1417.	*Citrullus colocynthis*	Cucurbitaceae	Weed
1418.	*Cleome viscosa*	Cleomaceae	Weed
1419.	*Commelina benghalensis*	Commelinaceae	Weed
1420.	*Commelina diffusa*	Commelinaceae	Weed
1421.	*Commelina erecta*	Commelinaceae	Weed
1422.	*Convolvulus arvensis*	Convolvulaceae	Weed
1423.	*Conyza stricta*	Poaceae	Weed
1424.	*Corchorus acutaangulus* L.	Malvaceae	Weed
1425.	*Corchorus tridens*	Tiliaceae	Weed
1426.	*Coronopus didymus*	Brassicaceae	Weed
1427.	*Crotalaria medicaginea*	Fabaceae	Weed
1428.	*Cucumis hardwickii*	Cucurbitaceae	Weed
1429.	*Cuscuta reflexa*	Convolvulaceae	Weed
1430.	*Cynodon dactylon*	Poaceae	Weed
1431.	*Cyperus rotandus*	Poaceae	Weed
1432.	*Cyperus compressus*	Poaceae	Weed
1433.	*Cyperus iria*	Poaceae	Weed
1434.	*Dactyloctenium aegyptium*	Poaceae	Weed
1435.	*Datura stramonium*	Solanaceae	Weed
1436.	*Desmodium gangetium* DC Pennel	Fabaceae	Weed
1437.	*Digera arvensis*	Amaranthaceae	Weed
1438.	*Digera muricata*	Amaranthaceae	Weed
1439.	*Digitaria sanguinalis*	Poaceae	Weed
1440.	*Echinochloa colona*	Poaceae	Weed
1441.	*Echinochloa crusgalli*	Poaceae	Weed
1442.	*Eclipta alba*	Asteraceae	Weed
1443.	*Eichhornia crassipes*	Pontederiaceae	Weed
1444.	*Eleusine indica*	Poaceae	Weed
1445.	*Eragrostis pilosa*	Poaceae	Weed
1446.	*Eragrostis tenella*	Poaceae	Weed
1447.	*Euphorbia heterophylla*	Euphorbiaceae	Weed
1448.	*Euphorbia hirta*	Euphorbiaceae	Weed
1449.	*Euphorbia microphylla*	Euphorbiaceae	Weed

	Name	Family	Purpose
1450.	*Euphorbia prostrata*	Euphorbiaceae	Weed
1451.	*Fumaria parviflora*	Papaveraceae	Weed
1452.	*Gnaphalium purpureum*	Asteraceae	Weed
1453.	*Heliotropium indicum*	Boraginaceae	Weed
1454.	*Hibiscus ovalifolius*	Malvaceae	Weed
1455.	*Hydrilla verticillata*	Hydrocharitaceae	Weed
1456.	*Indigofera glandulosa* J.C. Wendl	Fabaceae	Weed
1457.	*Indigofera linnaei*	Fabaceae	Weed
1458.	*Ipomoea carnea*	Convolvulaceae	Weed
1459.	*Ipomoea nil*	Convolvulaceae	Weed
1460.	*Ipomoea pestigridis*	Convolvulaceae	Weed
1461.	*Ischaemum rugosum*	Poaceae	Weed
1462.	*Lactuca scariola*	Asteraceae	Weed
1463.	*Lantana camara*	Verbenaceae	Weed
1464.	*Lathyru sativus*	Fabaceae	Weed
1465.	*Lathyrus aphacaora*	Fabaceae	Weed
1466.	*Leptochloa chinensis*	Poaceae	Weed
1467.	*Lolium temulentum*	Poaceae	Weed
1468.	*Loranthus micranthus*	Orobanchaceae	Weed
1469.	*Lagascea mollis*	Asteraceae	Weed
1470.	*Malva parviflora*	Malvaceae	Weed
1471.	*Marsilea hirsuta*	Marsileaceae	Weed
1472.	*Medicago denticulata*	Fabaceae	Weed
1473.	*Medicago polymorpha*	Fabaceae	Weed
1474.	*Medicago truncatula*	Fabaceae	Weed
1475.	*Melilotus alba*	Fabaceae	Weed
1476.	*Melilotus indicus*	Fabaceae	Weed
1477.	*Mollugo pentaphylla*	*Molluginaceae*	Weed
1478.	*Mollugo verticillata*	*Molluginaceae*	Weed
1479.	*Momordica blasamina*	Cucurbitaceae	Weed
1480.	*Momordica cochinchinesis*	Cucurbitaceae	Weed
1481.	*Momordica diocia*	Cucurbitaceae	Weed
1482.	*Myriophyllum aquaticum*	Haloragaceae	Weed
1483.	*Nymphaea alba*	Nymphaeaceae	Weed
1484.	*Nymphoides cristata*	Menyanthaceae	Weed
1485.	*Oenothera laciniata*	Onagraceae	Weed
1486.	*Opuntia dillenii*	Cactaceae	Weed
1487.	*Orobanche aegyptiaca*	Orobanchaceae	Weed
1488.	*Oryza sativa var. fatua*	Poaceae	Weed
1489.	*Oxalis corniculata*	Oxalidaceae	Weed
1490.	*Parthenium hysterophorus*	Asteraceae	Weed
1491.	*Paspalum distichum*	Poaceae	Weed
1492.	*Paspalum vaginatum*	Poaceae	Weed
1493.	*Phalaris minor*	Poaceae	Weed
1494.	*Phragmites australis*	Poaceae	Weed
1495.	*Phyllanthus niruri*	Phyllanthaceae	Weed
1496.	*Physalis minima*	Solanaceae	Weed
1497.	*Pistia stratiotes*	Araceae	Weed
1498.	*Poa annua*	Poaceae	Weed
1499.	*Polypogon monspeliensis*	Poaceae	Weed

	Name	Family	Purpose
1500.	*Portulaca oleracea*	Portulacaceae	Weed
1501.	*Potamogeton perfoliatus*	Potamogetonaceae	Weed
1502.	*Ranunculus muricatus*	Ranunculaceae	Weed
1503.	*Ranunculus sceleratus*	Synonyms	Weed
1504.	*Rumex dentatus*	Polygonaceae	Weed
1505.	*Rumex hymenosepalus*	Polygonaceae	Weed
1506.	*Rumex spinosus*	Polygonaceae	Weed
1507.	*Saccharum spontaneum*	Poaceae	Weed
1508.	*Sagittaria subulata*	Alismataceae	Weed
1509.	*Salvinia minima*	Salviniaceae	Weed
1510.	*Saponaria vaccaria*	Caryophyllaceae	Weed
1511.	*Setaria verticillata*	Poaceae	Weed
1512.	*Setaria viridis*	Poaceae	Weed
1513.	*Sida acuta*	Malvaceae	Weed
1514.	*Silene conoidea*	Caryophyllaceae	Weed
1515.	*Sisymbrium irio*	Brassicaceae	Weed
1516.	*Solanum dulcamara*	Solanaceae	Weed
1517.	*Solanum nigrum*	Solanaceae	Weed
1518.	*Sonchus arvensis*	Asteraceae	Weed
1519.	*Sorghum halepense*	Poaceae	Weed
1520.	*Spergula arvensis*	Caryophyllaceae	Weed
1521.	*Striga lutea*	Loranthaceae	Weed
1522.	*Trianthema monogyna*	Aizoaceae	Weed
1523.	*Trianthema portulacastrum*	Aizoaceae	Weed
1524.	*Tribulus terrestris*	Zygophyllaceae	Weed
1525.	*Trichosanthes cucumerina*	Cucurbitaceae	Weed
1526.	*Trichosanthes lobata*	Cucurbitaceae	Weed
1527.	*Trichosanthes palmata*	Cucurbitaceae	Weed
1528.	*Tridax procumbens*	Asteraceae	Weed
1529.	*Trigonella polycerata*	Fabaceae	Weed
1530.	*Triumfetta rhombidea* Jacq.	Malvaceae	Weed
1531.	*Typha angustifolia*	Typhaceae	Weed
1532.	*Typha latifolia*	Typhaceae	Weed
1533.	*Veronica agrestis*	Plantaginaceae.	Weed
1534.	*Vicia hirsuta*	Fabaceae	Weed
1535.	*Vicia sativa*	Fabaceae	Weed
1536.	*Wild Helianthus*	Compositeae	Weed
1537.	*Wolffia arrhiza*	Araceae	Weed
1538.	*Xanthium strumarium*	Asteraceae	Weed
1539.	*Xanthium strumarium*L.	Asteraceae	Weed
1540.	*Acanthiophilus helianthi* Rossi	Tephritidae	Insect
1541.	*Achaea janata* (L.)	Noctuidae	Insect
1542.	*Acherontia styx* (Westwood)	Sphingidae	Insect
1543.	*Acheta domestica* (Linnaeus)	Gryllidae	Insect
1544.	*Acyrthosiphon pisum*	Aphididae	Insect
1545.	*Agrius convolvuli* (Linn.)	Sphingidae	Insect
1546.	*Agrotis ipsilon* Hufnagel	Noctuidae	Insect
1547.	*Aiolopus thalassinus* (Fab.)	Acrididae	Insect
1548.	*Aleurocanthus woglumi* Ashby	Aleyrodidae	Insect
1549.	*Altica oleracea* (Linnaeus)	Chrysomelidae	Insect

	Name	Family	Purpose
1550.	*Altica* sp.	Chrysomelidae	Insect
1551.	*Amata cyssea* (Cramer)	Erebidae	Insect
1552.	*Amata* sp.	Arctiidae	Insect
1553.	*Amerila astrea* Drury	Erebidae	Insect
1554.	*Amerila eugenia* Fab.	Erebidae	Insect
1555.	*Amrasca bigutulla bigutulla* (Ishida)	Cicadellidae	Insect
1556.	*Amrasca devastans* Dist.	Jassidae	Insect
1557.	*Amritodus atkinsoni* (Leth.)	Cicadellidae	Insect
1558.	*Amsacta lactinea* (Cramer)	Arctidae	Insect
1559.	*Amsacta moorei* Butler	Arctidae	Insect
1560.	*Anarsia ephbippias* May	Gelechiidae	Insect
1561.	*Anomis flava* (Fabricius)	Erebidae	Insect
1562.	*Anoplonemis phasiana* (F)	Pentatomidae	Insect
1563.	*Antestiopsis cruciata* (Fabricius)	Pentatomidae	Insect
1564.	*Antigastra catalaunalis* (Duponchel)	Pyralidae	Insect
1565.	*Apheliona maculosa* (Dist.)	Delphacidae	Insect
1566.	*Aphis craccivora* Koch	Aphididae	Insect
1567.	*Aphis gossypii* Glover	Aphididae	Insect
1568.	*Aphis punicae* Passerini	Aphididae	Insect
1569.	*Apis cerana indica* Fab.	Apidae	Insect
1570.	*Apis dorsata* Fab.	Apidae	Insect
1571.	*Apis florea* Fab.	Apidae	Insect
1572.	*Apis melipona* Illiger	Apidae	Insect
1573.	*Apis mellifera* Linn.	Apidae	Insect
1574.	*Arachnis picta* (Packard)	Erebidae	Insect
1575.	*Argina astrea* (Durry)	Erebidae	Insect
1576.	*Argina cribraria* (Clerck)	Arctiidae	Insect
1577.	*Asota caricaei* (Fabr.)	Erebidae	Insect
1578.	*Asota ficus* (Fabricius)	Noctuidae	Insect
1579.	*Aspondylia capsici* Barnes	Cecidomyiidae	Insect
1580.	*Aspondylia sesami* (Felt.)	Cecidomyiidae	Insect
1581.	*Athalia lugens proxima* (Klug)	Tenthridinidae	Insect
1582.	*Athalia* sp.	Tenthridinidae	Insect
1583.	*Atherigona bituberculata* Malloch	Muscidae	Insect
1584.	*Atherigona soccata* Rond.	Muscidae	Insect
1585.	*Atherigona* sp.	Muscidae	Insect
1586.	*Atractomorpha* sp.	Pyrgomorphidae	Insect
1587.	*Aulacophora foveicollis* (Lucas)	Chrysomelidae	Insect
1588.	*Bacterocera cucurbitae* (Coquillet)	Tephritidae	Insect
1589.	*Bacterocera dorsalis* Hendel	Tephritidae	Insect
1590.	*Bacterocera rufomaculata* DeGeer	Cerambycidae	Insect
1591.	*Bagrada cruciferarum* Kirkaldy	Pentatomidae	Insect
1592.	*Banisia fenestrifera* (Walker)	Thyrididae	Insect
1593.	*Bastilla crameri* Moore	Noctuidae	Insect
1594.	*Bathrips melanicornis* (Shumsher)	Thripidae	Insect
1595.	*Bemisia tabaci* Gennadius	Aleyrodidae	Insect
1596.	*Brevicoryne brassicae* (L.)	Aphididae	Insect
1597.	*Cadra cautella* Walker	Pyralidae	Insect
1598.	*Caliothrips indicus (Bagnall)*	Thripidae	Insect
1599.	*Calliteara grotei (Moore)*	Erebidae	Insect

	Name	Family	Purpose
1600.	*Callosobruchus analis*	Bruchidae	Insect
1601.	*Callosobruchus chinensis* (L)	Bruchidae	Insect
1602.	*Callosobruchus maculates* (Fabricius)	Bruchidae	Insect
1603.	*Callosobruchus phaseoli* (Gylh)	Bruchidae	Insect
1604.	*Calocoris angustatus* Leth.	Miridae	Insect
1605.	*Chilo infuscatellus* Snell.	Pyralidae	Insect
1606.	*Chilo partellus* (Swinhoe)	Pyralidae	Insect
1607.	*Chionaema peregrina* (Walker)	Erebidae	Insect
1608.	*Chrotogonus* sp.	Pyrgomorphidae	Insect
1609.	*Chrotogonus trachypterus* (Blan.)	Acrididae	Insect
1610.	*Chrysocraspeda* sp.	Geometridae	Insect
1611.	*Chrysodeixis acuta* (Walker)	Noctuidae	Insect
1612.	*Chrysodeixis chalcites* (Esper)	Noctuidae	Insect
1613.	*Chrysomya* sp.	Calliphoridae	Insect
1614.	*Cirrhochrista brizoalis* (Walker)	Crambidae	Insect
1615.	*Clavigralla gibbosa* Spinola	Coreidae	Insect
1616.	*Cnapholocris medinalis* (Guen)	Pyraustidae	Insect
1617.	*Cneorane* sp.	Chrysomelidae	Insect
1618.	*Cofana spectra distant* (Distant)	Cicadellidae	Insect
1619.	*Condica conducata*	Noctuidae	Insect
1620.	*Contarinia sorghicola* Coq.	Cecidomyiidae	Insect
1621.	*Corcyra cephalonica* Stainton	Pyralidae	Insect
1622.	*Creatonotos gangis* (Linnaeus)	Arctiidae	Insect
1623.	*Cretonotos gangis* Linn.	Arctiidae	Insect
1624.	*Cretonotus jangia* Linn.	Arctiidae	Insect
1625.	*Crocidolomia binotalis* Zell	Pyralidae	Insect
1626.	*Crytopeltis tenuis* (Reuter)	Miridae	Insect
1627.	*Cyana peregrina* (Walker)	Erebidae	Insect
1628.	*Cyrtacanthacris* sp.	Acrididae	Insect
1629.	*Cyrtacanthacris tatarica* (Linn)	Acrididae	Insect
1630.	*Dactynotus carthami* HRL	Aphididae	Insect
1631.	*Dacus cucurbitae* (Coquillett)	Tephritidae	Insect
1632.	*Daphnis nerii* (Linnaeus)	Sphingidae	Insect
1633.	*Daphnis nerii* (Linnaeus)	Sphingidae	Insect
1634.	*Dasyneura lini* Barnes	Cecidomyiidae	Insect
1635.	*Dialeurodes citri* (Ashmead)	Aleyrodidae	Insect
1636.	*Diaphorina citri* Kuwayama	Psyllidae	Insect
1637.	*Dichocrosis punctiferalis* Gn.	Crambidae	Insect
1638.	*Dicladispa armigera* (Oliv.)	Chrysomelidae	Insect
1639.	*Digama hearseyana similis* Moore	Erebidae	Insect
1640.	*Dioxyna sororcula* (Wiedemann)	Tephritidae	Insect
1641.	*Drosicha mangiferae* (Green)	Margarodidae	Insect
1642.	*Dudua aprobola* (Meyrick)	Tortricidae	Insect
1643.	*Dysdercus cingulatus* (Fab.)	Pyrrhocoridae	Insect
1644.	*Dysdercus koengii* (Fab.)	Pyrrhocoridae	Insect
1645.	*Earias fabia* Cram.	Nolidae	Insect
1646.	*Earias insulana* (Boisd.)	Nolidae	Insect
1647.	*Earias jabia* Stoll	Nolidae	Insect
1648.	*Earias vitella* (Fab.)	Nolidae	Insect
1649.	*Ectatops* sp.	Pyrrhocoridae	Insect

	Name	Family	Purpose
1650.	*Edwardsiana rosae* L.	Cicadellidae	Insect
1651.	*Eilema antica* (Walker)	Erebidae	Insect
1652.	*Emmalocera depresella* Swinh.	Pyralidae	Insect
1653.	*Empoasca fabae* (Harris)	Cicadellidae	Insect
1654.	*Empoasca devastans* Dist.	Jassidae	Insect
1655.	*Empoasca kerri* Pruthi	Jassidae	Insect
1656.	*Epilachna vigintioctopunctata* (Fab.)	Coccinellidae	Insect
1657.	*Etiella zinckenella* Treitschke	Pyralidae	Insect
1658.	*Euconocephalus* sp.	Tettigoniidae	Insect
1659.	*Eudocima aurantia (Moore)*	Erebidae	Insect
1660.	*Eupeodes confrater* (Wiedemann)	Syrphidae	Insect
1661.	*Euploea core core* Cramer	Nymphalidae	Insect
1662.	*Euproctis lunata* (Walker)	Lymantriidae	Insect
1663.	*Euproctis similis* (Moore)	Lymantriidae	Insect
1664.	*Euproctis subnotata* (Walker)	Lymantridae	Insect
1665.	*Eupterote* sp.	Eupterotidae	Insect
1666.	*Eureama hecabe var. contubernalis*	Pieridae	Insect
1667.	*Euscyrtus concinnus* (de Haan)	Gryllidae	Insect
1668.	*Exelastis atomosa* Walsh.	Pterophoridae	Insect
1669.	*Eysarcoris* sp.	Pentatomidae	Insect
1670.	*Ferrisia virgata* Ckll.	Pseudococcidae	Insect
1671.	*Flata* sps.	Flatidae	Insect
1672.	*Forficula auricularia* (Linnaeus)	Forficulidae	Insect
1673.	*Frankliniella sulphurea* Schmutz	Thripidae	Insect
1674.	*Gastropacha pardale* (Walker)	Lasiocampidae	Insect
1675.	*Gesonia gemma* (Swinhoe)	Noctuidae	Insect
1676.	*Glyphodes canthusalis* (Walker)	Crambidae	Insect
1677.	*Glyphodes* sp.	Crambidae	Insect
1678.	*Grammodes geometrica* (Fab.)	Erebidae	Insect
1679.	*Grapholita critica* (Meyr)	Tortricidae	Insect
1680.	*Gryllotalpa africana* Palisot de Beauvois	Gryllotalpidae	Insect
1681.	*Gryllotalpa orientalis* (Burmeister)	Gryllotalpidae	Insect
1682.	*Gryllus bimaculatus* (De Geer)	Gryllidae	Insect
1683.	*Gryllus* sp.	Gryllidae	Insect
1684.	*Halys serrigera* (Westwood)	Pentatomidae	Insect
1685.	*Haplothrips gowdeyi* (Franklin)	Thripidae	Insect
1686.	*Haptoncus reflexicollis* Motschulskyi	Nitidulidae	Insect
1687.	*Helicoverpa armigera* (Hub.)	Noctuidae	Insect
1688.	*Hellula undalis* Fab.	Pyralidae	Insect
1689.	*Hexacentrus* sp.	Tettigoniidae	Insect
1690.	*Hieroglyphus banian* Fab.	Acrididae	Insect
1691.	*Hieroglyphus oryzivorous* Carl	Acrididae	Insect
1692.	*Hippotion celerio* (Linnaeus)	Sphingidae	Insect
1693.	*Holotrichia consanguinea* Blanchard	Scarabidae	Insect
1694.	*Hyblaea puera* Cramer	Hyblaeidae	Insect
1695.	*Hydaphis coriandri* (Das)	Aphididae	Insect
1696.	*Hylemya cilicrura* (Rond.)	Anthomyiidae	Insect
1697.	*Hymenia fasciolis* (Balsam)	Pyrallidae	Insect
1698.	*Hypomecis umbrosaria* (Hubner)	Geometridae	Insect
1699.	*Hyposidra talaca* (Walker)	Geometridae	Insect

	Name	Family	Purpose
1700.	*Idioscopus niverosparsus* (Leth)	Cicadellidae	Insect
1701.	*Idioscopus* sp.	Cicadellidae	Insect
1702.	*Indarbela tetraonis* (Moore)	Cossidae	Insect
1703.	*Lampides boeticus* (Lin.)	Lycaenidae	Insect
1704.	*Lamprosema dieminalis* Guen.	Pyralidae	Insect
1705.	*Lamprosema indicata* (Fab.)	Pyralidae	Insect
1706.	*Lasioptera falcata* Felt	Cecidomyidae	Insect
1707.	*Laspeyresia pseudonectis* Meyr	Euscosmidae	Insect
1708.	*Leptispa pygmaea* (Baly)	Chrysomelidae	Insect
1709.	*Leptocorisa acuta* (Thumb)	Alydidae	Insect
1710.	*Leptocorisa oratorius* (Guen.)	Coreidae	Insect
1711.	*Leucinodes orbonalis* (Guen.)	Crambidae	Insect
1712.	*Lipaphis erisymi* (Kalt.)	Aphididae	Insect
1713.	*Liriomyza trifolii* Burgess	Agromyzidae	Insect
1714.	*Longitarsus belgaumensis* Jacoby	Chrysomelidae	Insect
1715.	*Lucilia sericata* Meigen	Calliphoridae	Insect
1716.	*Macroglossum stellatarum* (Linn)	Sphingidae	Insect
1717.	*Macrosiphum rosaeiformis* Das	Aphididae	Insect
1718.	*Madurasia obscurella* Jacoby	Olyrocidae	Insect
1719.	*Marsamia* sp.	Crambidae	Insect
1720.	*Maruca testulalis* Geyer	Pyraustidae	Insect
1721.	*Mecopoda* sp.	Tettigoniidae	Insect
1722.	*Megachile anthracina*	Megachilidae	Insect
1723.	*Megalurothrips usitatus* (Bagnall)	Thripidae	Insect
1724.	*Melanagromyza hibisci* Spencer	Agromyzidae	Insect
1725.	*Melanagromyza obtusa* Malloch	Agromyzidae	Insect
1726.	*Melanagromyza sojae* Zehntner	Agromyzidae	Insect
1727.	*Melanapsis glomerata* (Green)	Diaspididae	Insect
1728.	*Melanitis leda ismene* (Cramer)	Nymphalidae	Insect
1729.	*Microtermes obesi* Holmgr.	Termitidae	Insect
1730.	*Mocis undata* Fab.	Noctuidae	Insect
1731.	*Monomorium destructor* Jard	Formicidae	Insect
1732.	*Mylabris pustulata* Thunberg	Meloidae	Insect
1733.	*Mylabris vitrata Fab.*	Meloidae	Insect
1734.	*Myllocerus maculosus* Desbr.	Curculionidae	Insect
1735.	*Myllocerus* sp.	Curculionidae	Insect
1736.	*Mythimna separata* Wlk.	Noctuidae	Insect
1737.	*Mythimna unipunctata* (Haworth)	Noctuidae	Insect
1738.	*Myzus persicae* (Sulzer)	Aphididae	Insect
1739.	*Necroscia sparaxes* Westwood	Lonchodidae	Insect
1740.	*Nephotettix nigropictus* (Stal.)	Cicadellidae	Insect
1741.	*Nephotettix virescens* Dist.	Cicadellidae	Insect
1742.	*Nesidiocoris tenuis* (Rent.)	Miridae	Insect
1743.	*Nezara viridula* L.	Pentatomidae	Insect
1744.	*Nilaparvata lugens* Stal.	Delphacidae	Insect
1745.	*Nymphula depunctalis* Guen.	Pyraustidae	Insect
1746.	*Obereopsis (= Oberea) brevis* Swed.	Cerambycidae	Insect
1747.	*Odontotermes obesus* Rambur	Termitidae	Insect
1748.	*Oecanthus* sp.	Tettigoniidae	Insect
1749.	*Oecophylla smaragdina* (F)	Formicidae	Insect

	Name	Family	Purpose
1750.	*Oocassida pudibunda* Boh.	Chrysomelidae	Insect
1751.	*Ophiomyia phaseoli* (Tryon)	Agromyzidae	Insect
1752.	*Orgyia leucostigma* (Smith)	Erebidae	Insect
1753.	*Orocious albicinctus* (Distant)	Jassidae	Insect
1754.	*Orseolia oryzae* (WM) Mani	Cecidomyiidae	Insect
1755.	*Othreis fulonica* (Clerck)	Noctuidae	Insect
1756.	*Otinotus oneratus* W.	Membracidae	Insect
1757.	*Oyxa* sp.	Acrididae	Insect
1758.	*Pagria kamaraesis* Baly	Alticidae	Insect
1759.	*Palpita quadristigmalis* (Guenée)	Crambidae	Insect
1760.	*Palpita vitrealis* (Rossi)	Crambidae	Insect
1761.	*Papilio demoleus*	Papilionidae	Insect
1762.	*Parapoynx stagnalis* (Zeller)	Crambidae	Insect
1763.	*Parnara mathias* Fab.	Hesperidae	Insect
1764.	*Parotis marginata* (Hampson)	Crambidae	Insect
1765.	*Pectinophora gossypiella (Saund)*	Gelechiidae	Insect
1766.	*Pempherelus affinis* (Fst.)	Curculionidae	Insect
1767.	*Peregrinus maidis* D	Delphacidae	Insect
1768.	*Pericallia ricini* (F.)	Arctiidae	Insect
1769.	*Perina nuda* (Fab.)	Erebidae	Insect
1770.	*Phenacoccus solenopsis* Tinsley	Pseudococcidae	Insect
1771.	*Phengaris arion* L.	Lycaenidae	Insect
1772.	*Phthorimaea operculella* Zeller	Gelechiidae	Insect
1773.	*Phyllocnisti citrella* Stainton	Gracillaridae	Insect
1774.	*Phyllotreta chotanica* Duv.	Chrysomelidae	Insect
1775.	*Phyllotreta cruciferae* G.	Chrysomelidae	Insect
1776.	*Phytomyza atricornis M*eigen	Agromyzidae	Insect
1777.	*Pieris brassicae* (L)	Pieridae	Insect
1778.	*Plecia amplipennis* (Skuse)	Bibionidae	Insect
1779.	*Plusia acuta* Wlk.	Noctuidae	Insect
1780.	*Plusia chalcites*	Noctuidae	Insect
1781.	*Plusia orichalcea* (Fab.)	Noctuidae	Insect
1782.	*Plutella maculipennis* (Curt.)	Yponomeutidae	Insect
1783.	*Plutella xylostella* (L.)	Yponomeutidae	Insect
1784.	*Polyphagotarsonemus* latus Banks	Acarina	Insect
1785.	*Psalis pennatula* (Fab.)	Erebidae	Insect
1786.	*Psalis securis* Hb.	Lymantriidae	Insect
1787.	*Pulvinaria psidii* (Maskell)	Coccidae	Insect
1788.	*Pyhhria umbra* Huf.	Noctuidae	Insect
1789.	*Pyrausta machaeralis* Walker	Pyralidae	Insect
1790.	*Pyrilla perpusilla* Wlk.	Lophopidae	Insect
1791.	*Ragmus importunitas* Dist.	Miridae	Insect
1792.	*Raphidopalpa foveicollis* Lucas	Chrysomelidae	Insect
1793.	*Rastrococcus (Phenacoccusi) iceryoides*	Coccidae	Insect
1794.	*Rhipiphorothrips cruentatus* Hood	Thripidae	Insect
1795.	*Rhizopertha dominica* Fab.	Bostrichidae	Insect
1796.	*Rhopalosiphum maidis* (Fitch)	Aphididae	Insect
1797.	*Riptortus pedestris* Fab.	Coreidae	Insect
1798.	*Riptortus* sp.	Coreidae	Insect
1799.	*Riptortus strenuus* (Fabricius)	Coreidae	Insect

	Name	Family	Purpose
1800.	*Saccaricoccus sacchari* (Cockerell)	Pseudococcidae	Insect
1801.	*Sameodes cancellalis* Zeller	Crambidae	Insect
1802.	*Scirpophaga incertulas* Wlk.	Pyralidae	Insect
1803.	*Scirpophaga ninclle* Fab.	Pyraustidae	Insect
1804.	*Scirpophaga novella* Fab.	Pyraustidae	Insect
1805.	*Scirtothrips dorsalis* Hood	Thripidae	Insect
1806.	*Sesamia inferens* Walker	Noctuidae	Insect
1807.	*Sitophilus oryzae* (Linn.)	Curculionidae	Insect
1808.	*Sitotroga cerealella* (Olivier)	Gelechiidae	Insect
1809.	*Sogatella furcifera* Horvath	Delphacidae	Insect
1810.	*Spilarctia* sp.	Erebidae	Insect
1811.	*Spilarctica (= Spilosoma)*	Arctiidae	Insect
1812.	*Spodoptera (=Laphygma) exigua* (Hub.)	Noctuidae	Insect
1813.	*Spodoptera frugiperda* J.E. Smith	Noctuidae	Insect
1814.	*Spodoptera litura* (Fab.)	Noctuidae	Insect
1815.	*Spoladea recurvalis* (Fab.)	Crambidae	Insect
1816.	*Sternochetus mangiferae* (F)	Curculionidae	Insect
1817.	*Stomopteryx subsecivella* Zeller	Gelechidae	Insect
1818.	*Stromatium barbatum* (Fabricius)	Cerambycidae	Insect
1819.	*Sylepta derogata* Fabr.	Pyralidae	Insect
1820.	*Taeniothrips distalis* Karny	Thripidae	Insect
1821.	*Taeniothrips nigricornis* Schmutz	Thripidae	Insect
1822.	*Taeniothrips sp.*	Thripidae	Insect
1823.	*Taylorilygus pallidulus*	Miridae	Insect
1824.	*Tetranychus* sp.	Tetranychidae	Mites
1825.	*Tetranychus neocaledonicus* Andre	Tetranychidae	Mites
1826.	*Tetranychus telarius* K.	Tetranychidae	Mites
1827.	*Tetranychus urticae* Koch	Tetranychidae	Mites
1828.	*Tetrix subulata* Linn.	Tetrigidae	Mites
1829.	*Tettigonia viridissima* L.	Tettigoniidae	Insect
1830.	*Thrips florum* (Schmutz)	Thripidae	Insect
1831.	*Thrips palmi* (Karny)	Thripidae	Insect
1832.	*Thrips tabaci* Lindeman	Thripidae	Insect
1833.	*Thysanoplusia orichalcea* (Fab.)	Noctuidae	Insect
1834.	*Trabala vishnou* (Lefebve)	Lasiocampidae	Insect
1835.	*Trachys* sp.	Buprestidae	Insect
1836.	*Tribolium castaneum* (Herbst)	Tenebrionidae	Insect
1837.	*Tribolium confusum* (Herbst)	Tenebrionidae	Insect
1838.	*Trichoplusia ni* (Hub.)	Noctuidae	Insect
1839.	*Trigonidium* sp.	Orchidaceaetrigonidiidae	Insect
1840.	*Trigonodes hyppasia* (Cramer)	Noctuidae	Insect
1841.	*Trogoderma granarium* Everts	Dermestidae	Insect
1842.	*Tryporyza incertulas* Wlk	Crambidae	Insect
1843.	*Tuta absoluta* (Meyrick)	Gelechiidae	Insect
1844.	*Urentius sentis* Dist	Tingidae	Insect
1845.	*Uroleucon carthami*	Aphididae	Insect
1846.	*Uroleucon compositae*	Aphididae	Insect
1847.	*Utetheisa ornatrix* (Linnaeus)	Erebidae	Insect
1848.	*Utetheisa pulchella* (Linn.)	Arctiidae	Insect
1849.	*Virachola isocrates* F.	Lycaenidae	Insect

	Name	Family	Purpose
1850.	*Zamarada excisa* (Hampson)	Geometridae	Insect
1851.	*Zanna* sp.	Fulgoridae	Insect
1852.	*Antheraea paphia* Linnaeus	Saturniidae	Insect
1853.	*Apis cerana indica* Fab.	Apidae	Insect
1854.	*Apis mellifera* Linn.	Apidae	Insect
1855.	*Bombyx moorei* F.	Bombycidae	Insect
1856.	*Kerria (=Laccifer) lacca* Kerr.	Lacciferidae	Insect
1857.	*Lycoriella ingénue* Dufour	Sciaridae	Insect
1858.	*Megaselia scalaris* (Loew)	Phoridae	Insect
1859.	*Adalia bipunctata* Linn.	Coccinellidae	Insect Predator
1860.	*Amantis* sp.	Mantidae	Insect Predator
1861.	*Anaciaeschna jaspidea* (Burmeister)	Aeshnidae	Insect Predator
1862.	*Anax* sp.	Aeshnidae	Insect Predator
1863.	*Antilochus cocqueberti* (Fab.)	Pyrrhocoridae	Insect Predator
1864.	*Archimantis latistyla,* (Serville)	Mantidae	Insect Predator
1865.	*Ascalaphus sinister* Walker	Ascalaphidae	Insect Predator
1866.	*Ascalaphus* sp.	Ascalaphidae	Insect Predator
1867.	*Brachinus longipalpis* (Wiedemann)	Carabidae	Insect Predator
1868.	*Brachinus sexmaculatus* Dejan	Carabidae	Insect Predator
1869.	*Brachythemis contaminate* Fab.	Libellulidae	Insect Predator
1870.	*Brumus suturalis* F.	Coccinellidae	Insect Predator
1871.	*Calosoma* sp.	Carabidae	Insect Predator
1872.	*Cantheconidia furcellata* Wolff	Pentatomidae	Insect Predator
1873.	*Chalenites circumdatus* (Bru.)	Carabidae	Insect Predator
1874.	*Cheilomenes sexmaculatus* (F.)	Coccinellidae	Insect Predator
1875.	*Chlaenius* sp.	Carabidae	Insect Predator
1876.	*Chlaenius circumdatus* (Dejean)	Carabidae	Insect Predator
1877.	*Chlaenius medioguttatis* (Chaudoir)	Carabidae	Insect Predator
1878.	*Chlaenius nigricans* (Riley and Clausen)	Carabidae	Insect Predator
1879.	*Chlaenius panagaeoides* Chaudoir	Carabidae	Insect Predator
1880.	*Chlaenius pictus* (Choudoir)	Carabidae	Insect Predator
1881.	*Chrysopa madestes* Banks	Chrysopidae	Insect Predator
1882.	*Chrysoperla* sp.	Chrysopidae	Insect Predator
1883.	*Chrysoperla carnea* (Stephens)	Chrysopidae	Insect Predator
1884.	*Chrysoperla rufilabris*	Chrysopidae	Insect Predator
1885.	*Chrysoperla sillemi* (Esben-petersen)	Chrysopidae	Insect Predator
1886.	*Cicindela cardoni* Flr.	Cicindellidae	Insect Predator
1887.	*Cicindela flexuosa* (Distant)	Carabidae	Insect Predator
1888.	*Cicindela melancholica* Fab.	Cicindellidae	Insect Predator
1889.	*Cicindela venosa* Kollar	Cicindellidae	Insect Predator
1890.	*Coccinella reparada* Thumb	Coccinellidae	Insect Predator
1891.	*Coccinella septempunctata* (L)	Coccinellidae	Insect Predator
1892.	*Coccinella transversalis* (Fab.)	Coccinellidae	Insect Predator
1893.	*Coenagrion* sp.	Coenagrionidae	Insect Predator
1894.	*Conocephalus* sp.	Tettigoniidae	Insect Predator
1895.	*Creobroter apicalis* Saussure	Hymenopodidae	Insect Predator
1896.	*Crocothemis servilia* Drury	Libellulidae	Insect Predator
1897.	*Crosopedophorus elegans* Dejean	Carabidae	Insect Predator
1898.	*Cyphon padi* Fleming	Scirtidae	Insect Predator
1899.	*Deserida lineola* (Macleay)	Carabidae	Insect Predator

	Name	Family	Purpose
1900.	*Didymocorypha lanceolata*Fab	Tarachodidae	Insect Predator
1901.	*Diplacodes trivialis* Rambur	Libellulidae	Insect Predator
1902.	*Diplocheila polita* (Fab.)	Carabidae	Insect Predator
1903.	*Diplonychus rusticus* (Fabricius)	Belostomatidae	Insect Predator
1904.	*Dytiscus marginalis* (Linnaeus)	Dytisadae	Insect Predator
1905.	*Ectomocoris cordiger* (Stal.)	Reduvidae	Insect Predator
1906.	*Ectomocoris ululans* (Rossi)	Reduviidae	Insect Predator
1907.	*Elaunon bipartitus* (Kirby)	Forficulidae	Insect Predator
1908.	*Eocanthecona furcellata* (Wolff)	Pentatomidae	Insect Predator
1909.	*Eublemma amabilis* Salmuller	Noctuidae	Insect Predator
1910.	*Eupeodes confrater* (Wiedemann)	Syrphidae	Insect Predator
1911.	*Geocoris tricolor* Fab.	Lygaeidae	Insect Predator
1912.	*Heteronychus* sp.	Scarabaeidae	Insect Predator
1913.	*Hydrochara caraboides* (Latreille)	Hydrophilidae	Insect Predator
1914.	*Hydropyche* sp.	Hydropsychidae	Insect Predator
1915.	*Hyperaspis maindroni Sci. sp.brumoides*	Coccinellidae	Insect Predator
1916.	*Illeis cincta* (Fabr.)	Coccinellidae	Insect Predator
1917.	*Libellua* sp.	Libellulidae	Insect Predator
1918.	*Macrochilustri pustulatus* (Dejean)	Carabidae	Insect Predator
1919.	*Mantis* sp.	Mantispidae	Insect Predator
1920.	*Mantis religiosa* (Linnaeus)	Mantidae	Insect Predator
1921.	*Menochilus sexmaculatus* Fabr.	Coccinellidae	Insect Predator
1922.	*Myrmicaria brunnea* Saunders	Formicidae	Insect Predator
1923.	*Myrmicaria sp.*	Formicidae	Insect Predator
1924.	*Nehalennia* spp.	Ceriogrionidae	Insect Predator
1925.	*Nephus regularis* Sic.	Coccinellidae	Insect Predator
1926.	*Oecophylla* sp.	Formicidae	Insect Predator
1927.	*Oligota* sp.	Staphylinidae	Insect Predator
1928.	*Oncocephalus* sp.	Reduviidae	Insect Predator
1929.	*Onitis falcutus* (Wulfen)	Scarabaeidae	Insect Predator
1930.	*Ophionea indica* (Thunberg)	Carabidae	Insect Predator
1931.	*Orthectrum cancellatum*	Aeshnidae	Insect Predator
1932.	*Orthetrum sabina* (Drury)	Libellulidae	Insect Predator
1933.	*Pantala flavescens* Fabricus	Rhyothemisidae	Insect Predator
1934.	*Podisus maculiventris* (Say)	Pentatomidae	Insect Predator
1935.	*Poecilus cupreus* (Linnaeus)	Carabidae	Insect Predator
1936.	*Prothyma proxima* (Chd.)	Cicindellidae	Insect Predator
1937.	*Prothyma* sp.	Carabidae	Insect Predator
1938.	*Pseudagrion microcephalum* Rambur	Ceriogrionidae	Insect Predator
1939.	*Pseudohypatopa pulverea Meyrick*	Blastobasidae	Insect Predator
1940.	*Pullus coccidivora* Ayyar	Coccinellidae	Insect Predator
1941.	*Rodolia fumida* Muls.	Coccinellidae	Insect Predator
1942.	*Scadra annulipes* Reuter	Reduviidae	Insect Predator
1943.	*Scymnus nubillus* Mils.	Coccinellidae	Insect Predator
1944.	*Sirthenea carinata* (Fabricius)	Reduviidae	Insect Predator
1945.	*Sirthenes* sp.	Reduvidae	Insect Predator
1946.	*Statilia maculata* (Thunberg)	Mantidae	Insect Predator
1947.	*Stethorus gilvifrons* Muls.	Coccinellidae	Insect Predator
1948.	*Stethorus pauperculus* Weise	Coccinellidae	Insect Predator
1949.	*Stratiolaelaps scimitus*	Laelapidae	Insect Predator

	Name	Family	Purpose
1950.	*Tholymis tillarga* Fab.	Libellulidae	Insect Predator
1951.	*Xanthogramma javanna* Wd.	Syrphidae	Insect Predator
1952.	*Zosteria* sp.	Asilidae	Insect Predator
1953.	*Zuphium olens* (Rossi)	Carabidae	Insect Predator
1954.	*Achrysocharella* sp.	Eulophidae	Insect parasite
1955.	*Aenasius advena* Compere	Encyrtidae	Insect parasite
1956.	*Apanteles* sp.	Braconidae	Insect parasite
1957.	*Apanteles* sp.(ultor group)	Braconidae	Insect parasite
1958.	*Apanteles obliquae* Wilkinson	Braconidae	Insect parasite
1959.	*Apanteles ruficrus*	Braconidae	Insect parasite
1960.	*Apanteles taragamae* Viereck	Braconidae	Insect parasite
1961.	*Asympiesella* sp.	Eulophidae	Insect parasite
1962.	*Avga* sp.	Hormiini	Insect parasite
1963.	*Brachymeria lasus* (Walker)	Chalcididae	Insect parasite
1964.	*Brachymeria* sp.	Chalcididae	Insect parasite
1965.	*Bracon gelechiae* Ashm.	Braconidae	Insect parasite
1966.	*Bracon hebetor* Say	Braconidae	Insect parasite
1967.	*Bracon* s.l.sp.	Braconidae	Insect parasite
1968.	*Bracon* spp	Braconidae	Insect parasite
1969.	*Brevicoryne brassicae* Linn	Aphididae	Insect parasite
1970.	*Bulborogas* sp.	Braconidae	Insect parasite
1971.	*Calcelias* sp.		Insect parasite
1972.	*Campoletis chloridae* (Uchida)	Ichneumonidae	Insect parasite
1973.	*Cantheconidia furcellata* Wolf.	Pentatomidae	Insect parasite
1974.	*Carcelia* sp.	Tachinidae	Insect parasite
1975.	*Chelonus* sp.	Braconidae	Insect parasite
1976.	*Chrysopa madestes Banks*	*Chrysopidae*	Insect parasite
1977.	*Chrysoperla sp.*	*Chrysopidae*	Insect parasite
1978.	*Cirrospilus* sp.	Eulophidae	Insect parasite
1979.	*Cotesia* (=*Apanteles*) sp.	Braconidae	Insect parasite
1980.	*Dairetiella rapae* (M'Intosh)	Braconidae	Insect parasite
1981.	*Dinarmus* sp.	Pteromalidae	Insect parasite
1982.	*Disophrys* sp.	Braconidae	Insect parasite
1983.	*Doryctobracon* sp.	Braconidae	Insect parasite
1984.	*Dorylus* sp.(Red ant)	Formicidae	Insect parasite
1985.	*Echthromorpha agrestoria* Swed.	Ichneumonidae	Insect parasite
1986.	*Ecphoropsis* (*Campoletis*) *perdistinctus*	Ichneumonidae	Insect parasite
1987.	*Elasmus brevicornis* Gab.	Elasmidae	Insect parasite
1988.	*Elasmus* sp.	Elasmidae	Insect parasite
1989.	*Elasmus* sp. nr. *fumipennis* Cameron	Elasmidae	Insect parasite
1990.	*Enicospilus capensis* (Thunberg)	Ichneumonidae	Insect parasite
1991.	*Enicospilus purgatus* (Say)	Ichneumonidae	Insect parasite
1992.	*Enicospilus xanthocephalus* Cameron	Ichneumonidae	Insect parasite
1993.	*Eriborus argentecupilosus*	Ichneumonidae	Insect parasite
1994.	*Eucelatoria bryani* Sabrosky	Tachinidae	Insect parasite
1995.	*Eucollidea* sp.	Eucoilidae	Insect parasite
1996.	*Euderus agromyzae* Gangrade	Eulophidae	Insect parasite
1997.	*Euplectrus* sp.	Eulophidae	Insect parasite
1998.	*Euplemus tachardiae Howard*	*Eupelmidae*	Insect parasite
1999.	*Eurytoma* sp.	Eurytomidae	Insect parasite

	Name	Family	Purpose
2000.	*Exorista sorbillans* Wield.	Tachinidae	Insect parasite
2001.	*Glabridorsum stokesii* (Cameron)	Ichneumonidae	Insect parasite
2002.	*Glyptapanteles* spp. (Wasp)	Braconidae	Insect parasite
2003.	*Habrobracon brevicornis* (Wesmael)	Braconidae	Insect parasite
2004.	*Homolobus* sp.	Braconidae	Insect parasite
2005.	*Leurinion* sp.	Hormiini	Insect parasite
2006.	*Macrocentrus* sp.	Braconidae	Insect parasite
2007.	*Microbracon hebetor*		Insect parasite
2008.	*Microplitis* sp.	Braconidae	Insect parasite
2009.	*Microplitis demolitor* Wilkinson	Braconidae	Insect parasite
2010.	*Microplitis maculipennis* (Szep.)	Braconidae	Insect parasite
2011.	*Misocyclops* sp.	Selionidae	Insect parasite
2012.	*Myrmicaria brunnea* Saunders	Formicidae	Insect parasite
2013.	*Netelia* sp.	Ichneumonidae	Insect parasite
2014.	*Palexorista solenis* Walk.	Tachinidae	Insect parasite
2015.	*Phanerotoma* sp.	Braconidae	Insect parasite
2016.	*Pseudohypatopa pulverea* Meyrick	*Blastobasidae*	Insect parasite
2017.	*Rhogas percunets* Lyle	Braconidae	Insect parasite
2018.	*Rhogas* sp.	Braconidae	Insect parasite
2019.	*Tachardiaephagus tachardiae* Howard	*Encyrtidae*	Insect parasite
2020.	*Temelucha* sp.	Braconidae	Insect parasite
2021.	*Tetrastichus* sp.	Eulophidae	Insect parasite
2022.	*Thelaira macropus* Wield.	Tachinidae	Insect parasite
2023.	*Utetes anastrephae*	Braconidae	Insect parasite
2024.	*Vespa orientalis* Linnaeus (Wasp)	Vespidae	Insect parasite
2025.	*Xanthopimpla predator* (Fab.)	Ichneumonidae	Insect parasite
2026.	*Xanthopimpla* sp.	Ichneumonidae	Insect parasite
2027.	*Yellicones* sp.	Braconidae	Insect parasite
2028.	*Zele chlorophthalama*	Braconidae	Insect parasite
2029.	*Zelomorpha* sp.	Braconidae	Insect parasite
2030.	*Brachymeria sp.*	Chalcididae	Hyper parasite
2031.	*Eurytoma sp.*	Eurytomidae	Hyper parasite
2032.	*Metaphycus* sp.	Encyrtidae	Hyper parasite
2033.	*Pediobus parvulus* Ferr.	Eulophidae	Hyper parasite
2034.	*Beauveria bassiana* (Bals.)	Cordiycipitaceae	Disease in insect
2035.	*Lecanicillium lecanii* (Zimm.)	Cordiycipitaceae	Disease in insect
2036.	*Metarhizium anisopilae* (Metsch.) Sorok	Clavicipitaceae	Disease in insect
2037.	*Agaricus bisporus*		Mushroom
2038.	*Auricularia polytricha*		Mushroom
2039.	*Calocybe indica*		Mushroom
2040.	*Coprinus* sp.		Mushroom
2041.	*Ganoderma applanatum*		Mushroom
2042.	*Ganoderma lucidum*		Mushroom
2043.	*Hericium erinaceus*		Mushroom
2044.	*Hypsizygus ulmarius*		Mushroom
2045.	*Lepiota cristata*		Mushroom
2046.	*Lycoperdon* sp.		Mushroom
2047.	*Pleurotus djamor*		Mushroom
2048.	*Pleurotus eous*		Mushroom
2049.	*Pleurotus flabellatus*		Mushroom

	Name	Family	Purpose
2050.	*Pleurotus membranaceus*		Mushroom
2051.	*Pleurotus ostreatus*		Mushroom
2052.	*Pleurotus sajor-caju*		Mushroom
2053.	*Pleurotus florida*		Mushroom
2054.	*Termitomyces clypeatus*		Mushroom
2055.	*Volvariella volvacea*		Mushroom
2056.	*Alternaria brassicae*		Fungus
2057.	*Alternaria brassicicola*		Fungus
2058.	*Alternaria carthami*		Fungus
2059.	*Alternaria cassiae*		Fungus
2060.	*Alternaria citri*		Fungus
2061.	*Alternaria gomphrenae*		Fungus
2062.	*Alternaria infectoria*		Fungus
2063.	*Alternaria longissima*		Fungus
2064.	*Alternaria padwickii*		Fungus
2065.	*Alternaria porri*		Fungus
2066.	*Alternaria rephani*		Fungus
2067.	*Alternaria sesami*		Fungus
2068.	*Alternaria* sp.		Fungus
2069.	*Alternaria tenuis*		Fungus
2070.	*Alternaria tenuissima*		Fungus
2071.	*Alternaria zinniae*		Fungus
2072.	*Alternaria alternata*		Fungus
2073.	*Alysidium resinae*		Fungus
2074.	*Aristastoma guttulosum*		Fungus
2075.	*Arthrinium phaeospermum*		Fungus
2076.	*Arthrinium sacchari*		Fungus
2077.	*Ascochyta oleandri*		Fungus
2078.	*Ascochyta rabiei*		Fungus
2079.	*Ascochyta sorghi*		Fungus
2080.	*Ascochyta* sp.		Fungus
2081.	*Aspergillus clavatus*		Fungus
2082.	*Aspergillus flavus*		Fungus
2083.	*Aspergillus niger*		Fungus
2084.	*Aspergillus ochraceus*		Fungus
2085.	*Aspergillus* sp.		Fungus
2086.	*Bartalinia robillardoides*		Fungus
2087.	*Beauveria bassiana*		Fungus
2088.	*Biharia smilacis*		Fungus
2089.	*Botryodiplodia* sp.		Fungus
2090.	*Botryodiplodia theobromae*		Fungus
2091.	*Botrytis cinerea*		Fungus
2092.	*Cephaliophora irregularis*		Fungus
2093.	*Cephalosporium* sp.		Fungus
2094.	*Ceratocystis major*		Fungus
2095.	*Cercoseptoria balsaminae*		Fungus
2096.	*Cercospora abelmoschi*		Fungus
2097.	*Cercospora agarwalii*		Fungus
2098.	*Cercospora arachidicola*		Fungus
2099.	*Cercospora avicennae*		Fungus

	Name	Family	Purpose
2100.	*Cercospora balsamina*		Fungus
2101.	*Cercospora canescens*		Fungus
2102.	*Cercospora cruenta*		Fungus
2103.	*Cercospora gossypina*		Fungus
2104.	*Cercospora hyalospora*		Fungus
2105.	*Cercospora indica*		Fungus
2106.	*Cercospora justiciidola*		Fungus
2107.	*Cercospora kikuchii*		Fungus
2108.	*Cercospora lentis*		Fungus
2109.	*Cercospora neriella*		Fungus
2110.	*Cercospora paramignyae*		Fungus
2111.	*Cercospora personata*		Fungus
2112.	*Cercospora pisi sativae* f. sp. *lathyri-sativae*		Fungus
2113.	*Cercospora rauwolfia*		Fungus
2114.	*Cercospora schreberae*		Fungus
2115.	*Cercospora sesami*		Fungus
2116.	*Cercospora sojina*		Fungus
2117.	*Cercospora sorghi*		Fungus
2118.	*Cercospora* sp.		Fungus
2119.	*Cercospora viachueli*		Fungus
2120.	*Cercospora withaniae*		Fungus
2121.	*Cercospora zinniae*		Fungus
2122.	*Cercospora canescens*		Fungus
2123.	*Cercospora zeae-maydis*		Fungus
2124.	*Cercosporella* sp.		Fungus
2125.	*Cerebella* sp.		Fungus
2126.	*Chaconia tectonae*		Fungus
2127.	*Chaetomium globosum*		Fungus
2128.	*Chaetomium* sp.		Fungus
2129.	*Choanephora* sp.		Fungus
2130.	*Ciliochorella mangiferae*		Fungus
2131.	*Cintractia axicola*		Fungus
2132.	*Cintractia clintonii*		Fungus
2133.	*Cintractia minor*		Fungus
2134.	*Cladobotryum* sp.		Fungus
2135.	*Cladosporium* sp.		Fungus
2136.	*Cladosporium tenuissium*		Fungus
2137.	*Clasterosporium mori*		Fungus
2138.	*Cochliobolus carbonum*		Fungus
2139.	*Colletotrichum capsici*		Fungus
2140.	*Colletotrichum cylindricum*		Fungus
2141.	*Colletotrichum dematium*		Fungus
2142.	*Colletotrichum falcatum*		Fungus
2143.	*Colletotrichum gloeosporioides*		Fungus
2144.	*Colletotrichum graminicola*		Fungus
2145.	*Colletotrichum graminicolum*		Fungus
2146.	*Colletotrichum lindemuthianum*		Fungus
2147.	*Colletotrichum papayae*		Fungus
2148.	*Colletotrichum piperis*		Fungus
2149.	*Colletotrichum pithecolobii*		Fungus

	Name	Family	Purpose
2150.	*Colletotrichum* sp.		Fungus
2151.	*Colletotrichum sublineolum*		Fungus
2152.	*Colletotrichum truncatum*		Fungus
2153.	*Colletotrichum graminicola*		Fungus
2154.	*Coniella noviae-zealanidiae*		Fungus
2155.	*Coniothyrium fuckelii*		Fungus
2156.	*Coprinus* sp.		Fungus
2157.	*Corynespora cassiicola*		Fungus
2158.	*Curvularia andropogonis*		Fungus
2159.	*Curvularia geniculata*		Fungus
2160.	*Curvularia intermedia*		Fungus
2161.	*Curvularia lunata*		Fungus
2162.	*Curvularia pallescens*		Fungus
2163.	*Curvularia* sp.		Fungus
2164.	*Curvularia trifoli*		Fungus
2165.	*Curvularia verruculosa*		Fungus
2166.	*Curvularia pallescens*		Fungus
2167.	*Cylindrocladium scoparium*		Fungus
2168.	*Cytospora rosarum*		Fungus
2169.	*Cytospora* sp.		Fungus
2170.	*Dasturella divina*		Fungus
2171.	*Deightoniella jabalpurensis*		Fungus
2172.	*Deightoniella torulosa*		Fungus
2173.	*Dendryphiella vinosa*		Fungus
2174.	*Diaporthe longicolla*		Fungus
2175.	*Diaporthe phaseolorum* var. *caulivora*		Fungus
2176.	*Dictyoarthrinium sacchari*		Fungus
2177.	*Dictyosporium intermedium*		Fungus
2178.	*Dinermasporium graminum*		Fungus
2179.	*Dinermasporium strigosum*		Fungus
2180.	*Diplodia sarmentorum*		Fungus
2181.	*Diplodia* sp.		Fungus
2182.	*Diplodia zizyphina*		Fungus
2183.	*Drechslera haveliensis*		Fungus
2184.	*Drechslera oryzae*		Fungus
2185.	*Drechslera spicifera*		Fungus
2186.	*Drechslera glycines*		Fungus
2187.	*Endophragmiella angustispora*		Fungus
2188.	*Entyloma oryzae*		Fungus
2189.	*Epicoccum* sp.		Fungus
2190.	*Erratomyces patelii*		Fungus
2191.	*Erysiphe alphitoides*		Fungus
2192.	*Erysiphe biocellata*		Fungus
2193.	*Erysiphe cichoracearum*		Fungus
2194.	*Erysiphe glycines*		Fungus
2195.	*Erysiphe pisi*		Fungus
2196.	*Erysiphe polygoni*		Fungus
2197.	*Eurotium* sp.		Fungus
2198.	*Excipularia narsapurensis*		Fungus
2199.	*Fusarium moniliforme*		Fungus

	Name	Family	Purpose
2200.	*Fusarium semitectum*		Fungus
2201.	*Fusarium* sp.		Fungus
2202.	*Fusarium equiseti*		Fungus
2203.	*Fusarium fusarioides*		Fungus
2204.	*Fusarium graminiarum*		Fungus
2205.	*Fusarium lateritium*		Fungus
2206.	*Fusarium oxysporum*		Fungus
2207.	*Fusarium oxysporum* f. sp. *glycines*		Fungus
2208.	*Fusarium oxysporum*f. sp. *lini*		Fungus
2209.	*Fusarium oxysporum* f. sp. *psidii*		Fungus
2210.	*Fusarium oxysporum* f. sp. *sesami*		Fungus
2211.	*Fusarium oxysporum* f. sp. *tracheiphilum*		Fungus
2212.	*Fusarium oxysporum* f. sp. *udum*		Fungus
2213.	*Fusarium oxysporum* f. sp. *vasinfectum*		Fungus
2214.	*Fusarium oxysporum* f.sp. *ciceris*		Fungus
2215.	*Fusarium oxysporum f*.sp. *lycopersici*		Fungus
2216.	*Fusarium oxysporum* f.sp. *tracheiphilum*		Fungus
2217.	*Fusarium oxysporum* f. sp. *melongenae*		Fungus
2218.	*Fusarium peniseti*		Fungus
2219.	*Fusarium semitectum*		Fungus
2220.	*Fusarium solani*		Fungus
2221.	*Fusarium solani* f. sp. *glycines*		Fungus
2222.	*Fusarium* sp.		Fungus
2223.	*Fusarium oxysporum* f. sp. *lentis*		Fungus
2224.	*Fusarium oxysporum* f. sp. *pisi*		Fungus
2225.	*Geotrichum candidum*		Fungus
2226.	*Gloeosporium musarum*		Fungus
2227.	*Gliocladium* sp.		Fungus
2228.	*Gloeosporium artocarpi*		Fungus
2229.	*Gloeosporium* sp.		Fungus
2230.	*Haplosporella beaumontiana*		Fungus
2231.	*Haplosporella rosarum*		Fungus
2232.	*Haplosporella* sp.		Fungus
2233.	*Helicosporium elinorage*		Fungus
2234.	*Helminthosporium bambus*		Fungus
2235.	*Helminthosporium erythrinae*		Fungus
2236.	*Helminthosporium oryzae*		Fungus
2237.	*Helminthosporium rostratum*		Fungus
2238.	*Helminthosporium sativum*		Fungus
2239.	*Helminthosporium* sp.		Fungus
2240.	*Helminthosporium turcicum*		Fungus
2241.	*Hermatomyces sphaericum*		Fungus
2242.	*Lasiodiplodia theobromae*		Fungus
2243.	*Lecanicillium lecanii*		Fungus
2244.	*Leveillula taurica*		Fungus
2245.	*Macrophoma musae*		Fungus
2246.	*Macrophomina phaseolina*		Fungus
2247.	*Macrospore levispora*		Fungus
2248.	*Melanopsichium pennsylvanicum*		Fungus
2249.	*Memnospora* sp.		Fungus

	Name	Family	Purpose
2250.	*Metarhizium anisopliae*		Fungus
2251.	*Microdiplodia purpusila*		Fungus
2252.	*Microsphaeropsis olivacea*		Fungus
2253.	*Monilia* sp.		Fungus
2254.	*Monochaetia jabalpurensis*		Fungus
2255.	*Monodictys* sp.		Fungus
2256.	*Mucor* sp.		Fungus
2257.	*Mycosphaerella populi*		Fungus
2258.	*Myrothecium roridum*		Fungus
2259.	*Myrothecium verrucaria*		Fungus
2260.	*Neovossia horrida*		Fungus
2261.	*Niesslia exilis*		Fungus
2262.	*Nigrospora* sp.		Fungus
2263.	*Nodulisporium gregarium*		Fungus
2264.	*Oedocephalum* sp.		Fungus
2265.	*Oidiopsis taurica*		Fungus
2266.	*Oidium* sp.		Fungus
2267.	*Oospora* sp.		Fungus
2268.	*Operculella padwickii*		Fungus
2269.	*Paathramaya sundara*		Fungus
2270.	*Penicillium digitatum*		Fungus
2271.	*Penicillium expansum*		Fungus
2272.	*Penicillium frequentans*		Fungus
2273.	*Penicillium italicum*		Fungus
2274.	*Penicillium* sp.		Fungus
2275.	*Periconia byssoides*		Fungus
2276.	*Peronospora pisi*		Fungus
2277.	*Pestalotia palmarum*		Fungus
2278.	*Pestalotia* sp.		Fungus
2279.	*Pestalotia suffocata*		Fungus
2280.	*Pestalotiopsis carbonacea*		Fungus
2281.	*Pestalotiopsis japonica*		Fungus
2282.	*Pestalotiopsis mangiferae*		Fungus
2283.	*Pestalotiopsis psidii*		Fungus
2284.	*Pestalotiopsis* sp.		Fungus
2285.	*Pestalotiopsis versicolor*		Fungus
2286.	*Pestalotiopsis woodfordeae*		Fungus
2287.	*Phaeodactylum uniseptata*		Fungus
2288.	*Phaeoisaria clematidis*		Fungus
2289.	*Phakopsora grewiae*		Fungus
2290.	*Phakopsora pachyrhizi*		Fungus
2291.	*Phloeospora cassiae*		Fungus
2292.	*Phoma bakeriana*		Fungus
2293.	*Phoma eucalyptica*		Fungus
2294.	*Phoma eucalypticola*		Fungus
2295.	*Phoma exigua*		Fungus
2296.	*Phoma glomerata*		Fungus
2297.	*Phoma medicaginis*		Fungus
2298.	*Phoma minutella*		Fungus
2299.	*Phoma sorghina*		Fungus

	Name	Family	Purpose
2300.	*Phoma* sp.		Fungus
2301.	*Phomopsis* sp.		Fungus
2302.	*Phomopsis vexans*		Fungus
2303.	*Phyllachora balansae*		Fungus
2304.	*Phyllosticta* sp.		Fungus
2305.	*Phyllosticta grevilleae*		Fungus
2306.	*Phyllosticta ipomoea*		Fungus
2307.	*Phyllosticta sesbaniae*		Fungus
2308.	*Phyllosticta sulata*		Fungus
2309.	*Phyllostictina artocarpina*		Fungus
2310.	*Phyllostictina murrayae*		Fungus
2311.	*Phytophthora boehmeriae*		Fungus
2312.	*Phytophthora capsici*		Fungus
2313.	*Phytophthora cinnamomi*		Fungus
2314.	*Phytophthora citrophthora*		Fungus
2315.	*Phytophthora cyperi*		Fungus
2316.	*Phytophthora infestans*		Fungus
2317.	*Phytophthora nicotianae* var. *nicotianae*		Fungus
2318.	*Phytophthora nicotianae* var. *parasitica*		Fungus
2319.	*Phytophthora parasitica*		Fungus
2320.	*Phytophthora parasitica* var. *sesami*		Fungus
2321.	*Phytophthora vignae*		Fungus
2322.	*Phytophthora colocasiae*		Fungus
2323.	*Phytophthora drechsleri* f.sp. *cajani*		Fungus
2324.	*Pithomyces ellisii*		Fungus
2325.	*Pithomyces* sp.		Fungus
2326.	*Plasmopara halstedii*		Fungus
2327.	*Pleospora infectoria*		Fungus
2328.	*Pseudopetrakia kambakkamensis*		Fungus
2329.	*Puccinia arachidis*		Fungus
2330.	*Puccinia calcitrapae* var. *centaureae*		Fungus
2331.	*Puccinia chloridis*		Fungus
2332.	*Puccinia hookeri*		Fungus
2333.	*Puccinia lateritia*		Fungus
2334.	*Puccinia purpurea*		Fungus
2335.	*Puccinia pusilla*		Fungus
2336.	*Puccinia romagnoliana*		Fungus
2337.	*Puccinia sorghi*		Fungus
2338.	*Puccinia sp.*		Fungus
2339.	*Pyrenochaeta* sp.		Fungus
2340.	*Pyrenochaeta tandonii*		Fungus
2341.	*Pyrenochaeta vexans*		Fungus
2342.	*Pyricularia oryzae*		Fungus
2343.	*Pythium aphanidermatum*		Fungus
2344.	*Pythium graminicolum*		Fungus
2345.	*Ramularia aromatici*		Fungus
2346.	*Ravenelia emblicae*		Fungus
2347.	*Ravenelia ornate*		Fungus
2348.	*Rhizoctonia bataticola*		Fungus
2349.	*Rhizoctonia solani*		Fungus

	Name	Family	Purpose
2350.	*Rhizopus nigricans*		Fungus
2351.	*Rhizopus* sp.		Fungus
2352.	*Sclerospora graminicola*		Fungus
2353.	*Sclerotinia sclerotiorum*		Fungus
2354.	*Sclerotium rolfsii*		Fungus
2355.	*Sclerotium* sp.		Fungus
2356.	*Septoria chrysanthemella*		Fungus
2357.	*Septoria leguminum*		Fungus
2358.	*Septoria lycopersici*		Fungus
2359.	*Sirosporium ziziphicola*		Fungus
2360.	*Spegazzinia tessarthra*		Fungus
2361.	*Sphacelotheca annulata*		Fungus
2362.	*Sphacelotheca apludae-aristatae*		Fungus
2363.	*Sphaerophragmium acaciae*		Fungus
2364.	*Sporisorium bursum (=Sphacelotheca bursa)*		Fungus
2365.	*Sporisorium cornutum*		Fungus
2366.	*Sporisorium sorghi*		Fungus
2367.	*Sporisorium inayati*		Fungus
2368.	*Sporisorium reilianum*		Fungus
2369.	*Stachybotrys* sp.		Fungus
2370.	*Stauronema sacchari*		Fungus
2371.	*Stemphylium sarciniformae*		Fungus
2372.	*Stemphylium* sp.		Fungus
2373.	*Syncephalastrum racemosum*		Fungus
2374.	*Synchytrium cookei*		Fungus
2375.	*Tilletia barclayana*		Fungus
2376.	*Tilletia dactyloctenii*		Fungus
2377.	*Tilletia sp.*		Fungus
2378.	*Tilletia sumati*		Fungus
2379.	*Tilletia vittata*		Fungus
2380.	*Tolyposporium penicillariae*		Fungus
2381.	*Torula herbarum*		Fungus
2382.	*Tretopileus sphaerophorus*		Fungus
2383.	*Trichoconiella padwickii*		Fungus
2384.	*Trichoderma hamatum*		Fungus
2385.	*Trichoderma harzianum*		Fungus
2386.	*Trichoderma longibrachiatum*		Fungus
2387.	*Trichoderma* sp.		Fungus
2388.	*Trichoderma viride*		Fungus
2389.	*Trichothecium* sp.		Fungus
2390.	*Triposporium myrti*		Fungus
2391.	*Ulocladium chartarum*		Fungus
2392.	*Uromyces fabae*		Fungus
2393.	*Uromyces hobsoni*		Fungus
2394.	*Uromyces orientalis*		Fungus
2395.	*Uromyces setariae-italicae*		Fungus
2396.	*Ustilaginoidea oplismenii*		Fungus
2397.	*Ustilaginoidea virens*		Fungus
2398.	*Ustilago coicis*		Fungus
2399.	*Ustilago cynodontis*		Fungus

	Name	Family	Purpose
2400.	*Ustilago hordei*		Fungus
2401.	*Ustilago neglecta*		Fungus
2402.	*Ustilago scitaminea*		Fungus
2403.	*Ustilago trichophora*		Fungus
2404.	*Ustilago tritici*		Fungus
2405.	*Ustilago vetiveriae*		Fungus
2406.	*Verticillium* sp.		Fungus
2407.	*Zygosporium oscheoides*		Fungus
2408.	*Erwinia carotovora*		Bacteria
2409.	*Bacillus* sp.		Bacteria
2410.	*Bacillus subtilis*		Bacteria
2411.	*Pseudomonas tolasii*		Bacteria
2412.	*Pseudomonas fluorescens*		Bacteria
2413.	*Pseudomonas* sp.		Bacteria
2414.	*Pseudomonas syringae pv. glycinea*		Bacteria
2415.	*Pseudomonas syringae* pv. *sesame*		Bacteria
2416.	*Ralstonia solanacearum (=Pseudomonassolanacearum)*		Bacteria
2417.	*Streptomyces* sp.		Bacteria
2418.	*Xanthomonas axonopodis* pv. *phaseoli*		Bacteria
2419.	*Xanthomonas axonopodis* pv. *glycines*		Bacteria
2420.	*Xanthomonas campestris* pv. *betlicola*		Bacteria
2421.	*Xanthomonas campestris* pv. *campestris*		Bacteria
2422.	*Xanthomonas oryzae* pv. *oryzae* (=*Xanthomonas campestris* pv. *oryzae*)		Bacteria
2423.	*Xanthomonas oryzae* pv. *oryzicola*		Bacteria
2424.	*Xanthomonas phaseoli* var. *fuscans*		Bacteria
2425.	*Xanthomonas sesami*		Bacteria
2426.	*Xanthomonas* sp.		Bacteria
2427.	*Xanthomonas vesicatoria*		Bacteria
2428.	Bhendi yellow vein mosaic virus (BYVMV)		Virus
2429.	Dasheen mosaic virus		Virus
2430.	Enation Leaf Curl Virus		Virus
2431.	Gemini virus		Virus
2432.	Mung bean yellow mosaic virus (MYMV)		Virus
2433.	Papaya leaf curl virus (PaLCuV)		Virus
2434.	Papaya ringspot virus (PRSV)		Virus
2435.	Pigeonpea sterility mosaic virus (PPSMV)		Virus
2436.	Pigeonpea Yellow Mosaic virus		Virus
2437.	Potato leaf roll virus		Virus
2438.	Potato virus X		Virus
2439.	Potato virus Y		Virus
2440.	Soybean mosaic virus (SMV)		Virus
2441.	Spotted wilt virus		Virus
2442.	Tobacco ringspot virus (TRSV)		Virus
2443.	Tobacco streak virus		Virus
2444.	Tomato leaf curl virus (TLCV)		Virus
2445.	Tomato mosaic virus (ToMV)		Virus
2446.	Tomato spotted wilt virus- Bud necrosis virus		Virus
2447.	*Aphelenchoides besseyi*		Nematode
2448.	*Heterodera cajani*		Nematode

	Name	Family	Purpose
2449.	*Hirschmanniella oryzae*		Nematode
2450.	*Meloidogyne graminicola*		Nematode
2451.	*Meloidogyne incognita*		Nematode
2452.	*Meloidogyne javanica*		Nematode
2453.	*Pratylenchus thornei*		Nematode
2454.	*Rotylenchulus reniformis*		Nematode
2455.	Phytoplasma		Phytoplasma
2456.	*Azospirillum* sp.		Bacteria benign
2457.	*Azotobacter* sp.		Bacteria benign
2458.	*Bacillus* sp.		Bacteria benign
2459.	*Bacillus* sp.		Bacteria benign
2460.	*Bradyrhizobium japonicum*		Bacteria benign
2461.	*Bradyrhizobium* sp.		Bacteria benign
2462.	Cyano bacteria hetero sp		Bacteria benign
2463.	*Fraturia aurentia*		Bacteria benign
2464.	*Pseudomonas fluorescence*		Bacteria benign
2465.	*Rhizobium cicerum*		Bacteria benign
2466.	*Rhizobium leguminosarum*		Bacteria benign
2467.	*Rhizobium leucaena*		Bacteria benign
2468.	*Rhizobium meliloti*		Bacteria benign
2469.	*Rhizobium phaseoli*		Bacteria benign
2470.	*Trichoderma viride*		Fungus benign
2471.	*Glomus* Sp.		Fungus benign
2472.	*Mermis* sp.	Mermithida	Nematode ento
2473.	*Strobiliomyia orbata*W.		Nematode ento
2474.	*Ariamnes simulans* (Cambridge)	Therididae	Predatory Spider
2475.	*Clubiona drassodes* Cambridge	Clubionidae	Predatory Spider
2476.	*Cyclosa* sp.	Araneidae	Predatory Spider
2477.	*Cyrtophora* sp.	Araneidae	Predatory Spider
2478.	*Dictyna* sp.	Dictynidae	Predatory Spider
2479.	*Drassodes* sp.	Gnaphosidae	Predatory Spider
2480.	*Gnaphosa poonaensis* Tikader	Gnaphosidae	Predatory Spider
2481.	*Gnaphosa* sp.	Gnaphosidae	Predatory Spider
2482.	*Guizygiella indica* (Tikader & Bal)	Tetragnathidae	Predatory Spider
2483.	*Guizygiella* sp.	Tetragnathidae	Predatory Spider
2484.	*Hogna lenta*Hentz	Lycosidae	Predatory Spider
2485.	*Hyllus semicupreus* (Simon)	Salticidae	Predatory Spider
2486.	*Larinia* sp.	Araneidae	Predatory Spider
2487.	*Leucauge decorata* (Blackwall)	Tetragnathidae	Predatory Spider
2488.	*Leucauge* sp.	Tetragnathidae	Predatory Spider
2489.	*Marpissa* sp.	Salticidae	Predatory Spider
2490.	*Neoscona ellipticus* (Tikader & Bal)	Araneidae	Predatory Spider
2491.	*Neoscona* sp.	Araneidae	Predatory Spider
2492.	*Neoscona vigilans* (Blackwall)	Araneidae	Predatory Spider
2493.	*Olios* sp.	Sparassidae	Predatory Spider
2494.	*Oxyopes birmanicus*Thorell	Oxyopidae	Predatory Spider
2495.	*Oxyopes javanus* Thorell	Oxyopidae	Predatory Spider
2496.	*Oxyopes salticus*Hentz	Oxyopidae	Predatory Spider
2497.	*Oxyopes* sp.	Oxyopidae	Predatory Spider
2498.	*Pardosa birmanica* Simon	Lycosidae	Predatory Spider

	Name	Family	Purpose
2499.	*Peucetia* sp.	Oxyopidae	Predatory Spider
2500.	*Peucetia viridans* (Hentz)	Oxyopidae	Predatory Spider
2501.	*Phidippus* sp.	Salticidae	Predatory Spider
2502.	*Plexippus* sp.	Salticidae	Predatory Spider
2503.	*Runcinia* sp.	Thomisidae	Predatory Spider
2504.	*Synema decoratum* Tikader	Thomisidae	Predatory Spider
2505.	*Synema* sp.	Thomisidae	Predatory Spider
2506.	*Telamonia dimediata* (Simon)	Salticidae	Predatory Spider
2507.	*Tetragnatha* sp.	Tetragnathidae	Predatory Spider
2508.	*Theridion* sp.	Therididae	Predatory Spider
2509.	*Thomisus* sp.	Thomisidae	Predatory Spider
2510.	*Uloborus jabalpurensis*	Uloboridae	Predatory Spider
2511.	*Uloborus* sp.	Uloboridae	Predatory Spider